Published for
OXFORD INTERNATIONAL AQA EXAMINATIONS

W0044039

International GCSE
CHEMISTRY

Lawrie Ryan
Patrick Fullick
Editor: Lawrie Ryan

OXFORD
UNIVERSITY PRESS

OXFORD
UNIVERSITY PRESS

Great Clarendon Street, Oxford, OX2 6DP, United Kingdom

Oxford University Press is a department of the University of Oxford.
It furthers the University's objective of excellence in research,
scholarship, and education by publishing worldwide. Oxford is a
registered trade mark of Oxford University Press in the UK and in
certain other countries

© Lawrie Ryan and Patrick Fullick 2016

The moral rights of the authors have been asserted

First published in 2016

All rights reserved. No part of this publication may be reproduced,
stored in a retrieval system, or transmitted, in any form or by any
means, without the prior permission in writing of Oxford University
Press, or as expressly permitted by law, by licence or under terms
agreed with the appropriate reprographics rights organization.
Enquiries concerning reproduction outside the scope of the above
should be sent to the Rights Department, Oxford University Press,
at the address above.

You must not circulate this work in any other form and you must
impose this same condition on any acquirer

British Library Cataloguing in Publication Data
Data available

978-0-19-837589-0

15

Paper used in the production of this book is a natural, recyclable
product made from wood grown in sustainable forests.
The manufacturing process conforms to the environmental
regulations of the country of origin.

Printed in China by Shanghai Offset Printing Products Ltd

Acknowledgements

The publishers would like to thank the following for permissions to
use copyright material:

p7: sciencephotos/Alamy; **p10:** luismmolina/iStockphoto; **p15:**
Martyn F. Chillmaid/Science Photo Library; **p24:** Leonard Lee Rue/
Science Photo Library; **p25:** Dirk Wiersma/Science Photo Library;
p26: Martyn F. Chillmaid; **p32:** Olga D. van de Veer/Fotolia.com;
p34: Bloomberg via Getty Images; **p35t:** Joe Gough/Fotolia.com;
p35m: iStockphoto; **p35b:** Pascal Goetgheluck/Science Photo
Library; **p36t:** Timothy A. Clary/Getty Images; **p36b:** berekin/iStock
photo; **p37t:** Laguna Design/Science Photo Library; **p37b:** Pasieka/
Science Photo Library; **p42:** Martyn F. Chillmaid/Science Photo
Library; **p43:** Martyn F. Chillmaid; **p45:** Dirk Wiersma/Science Photo
Library; **p52l:** RosaIreneBetancourt 1/Alamy Stock Photo; **p52r** &
p53r: iStockphoto; **p53l:** michellepix/fotolia.com; **p54:** Luis Veiga/
Getty Images; **p55t:** Ton Kinsbergen/Science Photo Library; **p55m:**
iStockphoto; **p60t:** Peter Arnold Images/Photolibrary/Getty Images;
p60b: DYK/iStockphoto; **p61:** Andrew Lambert Photography/
Science Photo Library; **p64–p72t:** iStockphoto **p72b:** aberenyi/
fotolia.com; **p78:** Martyn F. Chillmaid; **p84t:** David Taylor/Science
Photo Library; **p84b** & **p86:** Andrew Lambert Photography/Science
Photo Library; **p87:** Charles D. Winters/Science Photo Library; **p90t:**
Andrew Lambert Photography/Science Photo Library; **p90b:** Martyn
F. Chillmaid/Science Photo Library; **p94:** Adrian Sherratt/Alamy; **p95:**
iStockphoto; **p118t:** Michelle Lester/Fotolia.com; **p118b:** iStockphoto;
p120t: Photolibrary/Getty Images; **p120b:** iStockphoto; **p121:**
Phil Degginger/AlamyStock Photo; **p122:** Cordelia Molloy/Science
Photo Library; **p124l:** Dr Keith Wheeler/Science Photo Library;
p124r: Sheila Terry/Science Photo Library; **p128:** Andrew Lambert
Photography/Science Photo Library; **p129:** Sciencephotos/Alamy;
p134: ICI; **p136:** Maximilian Stock Ltd/Science Photo Library; **p142t:**
Nina Ryan; **p142b:** iStockphoto; **p143:** Photolibrary/Getty Images;
p144: Martyn F.Chillmaid; **p145:** Fuse/Getty Images; **p146:** Martyn F.
Chillmaid/Science Photo Library; **p147:** Andrew Lambert Photography/
Science Photo Library; **p158t:** Chuyu/123RF; **p162:** Tim Graham/Getty
Images; **p165:** iStockphoto; **p166:** John Millar/Getty Images; **p167:**
Photolibrary/Getty Images; **p168:** USDA; **p169:** Bloomberg/Getty
Images; **p172:** Paul Rapson/Science Photo Library; **p174t:** Cordelia
Molloy/Science Photo Library; **p174b:** Charles D. Winters/Science
Photo Library; **p176:** fotomatrix/Fotolia.com; **p177:** Innershadows/
Fotolia.com; **p178:** Image Source/Alamy; **p180:** iStockphoto; **p181:**
AP Photo/Josh Reynolds; **p185:** Martyn F. Chillmaid/Science Photo
Library; **p186:** OUP; **p187:** Andrew Lambert Photography/Science
Photo Library; **p188:** Craig Lovell/Agstockusa/Science Photo Library;

Cover: ALFRED PASIEKA/SCIENCE PHOTO LIBRARY

Although we have made every effort to trace and contact all copyright
holders before publication this has not been possible in all cases.
If notified, the publisher will rectify any errors or omissions at the
earliest opportunity.

Links to third party websites are provided by Oxford in good faith and
for information only. Oxford disclaims any responsibility for
the materials contained in any third party website referenced in
this work.

With special thanks to Mike Yates for his advice and guidance, and
Emma Gadsden for her editing and management of the project.

Chemistry   Contents

How to use this book

This book has been written for you by experienced teachers and subject experts. It covers the information you need to know for your exams and is packed full of features to help you achieve the very best that you can.

Figure 1 Many diagrams are as important for you to learn as the text, so make sure you revise them carefully

Key words are highlighted in the text. You can look them up in the glossary at the back of the book if you are not sure what they mean.

Required practical

This feature helps you become familiar with key practicals. It may be a simple introduction, a reminder or the basis for a practical in the laboratory.

Summary questions

These questions give you the chance to test whether you have learned and understood everything in the topic. If you get any wrong, go back and have another look. They are designed to be increasingly challenging.

And at the end of each chapter you will find …

Chapter summary questions

These will test you on what you have learned throughout the whole chapter, helping you to work out what you have understood and where you need to go back and revise.

Practice questions

These questions are examples of the types of questions you may encounter in your exam, so you can get lots of practice during your course.

Learning objectives

Each topic begins with key statements that you should know by the end of the lesson.

Study tip

Hints that give you important advice on things to remember and what to watch out for.

??? Did you know … ?

There are lots of interesting and often strange facts about science. This feature tells you about many of them.

⌾ links

Links will tell you where you can find more information about what you are learning and how different topics link up.

Activity

An activity is linked to a main lesson and could be a discussion or task in pairs, groups or by yourself.

Key points

At the end of each topic are the important points that you must remember. They can be used to help with revision and summarising your knowledge.

Practical skills

During this course, you will develop your understanding of the scientific process and the skills associated with scientific enquiry. Practical work is an important part of this as it develops these skills and also reinforces concepts and knowledge developed during the course.

As part of this course, all students are expected to undertake practical work in many topics and are required to carry out the five required practicals listed below:

Required Practicals

1. Investigate the products at the anode and cathode in the electrolysis of copper sulfate solution. (3.3.2) [Topic 5.6]

2. Identify the metal ion in an unknown compound using flame testing techniques. (3.4.3) [Topic 6.4]

3. Establish the concentration of an unknown strong acid through titration with a strong base. (3.6.4) [Topic 8.6]

4. Investigate factors affecting the rate of a reaction. (3.8.1) [Select from Chapter 9 practicals]

5. Test for the presence of a double bond in an unknown hydrocarbon. (3.10.1.3) [Topic 13.1]

In Paper 2, you will be assessed on aspects of the practical skills listed below, and may be required to read and interpret information from scales given in diagrams and charts, present data in appropriate formats, design investigations and evaluate information that is presented to them.

Designing a practical procedure

- Design a practical procedure to answer a question, solve a problem or test a hypothesis.
- Comment on/evaluate plans for practical procedures.
- Select suitable apparatus for carrying out experiments accurately and safely.

Control

- Appreciate that, unless certain variables are controlled, experimental results may not be valid.
- Recognise the need to choose appropriate sample sizes, and study control groups where necessary.

Risk assessment

- Identify possible hazards in practical situations, the risks associated with these hazards, and methods of minimising the risks.

Collecting data

- Make and record observations and measurements with appropriate precision and record data collected in an appropriate format (such as a table, chart or graph).

Analysing data

- Recognise and identify the cause of anomalous results and suggest what should be done about them.
- Appreciate when it is appropriate to calculate a mean, calculate a mean from a set of at least three results and recognise when it is appropriate to ignore anomalous results in calculating a mean.
- Recognise and identify the causes of random errors and systematic errors.
- Recognise patterns in data, form hypotheses and deduce relationships.
- Use and interpret tabular and graphical representations of data.

Making conclusions

- Draw conclusions that are consistent with the evidence obtained and support them with scientific explanations.

Evaluation

- Evaluate data, considering its repeatability, reproducibility and validity in presenting and justifying conclusions.
- Evaluate methods of data collection and appreciate that the evidence obtained may not allow a conclusion to be made with confidence.
- Suggest ways of improving an investigation or practical procedure to obtain extra evidence to allow a conclusion to be made.

1.1

States of matter

Learning objectives

After this topic, you should know:

- there are three states of matter

- the arrangement and motion of the particles in each state of matter

- the names of the processes and energy changes involved in changing state.

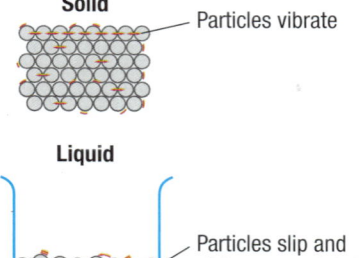

Solid — Particles vibrate

Liquid — Particles slip and slide over each other

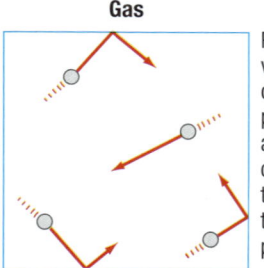

Gas

Particles move very quickly in all directions – as the particles bash against the walls of the container, they exert a force that causes pressure

Figure 1 *The particles in the three states of matter*

⊂⊃ **links**

To find out how to indicate a substance's state in a balanced symbol equation, see Topic 8.1 'Chemical equations'.

You can classify the majority of substances as solids, liquids, or gases. These are called the three **states of matter**.

Solids have a fixed shape and volume. They cannot be compressed. **Liquids** have a fixed volume, but they can flow and change their shape. Liquids occupy just slightly more space than when solid (water and ice are exceptions). **Gases** have no fixed shape or volume. They can be compressed easily.

To explain the properties of solids, liquids, and gases you use the kinetic theory of matter. It is based on the fact that all matter is made up of tiny particles and describes:

- the movement of the particles, and
- the average distance between particles

within each state of matter.

Look at the diagrams to the left that represent the three states of matter.

Each particle in a solid is touching its nearest neighbours and they remain in this fixed arrangement. They cannot move around, but they do vibrate constantly.

The particles in a liquid are also very close together but they can move past each other. This results in a constantly changing, random arrangement of particles.

The particles in a gas have much more space, on average, between them. They can move around at high speeds and in any direction. This means the particles have a random arrangement. The hotter the gas is, the faster the particles move. The pressure of a gas is caused by the particles colliding with the sides of the container. The more frequent and energetic the collisions are, the higher the pressure of the gas. So, in a sealed container, the pressure of the gas increases with temperature.

Changing state

If a solid is heated and changes directly to a gas without melting, that is, it does not pass through the liquid phase, the change of state is called sublimation.

Look at the changes of state that occur when water is heated and cooled:

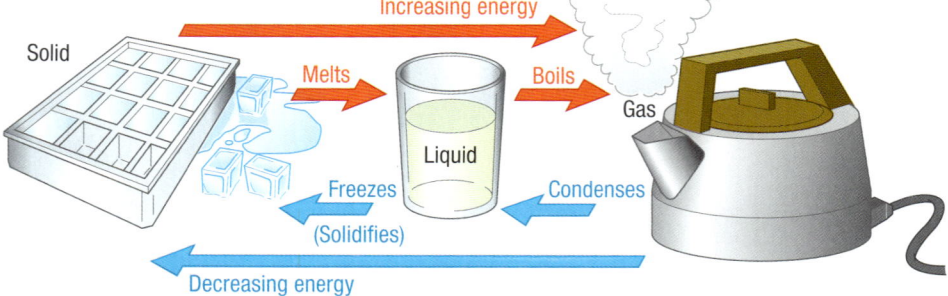

Increasing energy

Solid — Melts — Liquid — Boils — Gas

Freezes (Solidifies) — Condenses

Decreasing energy

Figure 2 *The changes of state in water*

The hotter a solid is, the faster its particles vibrate. Eventually, the vibrations will be so strong that the particles begin to break free from their neighbours. At this point the solid starts to melt and become a liquid.

The hotter a liquid is, the faster its particles move around. As the temperature rises, more and more particles gain enough energy to escape from the surface of the liquid. Its rate of evaporation increases. Eventually, the liquid boils and bubbles of gas rise and escape from within the liquid.

Each change of state is reversible. They are examples of physical changes. No new substances are formed in changes of state. For example, water molecules (H_2O) are the same in ice as they are in liquid water or in water vapour.

Energy changes during changes of state

When you monitor the temperature of a solid as you heat it to beyond its melting point, the results are surprising. The temperature stops rising at the solid's melting point. It remains constant until all the solid has melted, and only then starts to rise again.

At its melting point, the energy provided in heating the solid is being absorbed to break the forces between the particles in the solid. Once all the solid has melted, the energy from the heat source raises the temperature of the liquid as expected.

Changes of state which involve particles becoming closer together, that is, condensing and freezing (solidifying), transfer energy to the surroundings as stronger forces form between particles.

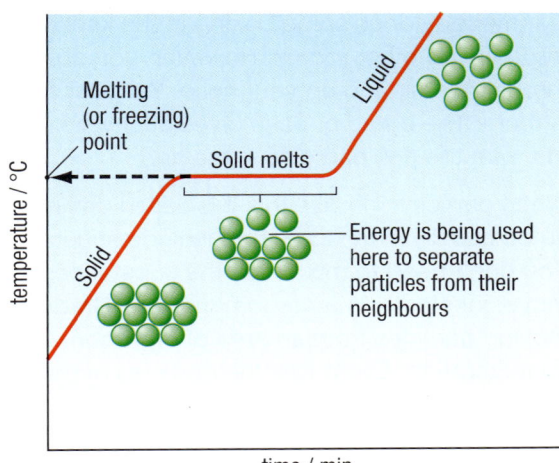

Figure 3 *The heating curve of a solid*

Practical

Cooling curve

Heat a test tube of stearic acid clamped in a water bath until its temperature reaches about 75 °C. Then remove the test tube from the hot water and monitor the temperature as it falls. Plot or print off a graph of the results.

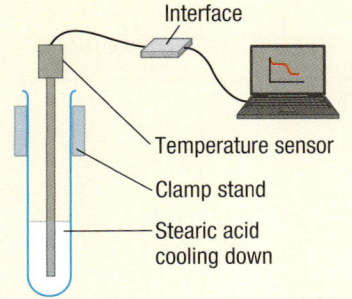

- What is the melting point of stearic acid?
- Explain the shape of the line on your graph.

Safety: Wear eye protection.

Summary questions

1 Draw a table to summarise the general properties of solids, liquids, and gases, as well as the average distance, arrangement, and movement of their particles.

2 Describe the changes that occur to the particles as a gas is cooled down to a temperature below its freezing point.

3 Name the following changes:
   a liquid → solid   b gas → liquid   c solid → liquid
   d liquid → gas   e solid → gas (in a single step)

4 Using the kinetic theory of matter, predict how temperature and pressure affect the density of a fixed mass of gas.

5 Explain why substances have different melting points in terms of their particles.

6 Evaporation is the change of state that occurs when a liquid changes to a gas below its boiling point. You can investigate the factors that affect the rate of evaporation using a wet paper towel on a high resolution electronic balance.
   Plan an investigation into one factor that might affect the rate of evaporation of water from the paper towel, writing a brief method.

Key points

- The three states of matter are solids, liquids, and gases.

- The particles in a solid are packed closely together, fixed in their positions and vibrate.

- The particles in a liquid are also close together but can slip and slide over each other in random motion.

- The particles in a gas have, on average, lots of space between them and zoom around randomly.

- Melting and boiling take in energy from the surroundings as they take place, whereas freezing and condensing transfer energy to the surroundings when they occur.

1.2

Evidence for particles

Learning objectives

After this topic, you should know:

- some evidence for the existence of particles in matter

- why diffusion takes place.

All the substances that make up our world are made of tiny particles. These particles are too small to see, although nowadays the latest powerful microscopes can show us images of particles. Scientists have developed special probes that can measure very small forces. These probes can be used to detect particles, and even position them where you want. However, long before the invention of these microscopes, people, such as the ancient Greeks, had proposed particle models to explain the properties of materials.

Diffusion

If somebody does some baking in the kitchen, the smell quickly spreads through into other rooms. However, you don't **see** any particles moving through the air and up your nose. Yet your sense of smell tells you that it really is there. The bread or cake gives off invisible particles of itself. These particles mix with the gas particles in the air.

When particles mix like this it is called **diffusion**. Diffusion happens automatically in mixtures that include substances which are liquids and gases. You don't need to mix or stir the substances. The particles in liquids and gases move constantly in a random manner. Gradually, this has the overall effect of moving particles from an area of **high** concentration to an area of **low** concentration. Eventually the particles of gas in a closed container will be evenly spread throughout the container.

Diffusion in liquids

Practical

Does temperature affect the rate of diffusion?

Plan an investigation to see how temperature affects the rate of diffusion through a liquid.

Let your teacher check your plan before you start any practical work.

Practical

Diffusion through a liquid

Use tweezers to pick up a few potassium manganate(VII) crystals.

Gently place them at the bottom of a beaker of water.

Observe the beaker for a few minutes.
- What do you see happen?

Leave your beaker until your next lesson.

Draw a diagram to show your results.
- Explain what you think happens to the purple particles.

Use the words **particles** and **diffuse** in your answer.

Safety: Wear eye protection. The crystals are harmful and will stain clothing and skin.

Purple colour starts spreading through the water

Potassium manganate(VII) crystals

Figure 1 *Diffusion of potassium manganate(VII) through water*

Diffusion in gases

As in liquids, the random motion of particles in a gas enables diffusion to take place.

You can show this using a coloured gas, just as you used coloured particles dissolved in water in the first practical to demonstrate diffusion in a liquid. A little bromine liquid is placed in a gas jar with a lid. Bromine vapour evaporates from the volatile liquid and fills the gas jar. Then a second gas jar of air is placed upside down on top of the original bromine-filled gas jar and the lid is removed, linking the jars together. Even though bromine vapour is much more dense than air, the bromine gas gets distributed evenly between the two gas jars after a time. The rapid, random movement of air and bromine particles (molecules) ensure complete mixing takes place.

Diffusion takes place more quickly through gases than through liquids. That is because the particles of gas are moving much more rapidly, on average, than the particles in a liquid.

You find that lighter particles of gas move faster, on average, than heavy ones at the same temperature. Therefore small, light particles diffuse faster than large, heavy ones.

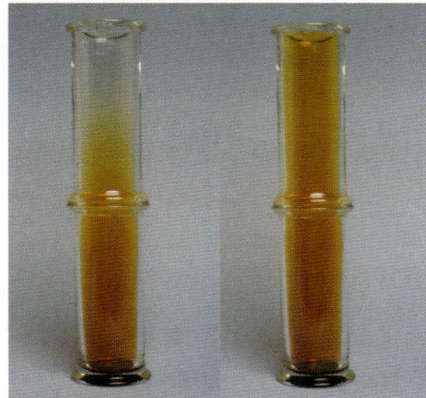

Figure 2 *Bromine vapour diffuses and mixes thoroughly with the air*

Demonstration

Diffusion through a gas

Watch what happens when the stoppers from bottles of concentrated ammonia solution and hydrochloric acid are brought near each other. The reaction is:

ammonia + hydrogen chloride → ammonium chloride

$$NH_3(g) + HCl(g) \rightarrow NH_4Cl(s)$$

(seen as white 'smoke')

Your teacher will set up the apparatus as shown below:

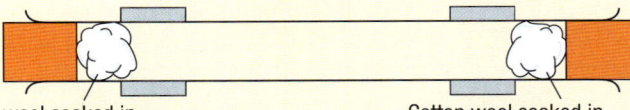

Cotton wool soaked in ammonia solution (concentrated)

Cotton wool soaked in hydrochloric acid (concentrated)

Figure 3 *Diffusion of NH₃(g) and HCl(g) through air*

- What do you see in the long tube where the HCl(g) and NH₃(g) particles meet?
- Which particles diffuse faster, NH₃(g) or HCl(g)? How can you tell?

In this experiment, the particles meet nearer the acid end of the tube.

- Which particles are lighter, the HCl(g) or NH₃(g)?

Safety: Do not inhale the gases. If you are asthmatic take particular care.

Summary questions

1 Explain each of the statements below in terms of particles:
   a You can smell a fish and chip shop from across the road.
   b Sugar dissolves faster in hot water than cold water.
   c Condensation forms on the inside of your bathroom windows in winter.
   d It is dangerous to smoke or strike a match in a petrol station.

2 a Define the word 'diffusion'.
   b Explain why diffusion takes place more quickly in a gas than in a liquid.

3 When pollen grains in water are viewed under a microscope, the grains are seen to be moving in a haphazard, jittery manner. This is called Brownian motion, after the scientist who first observed it. Explain how Brownian motion gives us evidence for the existence of particles in matter.

4 Hydrogen bromide (HBr) is a similar gas to hydrogen chloride (HCl) but its particles (molecules) are more than twice as heavy. If the demonstration 'Diffusion through a gas' in Figure 3 was done with HBr(g) instead of HCl(g), explain what you would expect to see happen.

Key points

- All substances are made up of particles.
- In liquids and gases, the random movement of particles mixes substances in a process called diffusion.
- Diffusion takes place faster in a gas than in a liquid.
- Small, light particles diffuse faster than large, heavy ones.

1.3

History of the atom

Learning objectives

After this topic, you should know:

- how and why the atomic model has changed over time

- that scientific theories are revised or replaced by new ones in the light of new evidence.

Early ideas about atoms

The ancient Greeks were the first to have ideas about particles and atoms. However, it was not until the early 1800s that these ideas became linked to strong experimental evidence when John Dalton put forward his ideas about atoms. From his experiments, he suggested that substances were made up of atoms that were like tiny, hard spheres. He also suggested that each chemical element had its own atoms that differed from others in their mass (Figure 1). Dalton believed that these atoms could not be divided or split. They were the fundamental building blocks of nature.

In chemical reactions, he suggested that the atoms re-arranged themselves and combined with other atoms in new ways. In many ways, Dalton's ideas are still useful today. For example, they help to visualise elements, compounds, and molecules, as well as the models still used to describe the different arrangement and movement of particles in solids, liquids, and gases.

Evidence for electrons in atoms

At the end of the 1800s, a scientist called J. J. Thomson discovered the **electron**. This is a tiny, negatively charged particle that was found to have a mass about 2000 times smaller than the lightest atom. Thomson was experimenting by applying high voltages to gases at low pressure (Figure 2).

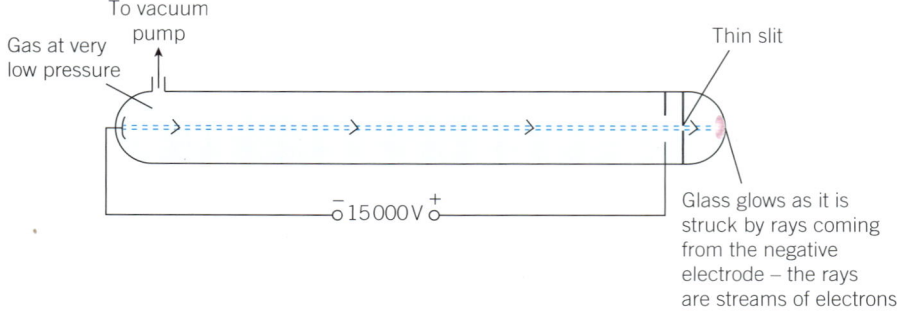

Figure 2 *Thomson's experimental evidence for the existence of electrons*

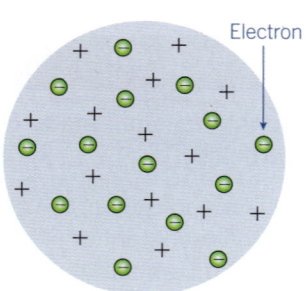

Figure 1 *Dalton made a list of substances that he believed were elements*

Thomson did experiments on the beams of particles. They were attracted to a positive charge, showing they must be negatively charged themselves. He called the tiny, negatively charged particles electrons. These electrons must have come from inside atoms in the tube. So Dalton's idea that atoms could not be divided or split had to be revised. Thomson proposed a different model for the atom. He said that the tiny, negatively charged electrons must be embedded in a cloud of positive charge. He knew that atoms themselves carry no overall charge so any charges in an atom must balance out. He imagined the electrons as the bits of plum in a plum pudding (Figure 3).

Figure 3 *Thomson's 'plum pudding' model of the atom*

Evidence for the nucleus

The next breakthrough in understanding the atom came about 10 years later. Two of Ernest Rutherford's students were doing an experiment with radioactive particles. They were firing dense, positively charged particles (called alpha particles) at the thinnest piece of gold foil they could make (see Figure 4). They expected the particles to pass straight through the gold atoms with their diffuse cloud of positive charge (as in Thomson's plum-pudding model). However, their results shocked them (Figure 4) as some particles came back towards the source.

In 1911, Ernest Rutherford interpreted their results and suggested a new model for the atom (Figure 5). He said that Thomson's atomic model was not possible. The positive charge must be concentrated at a tiny spot in the centre of the atom. Otherwise the large, positive particles fired at the foil could never be repelled back towards their source. He proposed that the electrons must be orbiting around this **nucleus** (centre of the atom), which contains very dense positively charged **protons** (Figure 5).

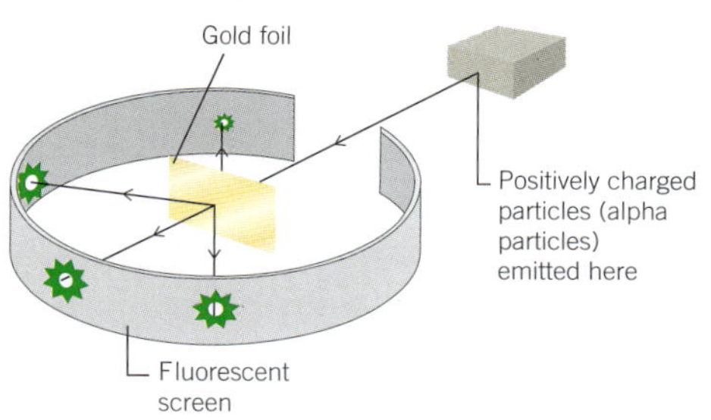

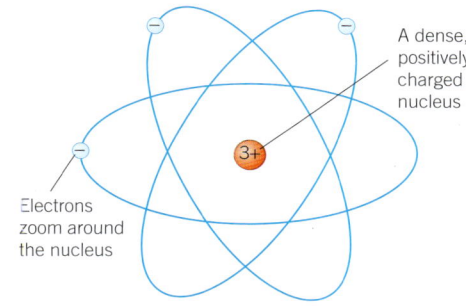

Figure 4 *The alpha particle scattering experiment that changed the 'plum pudding' theory*

Evidence for electrons in shells (energy levels)

The next important development came in 1914, when Niels Bohr revised the atomic model again (Figure 6). He noticed that the light given out when atoms were heated only had specific amounts of energy. He suggested that the electrons must be orbiting the nucleus at set distances, in certain fixed energy levels (or shells). The energy must be given out when 'excited' electrons fall from a high to a low energy level. Bohr matched his model to the energy values observed (see Figure 6).

Figure 5 *Rutherford's nuclear model of the atom*

Evidence for neutrons in the nucleus

Scientists at the time speculated that there were two types of sub-atomic particles inside the nucleus. They had evidence of protons, but a second sub-atomic particle in the nucleus was also proposed to explain the missing mass that had been noticed in atoms. These **neutrons** must have no charge and have the same mass as a proton.

Because neutrons have no charge, it was very difficult to detect them in experiments. It was not until 1932 that James Chadwick did an experiment that could only be explained by the existence of neutrons!

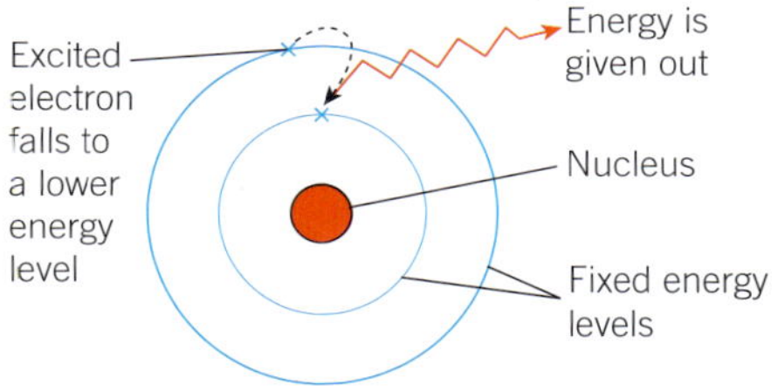

Figure 6 *Bohr's model of the atom*

Summary questions

1 **a** Which one of Dalton's ideas listed below about atoms do scientists no longer believe?
   A Elements contain only one type of atom.
   B Atoms get re-arranged in chemical reactions.
   C Atoms are solid spheres that cannot be split into simpler particles.
 b Which of the following substances from Dalton's list of elements (Figure 1) are not actually chemical elements?
      soda   oxygen   carbon   gold   lime

2 **a** Which sub-atomic particle did J. J. Thomson discover?
 b Describe J. J. Thomson's 'plum pudding' model of the atom.

3 State two ways in which Rutherford changed Thomson's model of the atom.

4 Explain why Bohr revised Rutherford's model of the atom.

Key points

- The ideas about atoms have changed over time.

- New evidence has been gathered from the experiments of scientists who have used their model of the atom to explain their observations and calculations.

- Key ideas were proposed successively by Dalton, Thomson, Rutherford, and Bohr, before arriving at the model of the atom you use at GCSE level today.

1.4 Atoms

Learning objectives

After this topic, you should know:

- the definition of an element
- that each type of atom has a chemical symbol
- the basic structure of the periodic table
- the basic structure of an atom.

Look at the things around you and the substances that they are made from. You will find wood, metal, plastic, glass ... the list is almost endless. Look further and the number of different substances is mind-boggling.

All substances are made of **atoms**. There are about 100 different types of atom found naturally on Earth. These can combine in a huge variety of ways, giving us all those different substances.

Some substances are made up of only one type of atom. These substances are called **elements**. It follows that as there are only about 100 different types of atom, there are only about 100 different elements.

Elements can have very different properties. Elements such as silver, copper, and gold are shiny, solid metals. Other elements such as oxygen, nitrogen, and chlorine are non-metals, and are gases at room temperature.

Chemical symbols

The name given to each different element depends on the language being spoken. For example, the element sulfur is called *schwefel* in German and *azufre* in Spanish. However, the scientific community communicates globally so it is important to have symbols for elements that all nationalities can understand. These symbols are shown in the **periodic table**.

- The symbols in the periodic table represent atoms. For example, O represents an atom of oxygen, Na represents an atom of sodium.

Figure 1 *An element contains only one type of atom – in this case gold*

Group numbers

1	2											3	4	5	6	7	0
			H 1 Hydrogen														**He** 2 Helium
Li 3 Lithium	**Be** 4 Beryllium											**B** 5 Boron	**C** 6 Carbon	**N** 7 Nitrogen	**O** 8 Oxygen	**F** 9 Fluorine	**Ne** 10 Neon
Na 11 Sodium	**Mg** 12 Magnesium											**Al** 13 Aluminium	**Si** 14 Silicon	**P** 15 Phosphorus	**S** 16 Sulfur	**Cl** 17 Chlorine	**Ar** 18 Argon
K 19 Potassium	**Ca** 20 Calcium	**Sc** 21 Scandium	**Ti** 22 Titanium	**V** 23 Vanadium	**Cr** 24 Chromium	**Mn** 25 Manganese	**Fe** 26 Iron	**Co** 27 Cobalt	**Ni** 28 Nickel	**Cu** 29 Copper	**Zn** 30 Zinc	**Ga** 31 Gallium	**Ge** 32 Germanium	**As** 33 Arsenic	**Se** 34 Selenium	**Br** 35 Bromine	**Kr** 36 Krypton
Rb 37 Rubidium	**Sr** 38 Strontium	**Y** 39 Yttrium	**Zr** 40 Zirconium	**Nb** 41 Niobium	**Mo** 42 Molybdenum	**Tc** 43 Technetium	**Ru** 44 Ruthenium	**Rh** 45 Rhodium	**Pd** 46 Palladium	**Ag** 47 Silver	**Cd** 48 Cadmium	**In** 49 Indium	**Sn** 50 Tin	**Sb** 51 Antimony	**Te** 52 Tellurium	**I** 53 Iodine	**Xe** 54 Xenon
Cs 55 Caesium	**Ba** 56 Barium	Lanthanum see below	**Hf** 72 Hafnium	**Ta** 73 Tantalum	**W** 74 Tungsten	**Re** 75 Rhenium	**Os** 76 Osmium	**Ir** 77 Iridium	**Pt** 78 Platinum	**Au** 79 Gold	**Hg** 80 Mercury	**Tl** 81 Thallium	**Pb** 82 Lead	**Bi** 83 Bismuth	**Po** 84 Polonium	**At** 85 Astatine	**Rn** 86 Radon
Fr 87 Francium	**Ra** 88 Radium	Actinium see below															

The transition metals

The alkali metals | The alkaline earth metals | The halogens | The noble gases

La 57 Lanthanium	**Ce** 58 Cerium	**Pr** 59 Praseodymium	**Nd** 60 Neodymium	**Pm** 61 Promethium	**Sm** 62 Samarium	**Eu** 63 Europium	**Gd** 64 Gadolinium	**Tb** 65 Terbium	**Dy** 66 Dysprosium	**Ho** 67 Holmium	**Er** 68 Erbium	**Tm** 69 Thulium	**Yb** 70 Ytterbium	**Lu** 71 Lutetium
Ac 89 Actinium	**Th** 90 Thorium	**Pa** 91 Protactinium	**U** 92 Uranium	**Np** 93 Neptunium	**Pu** 94 Plutonium	**Am** 95 Americium	**Cm** 96 Curium	**Bk** 97 Berkelium	**Cf** 98 Californium	**Es** 99 Einsteinium	**Fm** 100 Fermium	**Md** 101 Mendelevium	**No** 102 Nobelium	**Lr** 103 Lawrencium

Lanthanides
Actinides

Figure 2 *The periodic table shows the symbols for each type of atom*

- The elements in the table are arranged in columns, called **groups**. Each group contains elements with similar chemical properties.
- The 'staircase' drawn in bold black is the dividing line between metals and non-metals. The elements to the left of the line are metals. Those on the right of the line are non-metals. However, a few elements lying next to the dividing line are called metalloids or semi-metals as they have some metallic and some non-metallic characteristics. Examples include silicon (Si) and germanium (Ge) from Group 4.

Atoms, elements, and compounds

Most substances you encounter are not elements. They are made up of different types of atom joined together and are called **compounds**. Chemical bonds hold the atoms tightly together in compounds. Some compounds are made from just two types of atom (e.g., water, made from atoms of hydrogen and oxygen). However, most compounds consist of more than two different types of atom.

An atom is made up of a tiny central nucleus with electrons around it.

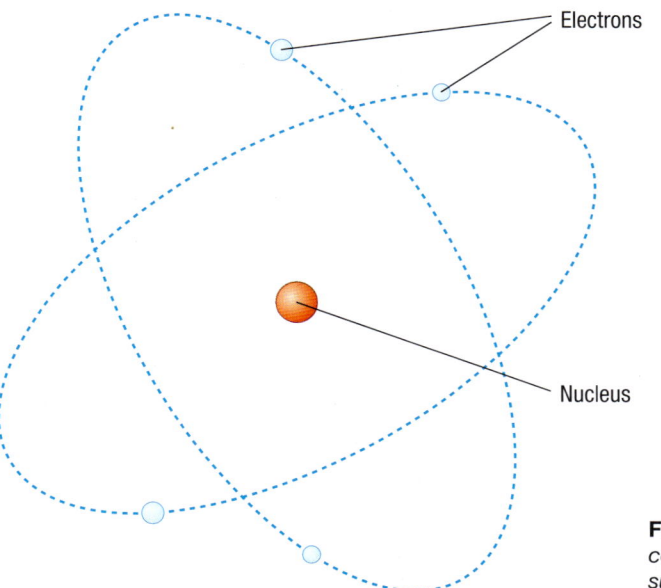

Electrons

Nucleus

Figure 3 *Each atom consists of a tiny nucleus surrounded by electrons*

links
For more information on the periodic table of elements, see 3.1 'The periodic table'.

??? Did you know … ?

Only 92 elements occur naturally on Earth. The other heavier elements in the periodic table have to be made artificially and might only exist for fractions of a second before they decay into other, lighter elements.

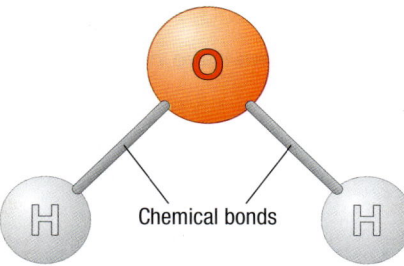

Chemical bonds

Figure 4 *A grouping of two or more atoms bonded together is called a* **molecule**. *Chemical bonds hold the hydrogen and oxygen atoms together in the water molecule. Water is an example of a compound.*

links
For more information on what is inside an atom, see 1.5 'Atomic structure' and 1.6 'The arrangement of electrons in atoms'.

Summary questions

1 **a** Arrange these elements into a table showing metals and non-metals: phosphorus (P), barium (Ba), vanadium (V), mercury (Hg), krypton (Kr), potassium (K), and uranium (U).
   b Would you classify hydrogen as a metal or a non-metal. Why?

2 Explain why, when you mix two elements together, you can often separate them again quite easily by physical means, but when two elements are chemically combined in a compound they are usually very difficult to separate.

3 Draw diagrams to explain the difference between an element and a compound.

4 Describe the basic structure of an atom.

5 Find out the Latin words from which the symbols of the following metallic elements are derived:
   a sodium, Na   **b** gold, Au   **c** lead, Pb   **d** potassium, K

Key points

- All substances are made up of atoms.
- The periodic table lists all the chemical elements, with eight main groups each containing elements with similar chemical properties.
- Elements contain only one type of atom.
- Compounds contain more than one type of atom.
- An atom has a tiny nucleus at its centre, surrounded by electrons.

1.5

Atomic structure

Learning objectives

After this topic, you should know:

- the location and relative charge of the protons, neutrons, and electrons in an atom

- what the 'atomic number' and 'mass number' of an atom represent

- why atoms have no overall charge

- that atoms of a particular element have the same number of protons.

??? Did you know …?

In 1808, a chemist called John Dalton published a theory of atoms. It explained how atoms joined together to form new substances (compounds). Not everyone liked his theory though – one person wrote 'Atoms are round bits of wood invented by Mr Dalton!'

In the centre of every atom there is a very small nucleus. This contains two types of sub-atomic particle, called protons and neutrons. A third type of sub-atomic particle orbits the nucleus. These really tiny particles are called electrons.

Protons have a positive charge. Neutrons have no charge, that is, they are neutral. So the nucleus itself has an overall positive charge. The electrons orbiting the nucleus are negatively charged. The relative charge on a proton is +1 and the relative charge on an electron is −1.

Type of sub-atomic particle	Relative charge
Proton	+1
Neutron	0
Electron	−1

Because every atom contains equal numbers of protons and electrons, the positive and negative charges cancel out and its charge is zero. So, there is no overall charge on any atom. For example, a carbon atom is neutral. It has six protons, so therefore it must have six electrons.

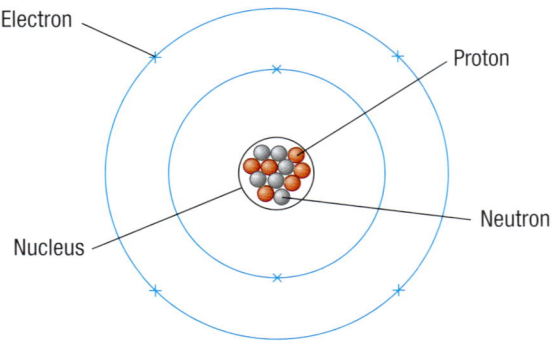

Figure 1 *Understanding the structure of an atom gives us important clues to the way chemicals react together. If you look carefully, this atom of carbon has six protons and six neutrons in its nucleus, and six electrons orbiting the nucleus. Note that although the number of protons and electrons are always equal in an atom, the number of neutrons can differ (but in this case they are the same).*

Atomic number and the periodic table of elements

All the atoms of a particular element have the same number of protons. For example, hydrogen has 1 proton in its nucleus, carbon has 6 protons in its nucleus, and sodium has 11 protons in its nucleus.

The **number of protons** in each atom of an element is called the **atomic number** of that element.

Figure 2 *You read the periodic table from left to right, and from the top down – just like reading a page of writing. This is an abbreviated version of the start of the periodic table, showing the first 18 elements. The elements in the periodic table are arranged in order of their atomic number.*

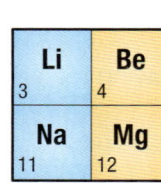

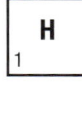

The elements in the periodic table are arranged in order of their atomic number (number of protons). If you are told that the atomic number of an element is 8, you can identify it using the periodic table. It will be the eighth element listed, that is, oxygen.

Mass number

The **number of protons plus the number of neutrons** in the nucleus of an atom is called its **mass number**. So, if an atom has four protons and five neutrons, its mass number will be $4 + 5 = 9$.

Given the atomic number and mass number, you can work out how many protons, electrons and neutrons are in an atom. For example, an argon atom has an atomic number of 18 and a mass number of 40.

- Its atomic number is 18 so it has **18 protons**. Remember that atoms have an equal number of protons and electrons. So argon also has **18 electrons**.
- The mass number is 40, so you know that:

    18 (the number of protons) + the number of neutrons = 40

- Therefore, argon must have **22 neutrons** (as $18 + 22 = 40$).

You can summarise the last part of the calculation as:

number of neutrons = mass number – atomic number

Study tip

In an atom, the number of protons is always equal to the number of electrons. You can find out the number of protons and electrons in an atom by looking up its atomic number in the periodic table.

∞ links

For more information on the patterns in the periodic table, see 1.6 'The arrangement of electrons in atoms'.

Summary questions

1 Draw a table showing the location and relative charge of the three sub-atomic particles.

2 Look at the abbreviated periodic table in Figure 2.
   a How does the number of electrons in each atom vary going across each row (called a period) in the table?
   b What pattern do you notice in the number of protons going down Group 0?
   c Sort out the elements in the table into metals, non-metals, and metalloids.

3 Explain why all atoms are neutral.

4 An atom has 27 protons and 32 neutrons. Use the periodic table in Topic 1.4 to name this element, and give its symbol, atomic number, and mass number.

5 How many protons, electrons, and neutrons do the following atoms contain?
   a A nitrogen atom whose atomic number is 7 and whose mass number is 14.
   b A chlorine atom whose atomic number is 17 and whose mass number is 35.
   c A silver atom whose atomic number is 47 and whose mass number is 108.
   d A uranium atom whose atomic number is 92 and whose mass number is 235.

Key points

- Atoms are made of protons, neutrons, and electrons.

- Protons have a relative charge of +1, and electrons have a relative charge of −1. Neutrons have no electric charge. They are neutral.

- Atoms contain an equal number of protons and electrons so carry no overall charge.

- Atomic number = number of protons (= number of electrons)

- Mass number = number of protons + number of neutrons

- Atoms of the same element have the same number of protons (and therefore electrons) in their atoms.

1.6

The arrangement of electrons in atoms

Learning objectives

After this topic, you should know:

- how the electrons are arranged in an atom

- the electronic structures of the first 20 elements in the periodic table

- how to represent electronic structures in diagrams and by using standard notation.

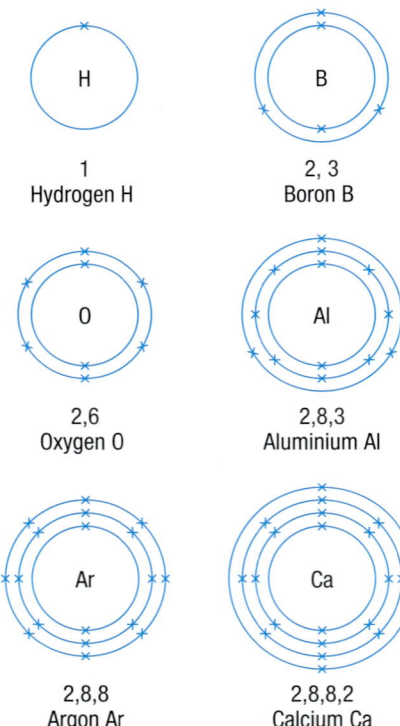

Figure 2 *Once you know the pattern, you should be able to draw the energy levels (shells) and electrons in the atoms of any of the first 20 elements (given their atomic number)*

The model of the atom which you use at this level has electrons arranged around the nucleus in **shells**, rather like the layers of an onion. Each shell represents a different **energy level**.

The lowest energy level is shown by the shell which is nearest to the nucleus. The electrons in an atom occupy the lowest available energy level (the shell closest to the nucleus).

Electron shell diagrams

An energy level (or shell) can only hold a certain number of electrons.

- The first, and lowest, energy level (nearest the nucleus) can hold up to 2 electrons.
- The second energy level can hold up to 8 electrons.
- Once there are 8 electrons in the third energy level, the fourth begins to fill up.

Beyond this point, that is, after the first 20 elements in the periodic table, the situation gets more complex so is covered in higher level courses. You only need to know the arrangement of electrons in atoms of the first 20 elements.

You can draw diagrams to show the arrangement of electrons in an atom. For example, a sodium atom has an atomic number of 11 so it has 11 protons, which means it also has 11 electrons. Figure 1 shows how you can represent an atom of sodium.

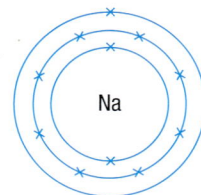

Figure 1 *A simple way of representing the arrangement of electrons in the energy levels (shells) of a sodium atom*

To save drawing atoms all the time, you can write down the number of electrons in each energy level. This is called an **electronic structure**. For example, the sodium atom in Figure 1 has an electronic structure of 2,8,1. You start at the lowest energy level (innermost or first shell), recording the numbers in each successive energy level or shell.

Silicon, whose atoms have 14 electrons, is in Group 4 of the periodic table. It has the electronic structure 2,8,4. This represents 2 electrons in the lowest energy level (first shell), then 8 in the next energy level or shell. There are 4 in the highest energy level (its outermost shell).

The best way to understand these arrangements is to look at the examples in Figure 2.

Study tip

Make sure that you can draw the electronic structure of the atoms for all of the first 20 elements. You will always be given their atomic number or their position in the periodic table (which tells you the number of electrons) – so you don't have to memorise these numbers.

Electrons and the periodic table

Look at the elements in any one of the main groups of the periodic table. Their atoms will all have the same number of electrons in their highest energy level. These electrons are often called the outer electrons because they are in the outermost shell. Therefore, all the elements in Group 1 have one electron in their highest energy level.

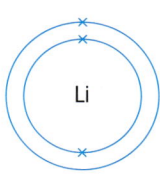 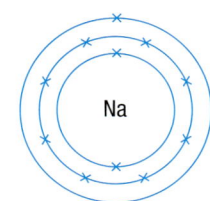 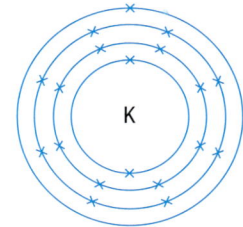

The chemical properties of an element depend on how many electrons it has. The way an element reacts is determined by the number of electrons in its highest energy level (or outermost shell). So because the elements in a particular group all have the same number of electrons in their highest energy level, they all react in a similar way.

For example:

lithium   + water → lithium hydroxide   + hydrogen
sodium   + water → sodium hydroxide   + hydrogen
potassium + water → potassium hydroxide + hydrogen

The elements in Group 0 of the periodic table are called the **noble gases** because they are unreactive. Their atoms have a very stable arrangement of electrons. All the Group 0 elements have eight electrons in their outermost shell, except for helium, which has only two electrons.

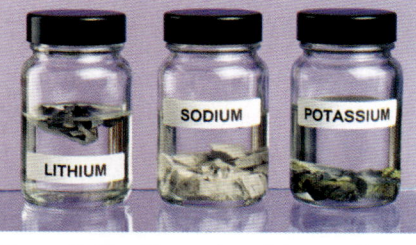

Figure 3 *The Group 1 metals are all reactive metals, shown here stored under oil*

∞ links

For more information on the reactions of elements and their electronic structures, see Chapter 3 'The periodic table'.

Summary questions

1 **a** Which shell represents the lowest energy level in an atom?
   b How many electrons can each of the lowest two energy levels hold?

2 Using the periodic table, draw the arrangement of electrons in the following atoms and label each one with its electronic structure.
   a He   **b** Be   **c** Cl   **d** Ar

3 **a** Write the electronic structure of potassium (atomic number 19).
   b How many electrons does a potassium atom have in its highest energy level (outermost shell)?

4 Give the name and symbol of the atom shown:

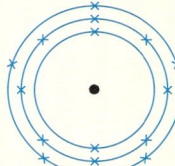

5 **a** Why do the Group 1 metals all react in a similar way with oxygen?
   b Write word equations for the reactions of lithium, sodium, and potassium with oxygen to form their oxides.

Key points

- The electrons in an atom are arranged in energy levels or shells.

- The lowest energy level (1st shell) can hold up to 2 electrons and the next energy level (the 2nd shell) can hold up 8 electrons.

- The 4th shell starts to fill after 8 electrons occupy the 3rd shell.

- The number of electrons in the outermost shell of an element's atoms determines the way in which that element reacts.

1.7

Atoms and isotopes

Learning objectives

After this topic, you should know:

- the relative masses of protons, neutrons, and electrons

- how to represent an atom's atomic number and mass number

- the definition of isotopes.

Did you know ... ?

The volume of the nucleus of an atom can be between 10 000 and 100 000 times smaller than the volume of the atom itself.

As you have seen, an atom consists of a tiny nucleus containing positively charged protons, together with neutrons which have no charge. The negatively charged electrons are arranged in energy levels (shells) around the nucleus.

The mass of a proton and a neutron is the same so you can say that the 'relative mass' of a neutron compared with a proton is 1. Electrons are much, much lighter than protons and neutrons. They have a mass of about $\frac{1}{2000}$ of a proton or neutron. Because of this, the mass of an atom is concentrated in its nucleus and you can ignore the tiny mass of the electrons when working out the relative mass of an atom.

Type of sub-atomic particle	Relative mass
Proton	1
Neutron	1
Electron	very small

Representing the atomic number and mass number

Remember that:

- the number of protons in an atom is its atomic number, and
- the total number of protons and neutrons in an atom is its mass number.

You can show the atomic number and mass number of an atom like this:

Mass number

$$^{12}_{6}\text{C (carbon)} \qquad ^{23}_{11}\text{Na (sodium)}$$

Atomic number

Given this information, you can work out the number of neutrons in the nucleus of an atom by subtracting its atomic number from its mass number. (See page 13).

number of neutrons = mass number – atomic number

For the two examples above:

Carbon has 6 protons and a mass number of 12.
So the number of neutrons in a carbon atom is (12 – 6) = 6.

Sodium has an atomic number of 11 and the mass number is 23.
So a sodium atom has (23 – 11) = 12 neutrons.
In its nucleus there are 11 protons and 12 neutrons.

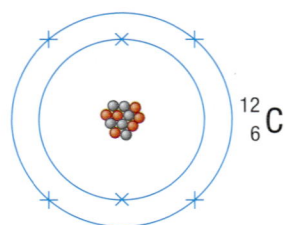

$^{12}_{6}\text{C}$

- ● Proton   Number of protons gives atomic number
- ● Neutron   Number of protons plus number of neutrons gives mass number

Figure 1 *An atom of carbon*

Isotopes

Atoms of the same element always have the same number of protons. However, they can have different numbers of neutrons.

The name **isotopes** is used to describe atoms of the same element with different numbers of neutrons.

Isotopes always have the same atomic number but different mass numbers. For example, two isotopes of carbon are $^{12}_{6}C$ (also written as carbon-12) and $^{14}_{6}C$ (carbon-14). The carbon-12 isotope has 6 protons and 6 neutrons in the nucleus. The carbon-14 isotope has 6 protons and 8 neutrons, that is, 2 more neutrons than carbon-12. Chemists use the carbon-12 isotope to compare masses of all the other elements listed in the periodic table. These 'relative atomic masses' take into account the proportions of an element's different isotopes (see page 100).

Sometimes extra neutrons make the nucleus unstable, so it is radioactive. However, not all isotopes are radioactive – they are simply atoms of the same element that have different masses.

Samples of different isotopes of an element have different *physical* properties. For example, they will have a different density and they may or may not be radioactive. However, they always have the same *chemical* properties because their reactions depend on their electronic structures. As their atoms will have the same number of protons, and therefore electrons, the electronic structure will be same for all isotopes of an element.

For example, look at the three isotopes of hydrogen in Figure 2. The three isotopes are called hydrogen (hydrogen-1), deuterium (or hydrogen-2), and tritium (or hydrogen-3). Each has a different mass and tritium is radioactive. However, all have identical chemical properties, for example, they all react with oxygen to make water:

$$2H_2(g) + O_2(g) \rightarrow 2H_2O(l)$$

Radioactive isotopes are used as tracers in medicine using relatively safe isotopes that can be tracked inside the body using radioactivity detectors. In industry their main use is in generating electricity in nuclear power stations.

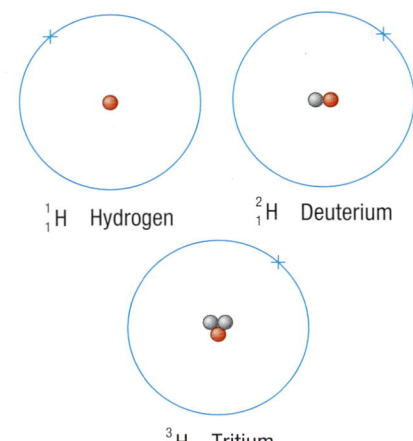

$^{1}_{1}H$ Hydrogen $^{2}_{1}H$ Deuterium

$^{3}_{1}H$ Tritium

Figure 2 *The isotopes of hydrogen – they have identical chemical properties but different physical properties, such as density*

∞ links

For more information on relative atomic mass, see Topic 8.2 'Relative masses and moles'

??? Did you know … ?

Cobalt-60 is a radioactive isotope used to kill cancerous cells using the gamma rays it emits. About 200 sources of the 'radio-isotope' can be focussed on a specific area e.g. a tumour in the brain.

Key points

- The relative mass of protons and neutrons is 1.

- You can represent the atomic number and mass number of an atom using the notation: $^{24}_{12}Mg$, where magnesium's atomic number is 12 and its mass number is 24.

- Isotopes are atoms of the same element with different numbers of neutrons. They have identical chemical properties, but their physical properties, such as density, can differ.

Summary questions

1 Draw a table to summarise the relative mass and charge of the three sub-atomic particles.

2 State how many protons and how many neutrons there are in the nucleus of each of the following elements:
   a $^{11}_{5}B$ b $^{14}_{7}N$ c $^{24}_{12}Mg$ d $^{37}_{17}Cl$ e $^{127}_{53}I$

3 Define the word 'isotopes'.

4 a Which physical property will always differ in pure samples of each isotope of the same element?
   b Explain why the isotopes of the same element have identical chemical properties?

5 a The same volume of any gas contains the same number of molecules. If the density of hydrogen gas containing only the hydrogen-1 isotope is 0.00008 g/cm³ at 25 °C, what will be the density of the gas containing only the hydrogen-3 isotope at 25 °C?
   b Name one hazard associated with both gases containing hydrogen-1 or hydrogen-3 isotopes.
   c Name a hazard associated with the gas containing hydrogen-3 which does not apply to the hydrogen-1 sample.

Chapter summary questions

1 Arsenic (As) is a solid element that sublimes at 613 °C.

a What is sublimation?

b Assuming arsenic exists as atoms, draw a diagram to show the arrangement of some of its particles in boxes like the ones below:

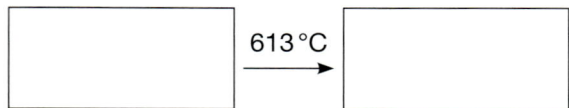

c Which of these processes take in energy from the surroundings?

boiling melting freezing condensation sublimation

2 Look at the experiment below:

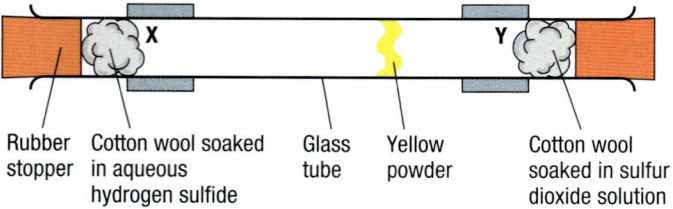

Rubber stopper   Cotton wool soaked in aqueous hydrogen sulfide   Glass tube   Yellow powder   Cotton wool soaked in sulfur dioxide solution

Hydrogen sulfide gas is released from the cotton wool at X and sulfur dioxide gas from the cotton wool at Y.

a The yellow powder inside the glass tube forms after 10 minutes. Explain how the yellow powder forms using the kinetic theory of matter.

b Name the physical process that this experiment demonstrates.

c Why is the yellow powder formed nearer to end Y of the tube?

d Draw a long tube and show the possible path of a hydrogen sulfide particle (molecule) travelling down the tube.

e The original experiment was carried out at 18 °C. Predict any difference you might notice if the experiment was repeated at 25 °C.

3 Look at the data in the table below:

Chemical element	Melting point (°C)	Boiling point (°C)	Density (g/cm³)
Bromine	−7	59	3.12
Caesium	29	669	1.88
Fluorine	−220	−188	0.00158
Strontium	769	1384	2.6
Xenon	−112	−108	0.0055

a What is the physical state of each element in the table at 25 °C?

b Which element exists as a liquid over the widest range of temperature?

c Explain the large differences in the densities of the elements in the table using the kinetic theory of matter.

d What is the chemical symbol for atoms of each element in the table?

e Classify each element in the table as a metal or a non-metal.

4 a i Which sub-atomic particles are found in the nucleus?

ii What is the maximum number of electrons that can occupy each of the first two energy levels or shells?

b i Explain the overall charge on any atom.

ii Define 'atomic number' and 'mass number'.

5 This question is about some of the elements in the periodic table.

You will need to use the periodic table to help you answer some parts of the question. See page 41.

a Neon (Ne) is the 10th element in the periodic table.

i Is neon a metal or a non-metal?

ii Are there more metals or non-metals in the periodic table?

iii How many protons does an neon atom contain?

iv The mass number of a neon atom is 20. How many neutrons does it contain?

v State the name and number of the group to which neon belongs.

vi Name two other elements in the same group as neon.

vii Write the electronic structure of a neon atom.

viii What is special about the electronic structure of neon and the other elements in its group?

b The element radium (Ra) has 88 electrons.

i How many protons are in the nucleus of each radium atom?

ii How many electrons does a radium atom have in its highest energy level (outermost shell)? How did you decide on your answer?

iii Is radium a metal or a non-metal?

iv Calcium is in the same group as radium. Its atomic number is 20. Write down its electronic structure.

v All the isotopes of radium are radioactive. Give one medical and one industrial use of radioactive isotopes.

Practice questions

1 Use words from the list to complete the table to show the state of matter described.

gas   liquid   solid

Description	State
particles close together but not moving around	
particles very far apart	
particles close together but continuously moving around	

(3)

2 a Explain the difference between an element and a compound. (2)

b Why is ammonia classified as a compound but air is classified as a mixture? (2)

c Name the change of state in each of these processes.
  i $H_2O(g) \rightarrow H_2O(l)$ (1)
  ii $H_2O(l) \rightarrow H_2O(g)$ (1)
  iii $I_2(s) \rightarrow I_2(g)$ (1)

3 A crystal of copper(II) sulfate is placed at the bottom of a beaker of water and left for a few days.

a Describe the appearance of the liquid in the beaker after a few days. (2)

b How does the appearance of the crystal change during the experiment? (1)

c Name **two** processes that occur during the experiment. (2)

4 The diagram shows the apparatus used in a teacher demonstration.

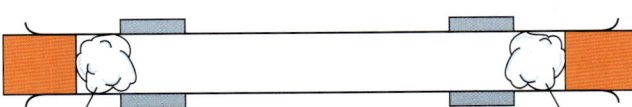

Cotton wool soaked in ammonia solution (concentrated)   Cotton wool soaked in hydrochloric acid (concentrated)

a Describe the observations you would make after a few minutes. (2)

b Use state symbols to complete the symbol equation for the reaction that occurs.

$NH_3(...) + HCl(...) \rightarrow NH_4Cl(...)$ (1)

c Explain how this experiment shows that ammonia molecules have a smaller mass than hydrogen chloride molecules. (1)

5 The atoms in a sample of neon can be represented as $^{20}_{10}Ne$ and $^{22}_{10}Ne$.

These atoms are isotopes.

a Explain the term 'isotopes'. (2)

b Explain why isotopes have identical chemical properties. (1)

c Copy these statements and use numbers to complete them.
  i   The number of protons in an atom of $^{20}_{10}Ne$ is .... (1)
  ii The number of electrons in an atom of $^{22}_{10}Ne$ is .... (1)
  iii The number of neutrons in an atom of $^{20}_{10}Ne$ is .... (1)
  iv The mass number of $^{22}_{10}Ne$ is .... (1)

d The relative atomic mass of neon is 20.2.

Which of the two isotopes is more abundant? (1)

6 Electrons in atoms are arranged in energy levels (or shells) at different distances from the nucleus. Copy these statements and use numbers to complete them.

a The maximum number of electrons in the first shell is .... (1)

b The maximum number of electrons in the second shell is .... (1)

c Atoms of sodium have electrons in .... shells. (1)

d The number of electrons in the third shell of an atom of phosphorus is .... (1)

e The atomic number of an atom with the electronic structure 2,8,6 is .... (1)

2.1 Atoms into ions

Learning objectives

After this topic, you should know:

- what a chemical compound is
- how elements form compounds
- how atoms can form either positive or negative ions
- how the elements in Group 1 bond with the elements in Group 7.

You already know that it is possible to mix two substances together without either of them changing. For example, you can mix sand and salt together and then separate them again. No change will have taken place. But in chemical reactions the situation is very different.

When the atoms of two or more elements react they make a compound.

A compound contains two or more elements which are chemically combined.

The compound formed is different from the elements and you cannot get the elements back again easily. You can also react compounds together to form other compounds. However, the reaction of elements is easier to understand as a starting point.

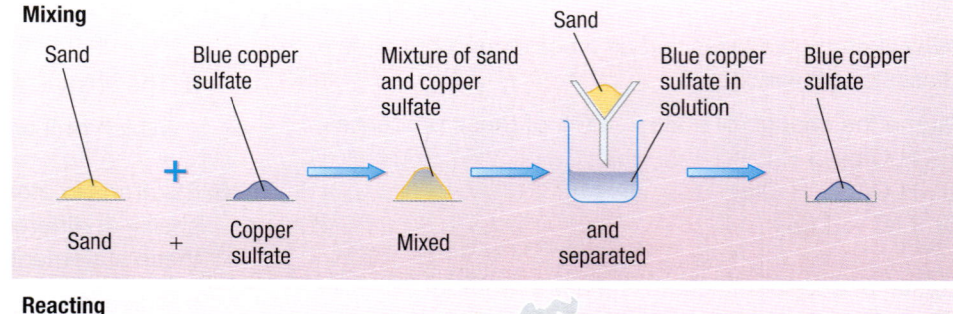
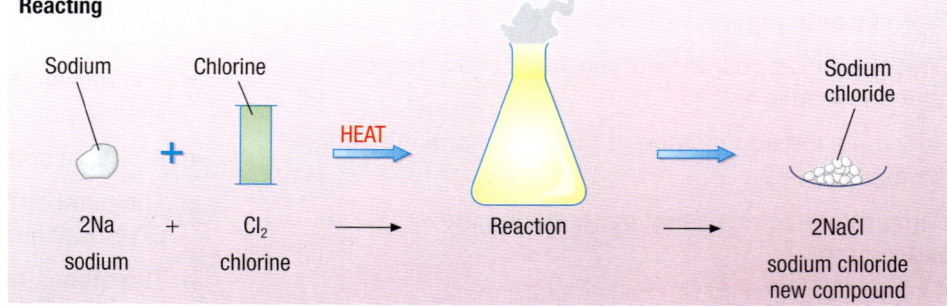

Figure 1 *The difference between mixing and reacting. Separating mixtures is usually quite easy, but separating the elements from compounds once they have reacted can be difficult.*

The atoms of the noble gases, in Group 0 of the periodic table, have an arrangement of electrons that make them stable and unreactive. However, most atoms do not have this electronic structure. When atoms react they take part in changes which give them a stable arrangement of electrons. They may do this either:

- by sharing electrons, which is called **covalent bonding**, or
- by transferring electrons, which is called **ionic bonding**.

Losing electrons to form positive ions

In ionic bonding the atoms involved lose or gain electrons to form charged particles called **ions**. The ions have the electronic structure of a noble gas. So, for example, if sodium (2,8,1), from Group 1 in the periodic table, loses one electron, it is left with the stable electronic structure of neon (2,8).

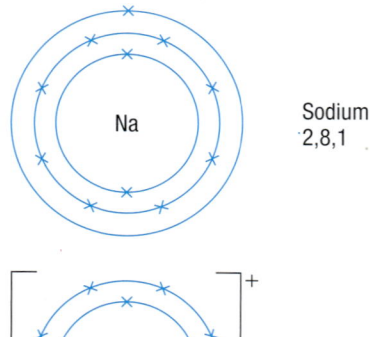
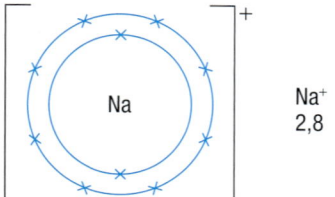

Figure 2 *A positive sodium ion (Na^+) is formed when a sodium atom loses an electron during ionic bonding*

However, it is also left with one more proton in its nucleus than there are electrons around the nucleus. The proton has a positive charge so the sodium atom has now become a positively charged ion. The sodium ion has a single positive charge. You write the formula of a sodium ion as Na^+. The electronic structure of the Na^+ ion is 2,8. This is shown in Figure 2.

Gaining electrons to form negative ions

When non-metals react with metals, the non-metal atoms gain electrons to achieve a stable noble gas structure. Chlorine, for example, has the electronic structure 2,8,7. It is in Group 7 of the periodic table. By gaining a single electron, it gets the stable electronic structure of argon (2,8,8).

In this case, there is now one more electron than there are positive protons in the nucleus. So the chlorine atom becomes a negatively charged ion. This carries a single negative charge. You write the formula of the chloride ion as Cl^-. Its electronic structure is 2,8,8. This is shown in Figure 3.

Representing ionic bonding

Metal atoms, which tend to lose electrons, react with non-metal atoms, which tend to gain electrons. So when sodium reacts with chlorine, each sodium atom loses an electron and each chlorine atom gains an electron. They both form stable ions. The electrostatic attraction between the oppositely charged Na^+ ions and Cl^- ions is called ionic bonding.

You can show what happens in a diagram. The electrons of one atom are represented by dots, and the electrons of the other atom are represented by crosses. This is shown in Figure 4.

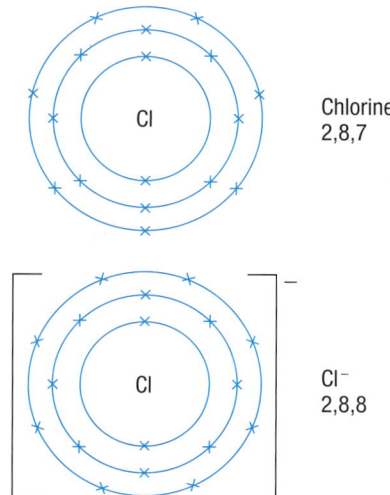

Figure 3 *A negative chloride ion (Cl^-) is formed when a chlorine atom gains an electron during ionic bonding*

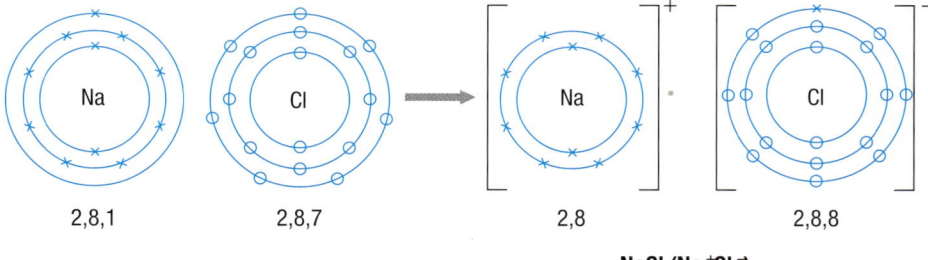

NaCl (Na^+Cl^-)

Figure 4 *The formation of sodium chloride (NaCl) – an example of ion formation by transferring an electron*

Summary questions

1 **a** When atoms bond together by *sharing* electrons, what type of bond is formed?
   b When ions bond together as a result of *gaining* or *losing* electrons, what type of bond is this?

2 Explain how and why atoms of Group 1 and Group 7 elements react with each other, in terms of their electronic structures.

3 Write electron structures to show the ions that would be formed when the following atoms are involved in ionic bonding. For each one, state how many electrons have been lost or gained and show the charge on the ions formed.
   a aluminium (Al)   **b** fluorine (F)
   c potassium (K)   **d** oxygen (O)

Key points

- Elements react together to form compounds by gaining or losing electrons or by sharing electrons.

- The elements in Group 1 react with the elements in Group 7. As they react, atoms of Group 1 elements can each lose one electron to gain the stable electronic structure of a noble gas. This electron can be given to an atom from Group 7, which then also achieves the stable electronic structure of a noble gas.

2.2 Ionic bonding

Learning objectives

After this topic, you should know:

- how ionic compounds are held together

- which elements, as well as those in Groups 1 and 7, form ions

- how the charges on ions are related to group numbers in the periodic table.

You have seen how positive and negative ions form during some reactions. Ionic compounds are usually formed when metals react with non-metals. It is the metals that form positive ions and the non-metals that form negative ions.

The ions formed are held next to each other by very strong forces of attraction between the oppositely charged ions. This electrostatic force of attraction, which acts in all directions, is called ionic bonding.

The **ionic bonds** between the charged particles result in an arrangement of ions that is known as a **giant structure** (or a **giant lattice**). If you could stand among the ions they would seem to go on in all directions forever.

Ions and the periodic table

You have seen how atoms in Group 1 of the periodic table have 1 electron in their outermost shell (or highest energy level) and form 1+ ions. Group 7 atoms have 7 electrons in their outermost shell and form 1− ions. The group number gives us the number of electrons in the outermost shell. So how does the group number relate to the charges on the ions formed from atoms?

Sometimes the atoms reacting need to gain or lose two electrons to gain the stable electronic structure of a noble gas. An example is when magnesium (2,8,2), from Group 2, reacts with oxygen (2,6), from Group 6. When these two elements react they form magnesium oxide (MgO). This is made up of magnesium ions with a double positive charge (Mg^{2+}) and oxide ions with a double negative charge (O^{2-}).

So you can say that when atoms form **ionic** bonds, atoms from

- Group 1 form 1+ ions
- Group 2 form 2+ ions
- Group 3 form 3+ ions, when they form ions as opposed to sharing electrons

- Group 4 do not form ions (apart from tin and lead at the bottom of the group)

- Group 5 form 3− ions, when they form ions as opposed to sharing electrons
- Group 6 form 2− ions, when they form ions as opposed to sharing electrons
- Group 7 form 1− ions, when they form ions as opposed to sharing electrons

- Group 0 never form ions in compounds.

Figure 1 shows how the electrons are transferred between a magnesium atom and an oxygen atom.

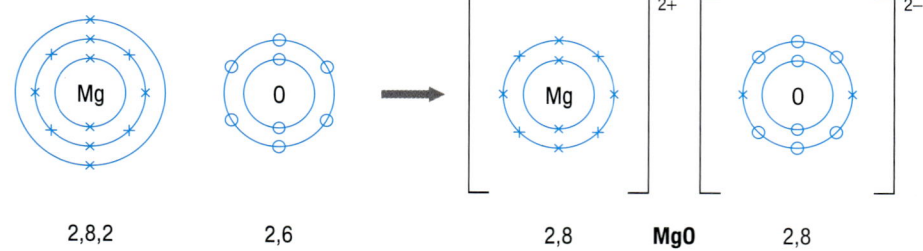

2,8,2   2,6   2,8   **MgO**   2,8

Figure 1 *When magnesium oxide (MgO) is formed, the reacting magnesium atoms lose two electrons and the oxygen atoms gain two electrons*

Another example of an ionic compound is calcium chloride. Each calcium atom (2,8,8,2) needs to lose two electrons but each chlorine atom (2,8,7) needs to gain only one electron. This means that two chlorine atoms react with every one calcium atom to form calcium chloride. So the formula of calcium chloride is $CaCl_2$.

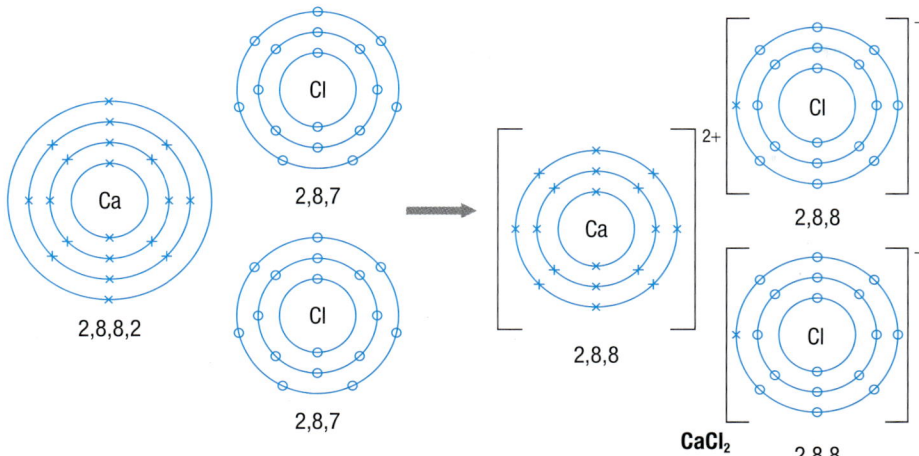

Figure 2 *The formation of calcium chloride ($CaCl_2$)*

Notice that the charges in the formula of an ionic compound cancel each other out as the overall charge on the compound is zero. So aluminium oxide, made up of aluminium ions, Al^{3+}, and oxide ions, O^{2-}, has the formula Al_2O_3 (total charges of 6+ on the aluminium ions cancelling out 6– on the oxide ions).

?? Did you know … ?

The structure of ionic lattices is investigated by passing X-rays through them.

Summary questions

1 a Copy and complete the table:

Atomic number	Atom	Electronic structure of atom	Ion	Electronic structure of ion
9	F			2,8
3		2,1	Li$^+$	
16	S		S^{2-}	
20		2,8,8,2		

2 Why do metal atoms form positively charged ions and non-metal atoms form negatively charged ions?

3 Think of rules to follow to remember the charge on any ions formed by elements in:
   a Groups 1, 2, and 3
   b Groups 5, 6, and 7.

4 a Explain why potassium bromide is KBr but potassium oxide is K_2O.
   b Explain why magnesium oxide is MgO but magnesium chloride is $MgCl_2$.

5 Draw diagrams to show how you would expect the following elements to form ions together:
   a potassium and oxygen
   b aluminium and fluorine

Key points

■ Ionic compounds are held together by strong forces of attraction between their oppositely charged ions. This is called ionic bonding.

■ Besides the elements in Groups 1 and 7, other elements that can form ionic compounds include those from Groups 2 and 6.

2.3 Covalent bonding

Reactions between metals and non-metals usually result in compounds with ionic bonding. However, many more compounds are formed in a very different way. When non-metals react together their atoms share pairs of electrons to form molecules. This is called covalent bonding.

Figure 1 *Most of the molecules in substances which make up living things are held together by covalent bonds between non-metal atoms*

Simple molecules

The atoms of non-metals generally need to gain electrons to achieve stable electron structures. When they react together neither atom can give away electrons. So they get the electronic structure of a noble gas by sharing electrons. The atoms in the molecules are then held together by the shared pairs of electrons. These strong bonds between the atoms are called **covalent bonds**.

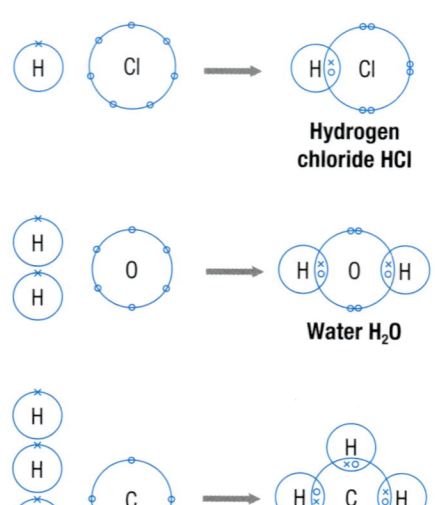

Hydrogen chloride HCl

Water H_2O

Methane CH_4

Figure 3 *The principles of covalent bonding remain the same however many atoms are involved*

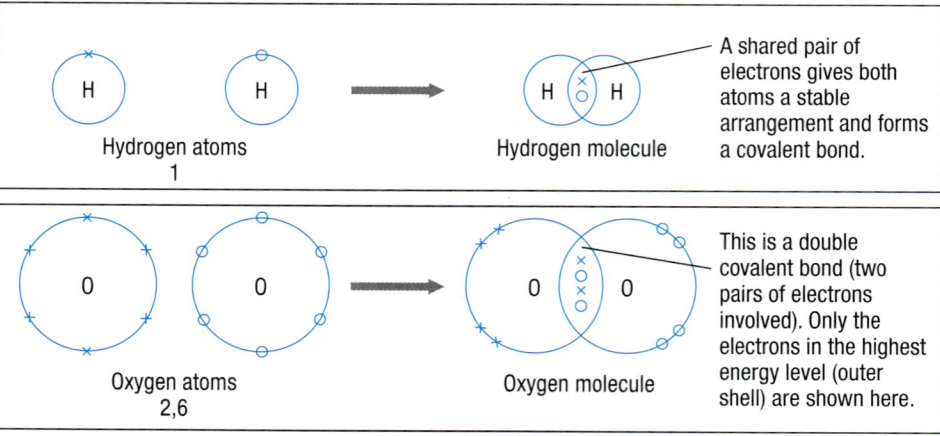

A shared pair of electrons gives both atoms a stable arrangement and forms a covalent bond.

Hydrogen atoms
1

Hydrogen molecule

This is a double covalent bond (two pairs of electrons involved). Only the electrons in the highest energy level (outer shell) are shown here.

Oxygen atoms
2,6

Oxygen molecule

Figure 2 *Atoms of hydrogen and oxygen join together to form stable molecules. The atoms in H_2 and O_2 molecules are held together by strong covalent bonds.*

Sometimes in covalent bonding each atom has the same number of electrons to share. But this is not always the case. Sometimes the atoms of one element will need several electrons, whilst those of the other element only needs one more electron for each atom to get a stable electron structure. In this case, more atoms become involved in forming the molecule, such as in water, H_2O, and methane, CH_4 (see Figure 3).

You can represent the covalent bonds in substances such as water, ammonia and methane in a number of ways. Each way represents the same thing. The method you should choose depends on what you want to show.

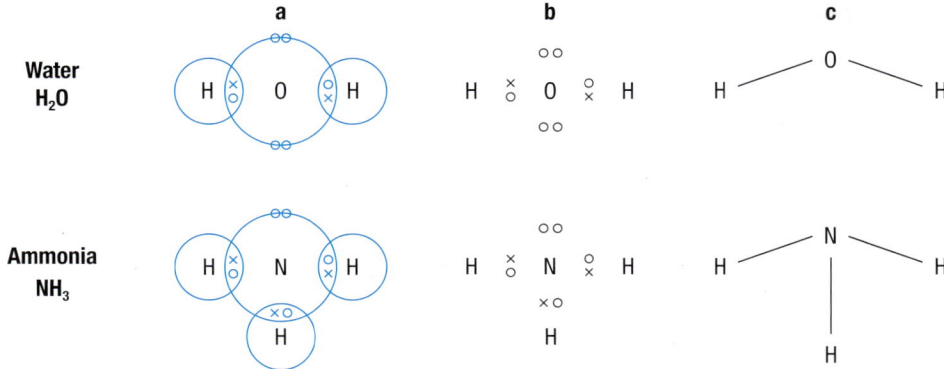

Figure 4 *You should represent a covalent compound by showing **a** the highest energy levels (or outer shells), **b** the outer electrons in a dot and cross diagram, or **c** the number of covalent bonds. Double bonds are shown by two lines, for example, O=O in an oxygen molecule*

Giant covalent structures

Many substances containing covalent bonds consist of small molecules, for example, H_2O. However, some covalently-bonded substances are very different. They have giant structures where huge numbers of atoms are held together by a network of covalent bonds. These are sometimes referred to as macromolecules.

Diamond has a giant covalent structure. In diamond, each carbon atom forms four covalent bonds with its neighbours. This results in a rigid giant covalent lattice.

Figure 5 *Diamonds owe their hardness to the way the carbon atoms are arranged in a giant covalent structure*

links

For more information about diamond, see 2.7 'Giant covalent structures'.

Summary questions

1 Which of these compounds will contain covalent bonds? How did you decide?

   hydrogen iodide, iron(II) chloride, lithium oxide, sulfur dioxide, nitrogen(III) chloride, magnesium nitride

2 Draw diagrams, showing all the electrons, to represent the covalent bonding between the following atoms:
   a two hydrogen atoms
   b two chlorine atoms
   c a hydrogen atom and a fluorine atom

3 Draw dot and cross diagrams to show the covalent bonds when:
   a a phosphorus atom bonds with three hydrogen atoms
   b a carbon atom bonds with two oxygen atoms.

4 Which noble gas electron structures do the atoms in a molecule of hydrogen chloride attain?

5 A covalent bond consists of a pair of electrons which occupy a space mainly between the nuclei of the two atoms bonded together. Explain how this pair of electrons bonds the atoms to each other. (Hint: think of the electrostatic forces of attraction between oppositely charged particles.)

Key points

- Covalent bonds are formed when atoms of non-metals share pairs of electrons with each other.

- Each shared pair of electrons is a covalent bond.

- Many substances containing covalent bonds consist of simple molecules, but some have giant covalent structures.

2.4 Bonding in metals

Learning objectives

After this topic, you should know:

- how the atoms in metals are arranged
- how the atoms in metals are bonded to each other.

links

For more information on protecting iron from rusting, see Topic 5.5 'Electroplating'.

Metal crystals

The atoms in metals are built up layer upon layer in a regular pattern (see Figure 1).

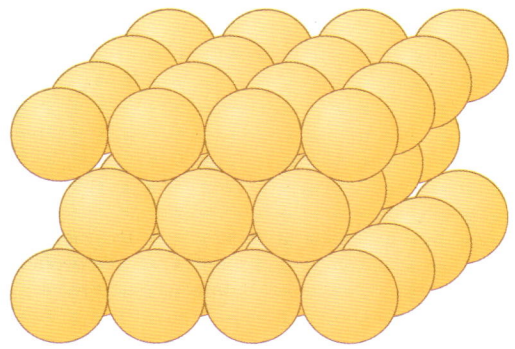

Figure 1 *The close-packed arrangement of copper atoms in copper metal*

This means that they form crystals, although these are not always obvious to the naked eye. However, sometimes you can see them. You can see zinc crystals on the surface of some steel that has been dipped into molten zinc to prevent it from rusting. This is called galvanised steel. For example, look at the surface of galvanised lamp posts and commercial wheelie bins (see Figure 3).

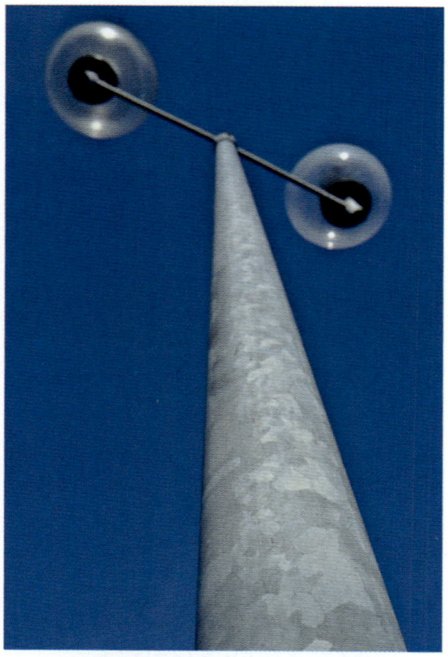

Practical

Growing silver crystals

You can grow crystals of silver metal by suspending a length of copper wire in silver nitrate solution. The crystals of silver will appear on the wire quite quickly. However, for the best results the experimental set-up needs to be left for several hours.

- Explain your observations.

Safety: Wear eye protection.

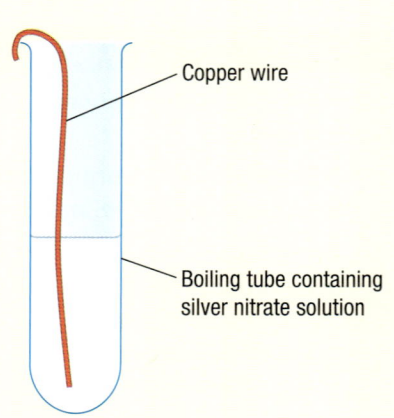

Copper wire

Boiling tube containing silver nitrate solution

Figure 2 Growing silver crystals

Figure 3 *Metal crystals, such as the zinc ones shown on this galvanised lamp post, give us evidence that metals are made up of atoms arranged in regular patterns*

Practical

Survey of metallic crystals

Take a look around your school or college to see if you can find any galvanised steel. See if you can spot the metal crystals. You can also look for crystals on brass fittings that have been left outside and not polished.

Metallic bonding

Metals are another example of giant structures. You can think of a metal as a lattice of positively charged ions. The metal ions are arranged in regular layers, one on top of another.

The outer electrons from each metal atom can easily move throughout the giant structure. The outer electrons (in the highest occupied energy level) form a 'sea' of free-moving electrons surrounding the positively charged metal ions. Strong electrostatic attraction between the negatively charged electrons and positively charged ions bonds the metal ions to each other.

The electrons in the 'sea' of free-moving electrons are called **delocalised electrons**. They are no longer linked with any particular metal ion in the giant metallic structure. These delocalised electrons help us explain the properties of metals. (See 2.8 'Giant metallic structures'.)

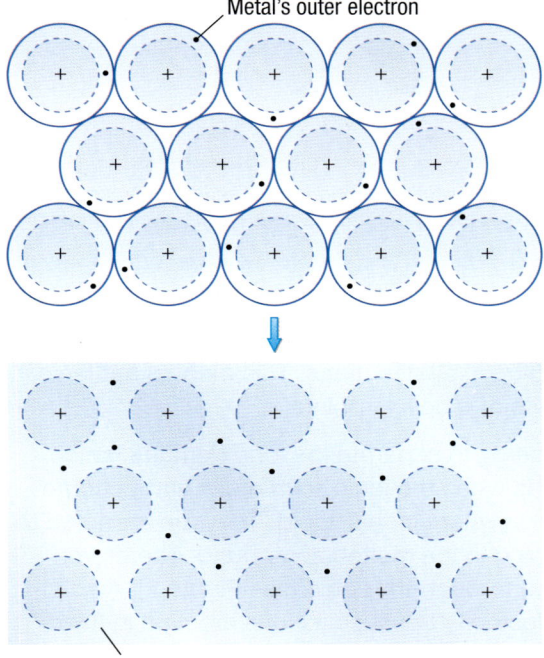

Metal's outer electron

The 'sea' of delocalised electrons

Figure 4 *A metal consists of positively charged metal ions surrounded by a 'sea' of delocalised electrons. This diagram shows us a model of metallic bonding.*

⚭ links

For more information about the properties of metals, see 2.8 'Giant metallic structures'.

Summary questions

1 What can you deduce about the arrangement of the particles in a metal from the fact that metals form crystals?

2 **a** Why are the particles that make up a metal described as positively charged ions?
b What are delocalised electrons?

3 Use the theory of metallic bonding to explain the bonding in magnesium metal. Make sure you mention delocalised electrons. (The atomic number of magnesium is 12.)

4 Using a model to explain metallic bonding, delocalised electrons could be thought of as a 'glue'.
Explain why thinking of delocalised electrons in a metal as a glue is a useful model but one which has a major drawback.

Key points

- The atoms in metals are closely packed together and arranged in regular layers.

- You can think of metallic bonding as positively charged metal ions which are held together by electrons from the outermost shell of each metal atom. These delocalised electrons are free to move throughout the giant metallic lattice.

2.5

Giant ionic structures

Learning objectives

After this topic, you should know:

■ why ionic compounds have high melting points

■ why ionic compounds conduct electricity when they are melted or dissolved in water.

In Topic 2.2 *Ionic bonding*, you have already seen that an ionic compound consists of a giant structure of ions arranged in a lattice. The attractive electrostatic forces between the oppositely charged ions act in all directions and are very strong. This holds the ions in the lattice together very tightly. Look at Figure 1.

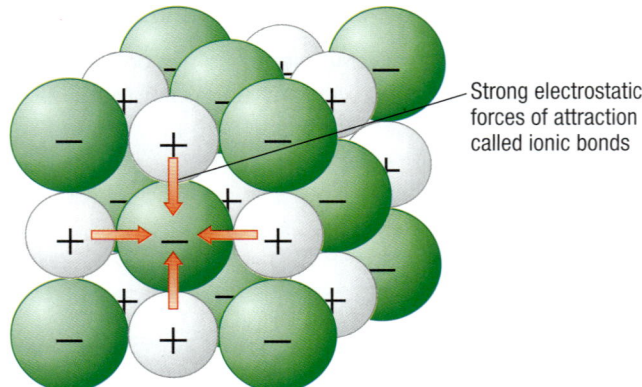

Strong electrostatic forces of attraction called ionic bonds

Figure 1 *The attractive forces between the oppositely charged ions in an ionic compound are very strong. The regular arrangement of ions in the giant lattice enables ionic compounds to form crystals.*

links

For more information on the movement and arrangement of particles in solids and liquids, look back to 1.1 'States of matter'.

Study tip

Remember that every ionic compound has a giant structure. The oppositely charged ions in these structures are held together by strong electrostatic forces of attraction. These act in all directions.

It takes a lot of energy to break up a giant ionic lattice. There are lots of strong ionic bonds to break. To separate the ions you have to overcome all those electrostatic forces of attraction acting in all directions. This means that ionic compounds have high melting points and boiling points.

Once you have supplied enough energy to separate the ions from the lattice, they are free to move around. That is when the ionic solid melts and becomes a liquid. The ions are free to move anywhere in this liquid. They are attracted to oppositely charged electrodes dipping in the molten compound. Therefore, they can carry their electrical charge through the liquid (see Figure 2). A solid ionic compound cannot conduct electricity because its ions are held in fixed positions in the lattice. They cannot move around. They can only vibrate 'on the spot' when in the solid state.

Bulb lights as current flows

Moving ions carry the electrical charge through the molten potassium chloride

Cl⁻ K⁺ Cl⁻ Cl⁻ K⁺
K⁺ K⁺ K⁺
K⁺ Cl⁻ K⁺ Cl⁻ Cl⁻ Cl⁻

Molten potassium chloride

Figure 2 *Because the ions are free to move, a molten ionic compound can conduct electricity*

Many ionic compounds will dissolve in water. When you dissolve an ionic compound in water, the lattice is split up by the water molecules. Then the ions are free to move around in the solution formed. They can carry their charge to oppositely charged electrodes in the solution. Just as molten ionic compounds will conduct electricity, solutions of ionic compounds will also conduct electricity. The ions are able to move to an oppositely charged electrode dipped in the solution.

Practical

Testing conductivity

Using a circuit as shown in Figure 2, dip a pair of electrodes into a 1 cm depth of sodium chloride crystals. What happens?

Now slowly add water.
- What happens to the bulb?

Repeat the experiment using potassium chloride.
- Explain your observations.

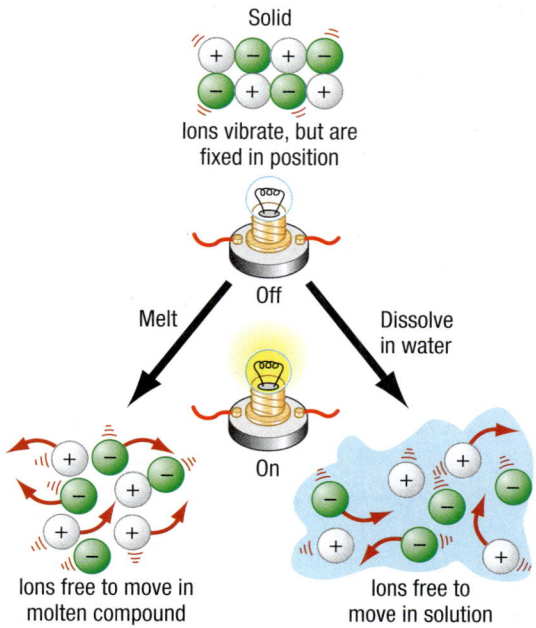

Figure 3 *Ionic compounds do not conduct electricity in the solid state but do when molten or when dissolved in water*

Ionic solid	Molten ionic compound	Ionic compound in solution
Ions are fixed in a giant lattice. They vibrate but cannot move around – it does not conduct electricity.	High temperature provides enough energy to overcome the many strong attractive forces between ions. Ions are free to move around within the molten compound – it does conduct electricity.	Water molecules separate ions from the lattice. Ions are free to move around within the solution – it does conduct electricity.

Summary questions

1 Explain why ionic compounds have high melting points.

2 Explain why ionic compounds conduct electricity only when they are molten or dissolved in water.

3 Why is seawater a better conductor of electricity than water from a freshwater lake?

4 Which of these ions will move to the positive electrode and which will move to the negative electrode in a circuit as shown in Figure 2?

   lithium ions, chloride ions, bromide ions, calcium ions, sodium ions, zinc ions, oxide ions, iodide ions, barium ions

5 Predict which of these two ionic compounds has the higher melting point – sodium oxide or aluminium oxide. Explain your reasoning.

Key points

- It takes a lot of energy to break the many strong ionic bonds that hold a giant ionic lattice together. So, ionic compounds have high melting points. They are all solids at room temperature.

- Ionic compounds will conduct electricity when they are melted or dissolved in water. That is because their ions can then move freely around and can carry charge through the liquid.

2.6 Simple molecules

Learning objectives

After this topic, you should know:

- why substances made of simple molecules have low melting points and boiling points

- why these substances do not conduct electricity.

Many substances made up of simple covalently-bonded molecules have low melting points and boiling points. Look at the graph in Figure 1.

This means that many substances made up of simple molecules are liquids or gases at room temperature. Others are solids with quite low melting points, such as iodine (I_2) and sulfur (S_8).

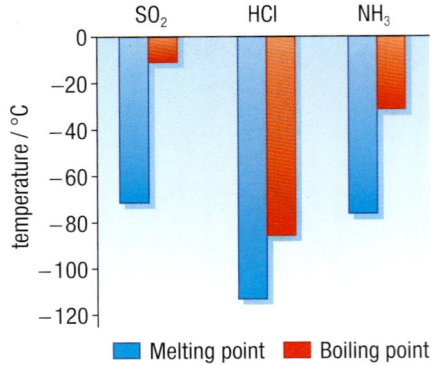

Figure 1 *Substances made of simple molecules usually have low melting points and boiling points*

Intermolecular forces

Covalent bonds are very strong so the atoms within each molecule are held very tightly together. However, each molecule tends to be quite separate from its neighbouring molecules. The force of attraction between the individual molecules in a covalent substance is relatively small. Scientists say that there are weak **intermolecular forces** between molecules. Overcoming these forces does not take much energy.

Look at the molecules in a sample of chlorine gas (Figure 2).

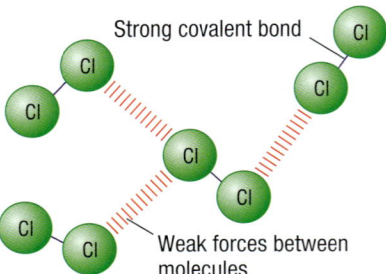

Figure 2 *Covalent bonds and the weak forces between molecules in chlorine gas. It is the weak intermolecular forces that are overcome when substances made of simple molecules melt or boil. The covalent bonds are not broken.*

You have seen that ionic compounds will conduct electricity when they are liquids. But although a substance that is made up of simple molecules may be a liquid at room temperature, it will not conduct electricity. Look at the demonstration below.

Study tip

Although the covalent bonds in molecules are strong, the forces between molecules are weak.

Demonstration

Conductivity

■ What happens?

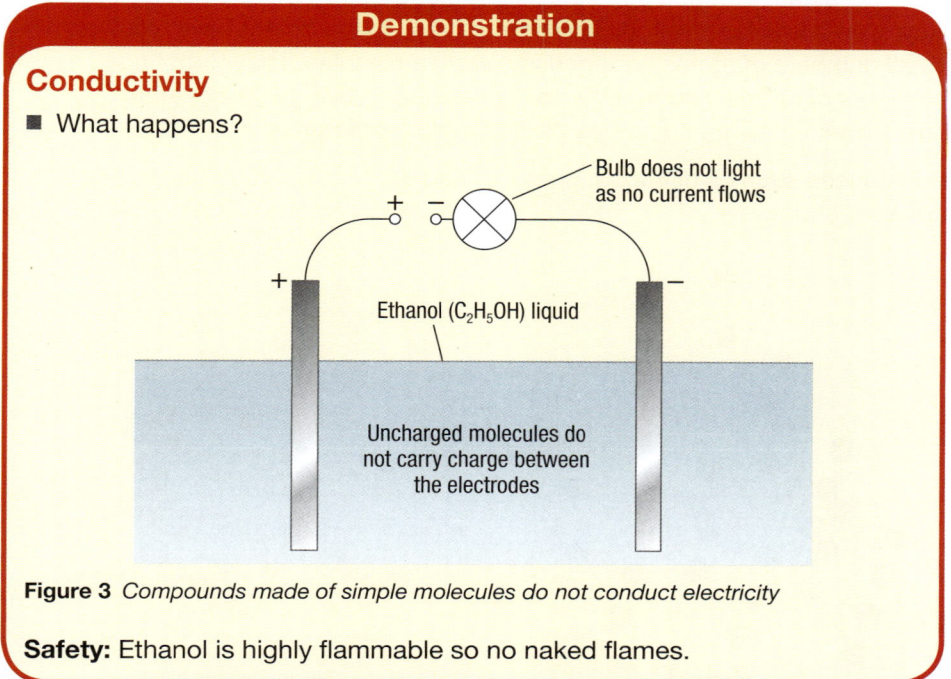

Bulb does not light as no current flows

Ethanol (C_2H_5OH) liquid

Uncharged molecules do not carry charge between the electrodes

Figure 3 *Compounds made of simple molecules do not conduct electricity*

Safety: Ethanol is highly flammable so no naked flames.

There is no overall charge on the simple molecules in a compound like ethanol. So their neutral molecules cannot carry electrical charge. This makes it impossible for substances made up of simple molecules to conduct electricity.

⬭⬭ **links**

For information on ionic compounds conducting electricity, look back at 2.5 'Giant ionic structures'.

Summary questions

1 Describe what is meant by the term 'intermolecular forces'?

2 Which of the following substances will be held together by weak intermolecular forces?
   a aluminium oxide, aluminium, oxygen
   b magnesium chloride, magnesium, chlorine
   c hydrogen sulfide, hydrogen, sulfur

3 a Diamond has a very high melting point. Explain why.
   b Nitrogen gas has a very strong triple covalent bond holding the nitrogen atoms together in diatomic molecules (N_2). Explain its boiling point of $-196\,°C$.

4 A compound called sulfur hexafluoride (SF_6) is used to stop sparks forming inside electrical switches designed to control large currents. Explain why the properties of this compound make it particularly useful in electrical switches.

5 Explain why the melting point of hydrogen chloride is $-115\,°C$ but the melting point of sodium chloride is $801\,°C$.

Key points

■ Substances made up of simple molecules have low melting points and boiling points.

■ The forces between simple molecules are weak. These weak intermolecular forces explain why substances made of simple molecules have low melting points and boiling points.

■ Simple molecules have no overall charge, so they cannot carry electrical charge. Therefore, substances made of simple molecules do not conduct electricity.

2.7 Giant covalent structures

Learning objectives

After this topic, you should know:

- the general properties of substances with giant covalent structures

- why diamond is hard and graphite is slippery

- why graphite can conduct electricity

- what fullerenes are made of and some of their uses.

Figure 2 *Hard, shiny, and transparent – diamonds make beautiful jewellery*

Study tip

Giant covalent structures are held together by covalent bonds throughout the lattice.

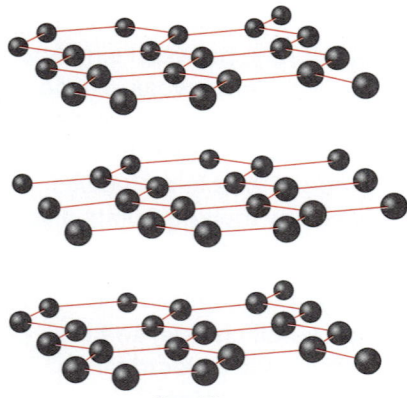

Graphite

Figure 3 *The giant structure of graphite. There are strong covalent bonds **within** its layers but only weak intermolecular forces **between** the layers.*

Most covalently-bonded substances are made up of individual molecules. However, a few form very different structures. These do not have relatively small numbers of atoms arranged in simple molecules. They form huge networks of atoms held together by covalent bonds in **giant covalent structures**. They are sometimes called **macromolecules**.

Substances such as diamond, graphite, and silicon dioxide (silica) have giant covalent structures.

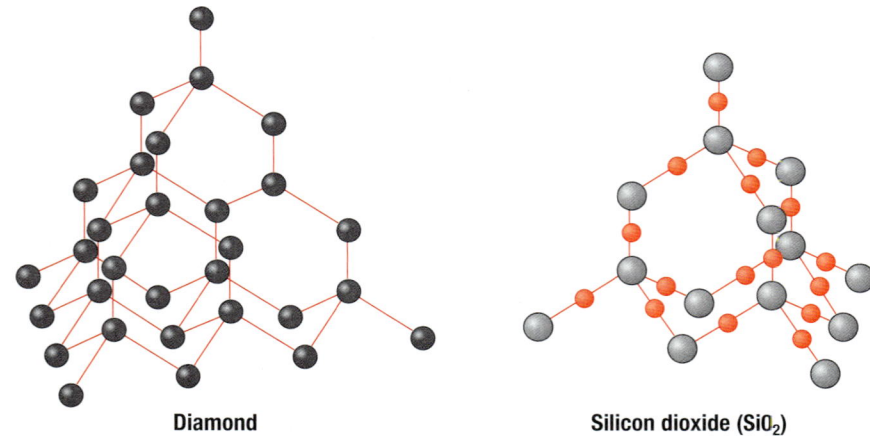

Diamond   Silicon dioxide (SiO$_2$)

Figure 1 *The structures of diamond and silicon dioxide (silica) continue on in all directions*

All the atoms in these giant lattices are held in position by strong covalent bonds, as in diamond and silicon dioxide shown above. This gives them some very special properties. They have very high melting points and boiling points and are insoluble in water. Apart from graphite, they are hard and do not conduct electricity. Diamond is exceptionally hard and boils at 4827 °C. All its carbon atoms each form four strong covalent bonds, arranged in a perfectly symmetrical giant lattice.

Carbon isn't only found in the form of diamonds. Another form is graphite (well known for its use in pencil 'lead'). In graphite, carbon atoms are only bonded to three other carbon atoms. They form hexagons which are arranged in giant layers. There are no covalent bonds between the layers, only weak intermolecular forces. So the layers can slide over each other easily. It is a bit like playing cards sliding off a pack of cards. This makes graphite a soft material that feels slippery.

  Did you know …?

Diamond is the hardest known natural substance. Artificial diamonds can be made by heating pure carbon to very high temperatures under enormous pressures. 'Industrial diamonds' made like this are embedded in the drill bits used by oil companies when drilling for oil.

Bonding in graphite

The carbon atoms in graphite's layers are arranged in hexagons so each carbon atom forms three strong covalent bonds (see Figure 3). Carbon atoms have four electrons in their outer shell available for bonding. This leaves one spare outer electron on each carbon atom.

This electron is free to move along the layers of carbon atoms. The free electrons found in graphite are called **delocalised electrons**. They behave rather like the electrons in a metallic structure.

These free electrons allow graphite to conduct electricity and heat along its layers of carbon atoms. Diamond – and most other covalently-bonded substances – cannot conduct electricity. This is because their atoms have no free electrons as all the outer shell electrons are involved in covalent bonding.

Fullerenes

Apart from diamond and graphite, there are other structures that carbon atoms can form. In these structures the carbon atoms join together to make large cages which can have all sorts of shapes. Chemists have made shapes looking like balls, onions, tubes, doughnuts, corkscrews, and cones! They are based on *hexagonal rings of carbon atoms*.

Chemists discovered carbon's ability to behave like this in 1985. The large carbon molecules containing these cage-like structures are called **fullerenes**. Scientists can now place other molecules inside these carbon cages. This has exciting possibilities, including the delivery of drugs to specific parts of the body. They are sure to become very important in nanoscience applications, for example, as catalysts and lubricants.

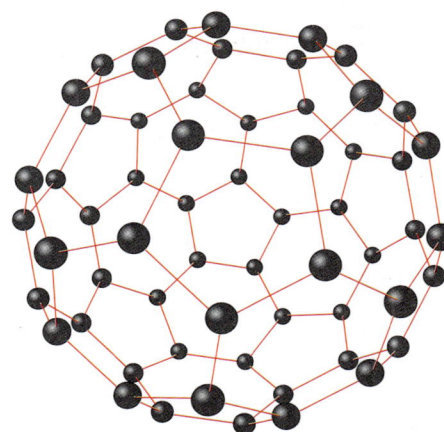

Figure 4 *The first fullerene to be discovered contained 60 carbon atoms, but chemists can now make giant fullerenes which contain many thousands of carbon atoms*

links

For information about delocalised electrons, look at 2.4 'Bonding in metals'.

links

For more information on nanoscience, see 2.9 'Nanoscience'.

Key points

- Some covalently-bonded substances have giant structures. These substances have very high melting points and boiling points.

- Graphite contains giant layers of covalently-bonded carbon atoms. However, there are no covalent bonds between the layers, just weak intermolecular forces. This means they can slide over each other, making graphite soft and slippery. The carbon atoms in diamond have a rigid giant covalent structure, making it a very hard substance.

- Graphite can conduct electricity and heat because of the delocalised electrons along its layers.

- As well as diamond and graphite, carbon also exists as fullerenes which can form large cage-like structures and tubes, based on hexagonal rings of carbon atoms.

Summary questions

1 a Name three forms of the element carbon.
  b Find out what chemists call different forms of the same element in the same state.
2 List the general properties of a substance with a typical giant covalent structure.
3 a State a possible medical application of fullerene cages.
  b Find out why chemists called a molecule of the new form of carbon discovered in 1985 a 'buckyball', and why related molecules became known as fullerenes.
4 Graphite is sometimes used to reduce the friction between two surfaces that are rubbing together. Explain how it does this.
5 Explain in detail why graphite can conduct electricity but diamond cannot.

2.8 Giant metallic structures

Learning objectives

After this topic, you should know:

- why metals can be bent and shaped without breaking
- why alloys are harder than pure metals
- why metals conduct electricity and heat
- what 'shape memory alloys' are.

Metals can be hammered and bent into different shapes, and drawn out into wires. This is because the layers of atoms in a pure metal are able to slide easily over each other.

The atoms in a pure metal, such as iron, are held together in a giant metallic structure. The atoms are arranged in closely-packed layers. This regular arrangement allows the atoms to slide over one another quite easily. This is why pure iron is relatively soft and easily bent and shaped.

Alloys are usually mixtures of two or more elements, at least one of which is a metal. However, most steels contain iron with controlled amounts of carbon, a non-metal, mixed in its structure. The carbon atoms are a different size to the iron atoms. This makes it more difficult for the layers in the metal's giant structure to slide over each other. So, alloys are harder than the pure metals used to make them. This is shown in Figure 2.

Figure 1 *Drawing copper out into wires depends on being able to make the layers of metal atoms slide easily over each other, without breaking the metal*

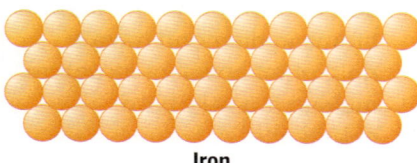

Iron

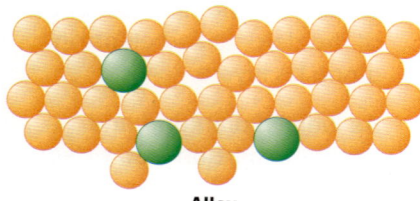

Alloy

Figure 2 *The atoms in pure iron are arranged in layers which can easily slide over each other. In alloys the layers cannot slide so easily because atoms of other elements distort the layers.*

Practical

Making models of metals

Tube connected to gas tap

Fine-pointed tube

Plastic container with soap solution

Figure 3 *Making a bubble raft to model the structure of a metal*

A regular arrangement of bubble 'atoms'

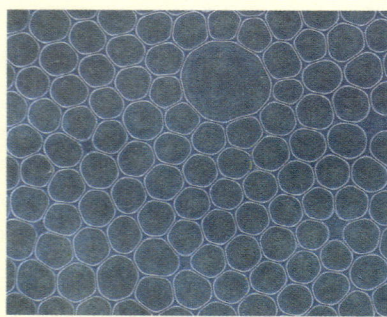

A larger bubble 'atom' disrupts the regular arrangement around it

You can make a model of the structure of a metal by blowing small bubbles of gas on the surface of soap solution to represent atoms.

- Why are models useful in science?

- Metal cooking utensils are used all over the world, because metals are good conductors of heat.
- Wherever electricity is generated, it passes through metal wires (usually made of copper) to get to where it is needed because metals are also good conductors of electricity.

Explaining the properties of metals

The positive ions in a metal's giant structure are held together by a sea of delocalised electrons. These electrons are a bit like 'glue'. Their negative charge between the positively charged ions holds the metal ions in position.

However, unlike glue, the electrons are able to move throughout the whole giant lattice. Because they can move around and hold the metal ions together at the same time, the delocalised electrons enable the lattice to distort. When struck, the metal atoms can slip past one another without breaking up the metal's structure.

Metals are good conductors of heat and electricity because the delocalised electrons can readily flow through the giant metallic lattice. The heat energy and electrical current are transferred quickly through the metal by the free-moving electrons.

⃝⃝ links

For more about the bonding in metals, look back at 2.4 'Bonding in metals'.

Shape memory alloys

Some alloys have a very special property. Like all metals they can be bent (or **deformed**) into different shapes. The difference comes when you heat them up and they return to their original shape automatically. These metals are called **shape memory alloys**, which describes the way they behave. They seem to 'remember' their original shape!

Shape memory alloys are used in many ways, for example, in health-care. Doctors treating a badly broken bone can use alloys to hold the bones in place whilst they heal. They cool the alloy before it is wrapped around the broken bone. When it heats up again the alloy goes back to its original shape. This pulls the bones together and holds them in place whilst they heal.

Dentists can also make braces to pull teeth into the right position using this technique (see Figure 5).

Figure 4 *Metals are essential in our lives – the delocalised electrons mean that they are good conductors of both heat and electricity*

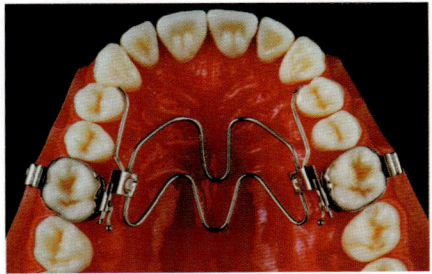

Figure 5 *This dental brace pulls the teeth into the right position as it warms up. It is made of a shape memory alloy called nitinol. It is an alloy of nickel and titanium.*

Key points

- Metals can be bent and shaped because the layers of atoms (or positively charged ions) in a giant metallic structure can slide over each other.

- Alloys are harder than pure metals because the regular layers in a pure metal are distorted by differently-sized atoms in an alloy.

- Delocalised electrons in metals enable electricity and heat to pass through a metal easily.

- If a shape memory alloy is deformed, it can return to its original shape on heating.

Summary questions

1 a Why can metals be bent, shaped and pulled out into wires when forces are applied?
   b Find out the word used to describe a material, such as a metal, that can be:
   i hammered into shapes   ii drawn out into a wire.

2 a Use your knowledge of metal structures to explain why alloying a metal can make the metal harder.
   b What is a shape memory alloy?

3 Explain how a dental brace made out of nitinol is more effective than a brace made out of a traditional alloy.

4 Explain why metals are good conductors of heat and electricity in terms of their structure and bonding.

2.9 Nanoscience

Learning objectives

After this topic, you should know:

- what nanoscience is
- some possible uses and potential benefits and risks associated with nanoscience.

Did you know ... ?

A pin-head measures about 1 million nanometres, and a human hair is about 80 000 nm wide.

Figure 1 *Nanoparticles will save many people from damaged skin and cancers caused by too much UV light*

Figure 2 *Nanoparticles in cosmetic products can work deeper in the skin*

Nanoscience is a new and exciting area of science. 'Nano' is a prefix like 'milli' or 'mega'. Whilst 'milli' means 'one-thousandth', 'nano' means 'one thousand-millionth'.

$$1 \text{ nanometre (1 nm)} = 1 \times 10^{-9} \text{ metres}$$
$$(= 0.000000001 \text{ m or a billionth of a metre})$$

So nanoscience is the science of really tiny things. It deals with structures that are just a few hundred atoms in size or even smaller (between 1 and 100 nm in size).

Scientists have discovered that materials behave very differently at a very tiny scale. Nanoparticles are so tiny that they have a huge surface area for a small volume of material. When atoms and molecules are arranged on a nanoscale, their properties can be truly remarkable.

Nanoscience at work

Here are some uses of nanoscience.

- Glass can be coated with titanium oxide nanoparticles. Sunshine triggers a chemical reaction which breaks down dirt that lands on the window. When it rains the water spreads evenly over the surface of the glass, washing off the broken down dirt.
- Titanium oxide and zinc oxide nanoparticles are also used in modern sunscreens. Scientists can coat nanoparticles of the metal oxide with a coating of silica. The thickness of the silica coating can be adjusted at an atomic level. These coated nanoparticles seem more effective at blocking the Sun's rays than conventional UV absorbers.
- The cosmetics industry is one of the biggest users of this new technology. The nanoparticles in face creams are absorbed deeper into the skin. They are also used in sun-tan creams and deodorants.

The delivery of active ingredients in cosmetics can also be applied to medicines. The latest techniques being developed use nanocages of gold to deliver drugs where they need to go in the body. Researchers have found that the tiny gold particles can be injected and absorbed by tumours. Tumours have thin, leaky blood vessels with holes large enough for the gold nanoparticles to pass into. However, they cannot get into healthy blood vessels. When a laser is directed at the tumour the gold nanoparticles absorb energy and warm up. The temperature of the tumour increases enough to change the properties of its proteins but barely warms the surrounding tissue. This destroys the tumour cells without damaging healthy cells.

There is potential to use the gold nanocages to carry cancer-fighting drugs to the tumour at the same time. The carbon nanocages (called fullerenes) that you met in Topic 2.7 can also be used to deliver drugs in the body. Incredibly strong, yet light, fullerene nanotubes are already being used to reinforce materials (see Figure 3). The new materials are finding uses in sport, such as making very strong but light tennis racquets.

Silver nanoparticles are antibacterial. They are used inside fridges to inhibit the growth of microorganisms. The silver particles also act against viruses and fungi. They are put in the sprays used to clean operating theatres in hospitals.

Future developments?

Nanotubes are now being developed that can be used as nanowires. This will make it possible to construct incredibly small electronic circuits. Nanotubes can be used to make highly sensitive selective sensors. For example, nanotube sensors have been made that can detect tiny traces of a gas present in the breath of asthmatics before an attack. This will let patients monitor and treat their own condition without having to visit hospital to use expensive machines.

Nanowires would also help to make computers with vastly improved memory capacities and speeds.

Scientists in the US Army are developing nanotech suits – thin, or even spray-on, uniforms which are flexible and tough enough to withstand bullets and blasts. The uniforms would receive aerial views of the battlefield from satellites, transmitted directly to the soldier's brain. There would also be a built-in air conditioning system to keep the body temperature normal. Inside the suit there would be a full range of nanobiosensors that could send medical data back to a medical team.

Possible risks

The large surface area of nanoparticles would make them very effective as catalysts. However, their large surface area also makes them dangerous. If a spark is made by accident, they may cause a violent explosion.

If nanoparticles are used more and more there is also going to be more risk of them finding their way into the air around us. Breathing in tiny particles could damage the lungs. Nanoparticles could enter the bloodstream this way, or from their use in cosmetics, with unpredictable effects. More research needs to be done to find out their effects on health and the environment.

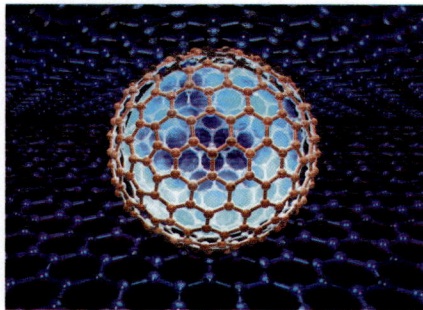

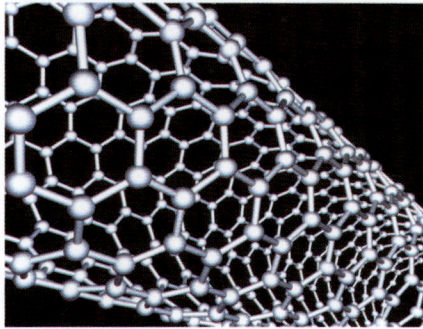

Figure 3 *Nanocages (top) can carry drugs inside them and nanotubes (below) can reinforce materials*

Summary questions

1 What is meant by 'nanoscience'?

2 Give two uses of silver nanoparticles.

3 a Give an advantage of using nanoparticles as catalysts.
  b Why are some people concerned about the use of nanoparticles as catalysts?

4 a Give two uses of nanoscience in cosmetic products for the skin.
  b How can nanoscience possibly help to fight cancers?

5 In his book *Engines of Creation,* K. Eric Drexler speculates that one day humans may invent a nanomachine that can reproduce itself. Then the world could be overrun by so-called 'grey goo'. Some people are so worried that they have called for a halt in nanoscience research. What are your views?

Key points

- Nanoscience is the study of small particles that are between 1 and 100 nanometres in size.

- Nanoparticles behave differently from the materials they are made from on a large scale.

- New developments in nanoscience are very exciting but will need more research into possible issues that might arise from their increased use.

Chapter summary questions

1 Copy each sentence and write a number to fill in the blanks in **a** to **e**:

a The elements in Group .... in the periodic table all form ions with a charge of 1+.

b The elements in Group .... in the periodic table all form ions with a charge of 2+.

c The elements in Group .... in the periodic table all form ions with a charge of 1−.

d The elements in Group .... in the periodic table all form ions with a charge of 2−.

e The elements in Group 4 in the periodic table have atoms which can form .... covalent bonds.

2 This table contains data about some different substances:

Substance	Melting point (°C)	Boiling point (°C)	Electrical conductor
ammonia	−78	−33	solid – poor liquid – poor
magnesium oxide	2852	3600	solid – poor liquid – good
lithium chloride	605	1340	solid – poor liquid – good
silicon dioxide	1610	2230	solid – poor liquid – poor
hydrogen bromide	−88	−67	solid – poor liquid – poor
graphite	3652	4827	solid – good liquid – good

a Make a table with the following headings:

Giant covalent, Giant ionic, Simple molecules

Now write the name of each substance from the table above in the correct column.

b Which substances are gases at 25 °C?

c One of these substances behaves in a surprising way. Which one and why?

d Draw a diagram to show the electron structures of a lithium atom and a chlorine atom and their ions in lithium chloride. (Atomic number of Li = 3, Cl = 17)

3 **a** Which of the following substances will have covalent bonding?

hydrogen iodide
chlorine(VII) oxide
silver chloride
phosphorus(V) fluoride
calcium bromide
silver nitrate

b Explain how you decided on your answers in part **a**.

c What type of bonding will the remaining substances in the list have?

d What is the formula of:
   i hydrogen iodide?
   ii calcium bromide?

4 Copy and complete the following table with the formula of each ionic compound formed.
(The first one is done for you.)

	chloride, Cl^-	oxide, O^{2-}	sulfate, SO_4^{2-}	phosphate(V), PO_4^{3-}
potassium, K^+	KCl			
magnesium, Mg^{2+}				
iron(III), Fe^{3+}				

5 Draw a diagram which shows the bonding in:

a hydrogen, H_2

b carbon dioxide, CO_2.

Practice questions

1 Lithium and oxygen combine to form lithium oxide, an ionic compound.

The diagrams show the electronic structures of a lithium atom and an oxygen atom.

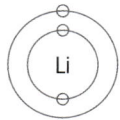 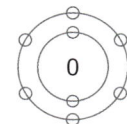

a Describe, in terms of electrons, what happens when a lithium atom combines with an oxygen atom. (3)

b The electronic structures of lithium and oxygen can be shown as follows:

Li 2,1   O 2,6

Write the charge of each of the ions formed in lithium oxide. (2)

c Explain why lithium oxide has a high melting point. (2)

d Explain why lithium oxide does not conduct electricity when solid but does when molten. (2)

2 Hydrogen and fluorine atoms combine to form the compound hydrogen fluoride.

The diagrams show the electronic structures of a hydrogen atom and a fluorine atom.

 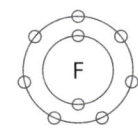

a Draw a diagram to show how the electrons are arranged in a hydrogen fluoride molecule. (2)

b What is the name of the bond in a hydrogen fluoride molecule? (1)

c The bond in a hydrogen fluoride molecule is strong.

Explain why hydrogen fluoride is a gas at room temperature. (2)

d Hydrogen bromide has the same type of bonding as hydrogen fluoride.

A bond in a molecule can be represented by a straight line.

Show the hydrogen bromide molecule in this way. (1)

3 Silicon dioxide has a giant structure.

a Explain why silicon dioxide has a very high melting point. (3)

b Explain why molten silicon dioxide does not conduct electricity. (2)

4 Fullerenes and nanoscience may become more important in the future.

a One type of fullerene exists as molecules containing sixty atoms.

State the element in this molecule and how the atoms are arranged. (2)

b Suggest **one** use of fullerenes in medicine. (1)

c What is the size of a structure called a nanoparticle? (1)

d Suggest a property of nanoparticles that makes them suitable as catalysts. (1)

5 Copper is a useful metal.

a The diagram represents the structure of copper.

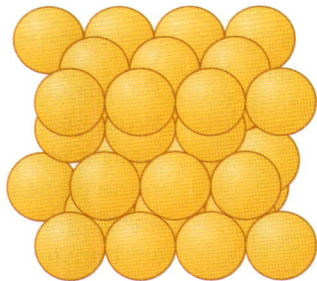

  i Use this diagram to explain why copper can be easily shaped. (2)

  ii Explain why copper is a good conductor of electricity. (2)

b Pure copper is very soft and is made harder by adding other metals.

  i What name is used for a mixture of metals? (1)

  ii Explain how the other metals make copper harder. (2)

3.1

The periodic table

Learning objectives

After this topic, you should know:

- how atomic structure is linked to the periodic table

- why the noble gases are so unreactive.

 Did you know ... ?

The alkali metal caesium reacts explosively with cold water, and will even react with ice at temperatures above −116 °C.

Mendeleev's predictions

When Mendeleev published his periodic table in 1869, not all scientists were convinced that he had it right. However, Mendeleev's brilliant idea was to leave gaps in his table for elements that had not yet been discovered. Then he used his table to predict what their properties should be, based on elements in the same group.

A few years later, new elements were discovered with properties that closely matched Mendeleev's predictions. Then there were few doubts that his table was a great breakthrough in scientific understanding.

 Did you know ... ?

Radon is a radioactive noble gas, produced in some rocks by the nuclear decay of radium. In some areas it is monitored as a health hazard as it can be released from the rocks and collect in the lower parts of buildings.

The periodic table was first developed by a Russian chemist called Dmitri Mendeleev and published in 1869. He put elements in order of their atomic masses and started new rows to produce groups of similar elements. The word 'periodic' means repeated at regular intervals.

However, not all elements fit in with the pattern. For example, argon atoms have a greater average mass than potassium atoms. Ordering by atomic mass would result in argon (a **noble gas**) being in the same group as reactive metals such as sodium and lithium, and would group potassium (an extremely reactive metal) with the unreactive noble gases. So argon must be put before potassium in the periodic table to maintain the periodic pattern, even though the average mass of its atoms is heavier than that of potassium's atoms. Look at Figure 1 to check this out.

At the start of the 20th century scientists began to find out more about the structure of the atom. Only then could they solve this problem of certain elements breaking the periodic pattern. The elements are now arranged ***in order of their atomic (proton) number***. This puts them all in exactly the right place in the periodic table. It arranges the elements so that they line up in groups (vertical columns) with similar properties.

Electronic structures and the periodic table

The periodic table also gives you an important summary of the electronic structures of all the elements. Elements in the same group of the periodic table react in similar ways because their atoms have the same number of electrons in the highest occupied energy level (outer shell). You saw in Topic 2.2 *Ionic bonding*, and Topic 2.3 *Covalent bonding*, how it is these outer electrons that are transferred (in ionic bonding) or shared (in covalent bonding) when atoms combine with each other.

In Chapter 2, you learnt that **the group number in the periodic table tells you the number of electrons in the outermost shell (highest occupied energy level) of an atom.**

For example, all the atoms of Group 2 elements have 2 electrons in their outermost shell (highest energy level) and those in Group 6 have 6 electrons in their outermost shell.

You also saw in Chapter 2 how atoms transfer or share electrons when bonding in order to attain the stable electronic structures of the noble gases. Look at the Group 0 noble gases, shaded purple in Figure 1. You can see how atoms of elements in Group 6 and 7 can gain electrons to form negative ions. These then have the electronic structure of the noble gas at the end of their row (called a period). Elements in Groups 1, 2, or 3 lose electrons and their positive ions attain the electronic structure of the noble gas at the end of the period one row above them.

The atoms of noble gases have eight electrons in their outermost shell, making the atoms very stable. The exception is the first of the noble gases, helium, which has just two electrons, but this complete first shell is also a very stable electronic structure.

The stable electronic structures of the noble gases explains why they exist as single atoms. They are monatomic gases. They have no tendency to react to form molecules. However, chemists have managed to make a few compounds of the larger noble gases. These contain the most reactive non-metallic elements, fluorine and oxygen, for instance in the compounds XeF_6 and XeO_4.

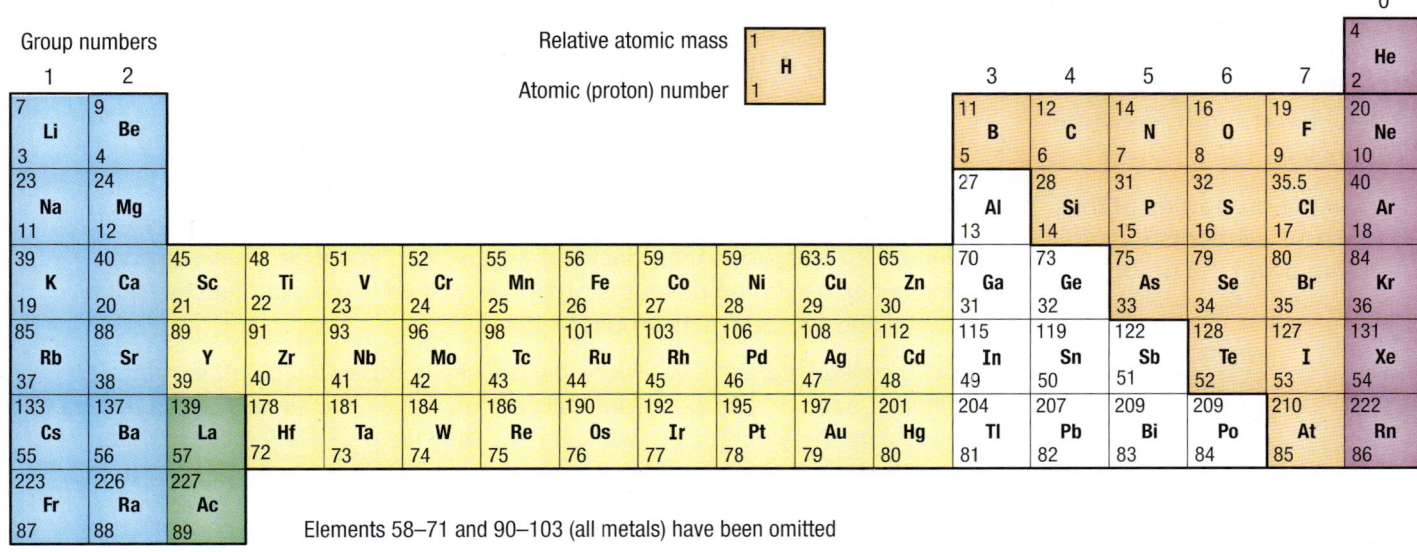

Elements 58–71 and 90–103 (all metals) have been omitted

Key

 Reactive metals These metals react vigorously with other elements like oxygen or chlorine, and with water. They are all soft – some of them can even be cut with a knife, like cheese!

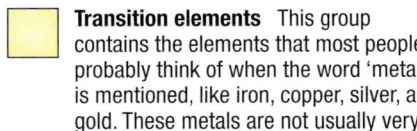 **Transition elements** This group contains the elements that most people probably think of when the word 'metal' is mentioned, like iron, copper, silver, and gold. These metals are not usually very reactive – some, like silver and gold, are very unreactive.

 Non-metals These elements have low melting and boiling points, and many are liquids or gases at room temperature and pressure.

Noble gases These (non-metal) elements are very unreactive, and it is very difficult to get them to combine with other elements.

Figure 1 *The modern periodic table. The upper number on the left of a symbol is the element's relative atomic mass. The lower number is its atomic number (or proton number) – the number of protons in the nucleus. The elements are arranged in order of atomic number.*

> ### Study tip
> Metals react by losing electrons. Non-metals react with metals by gaining electrons.

Summary questions

1 a What does 'periodic' mean in the term 'periodic table'?
   b In the periodic table, what is:
      i   a group?   ii   a period?

2 In the periodic table, how does the number of metallic elements compare with the number of non-metal elements?

3 Explain why elements in many groups of the periodic table have similar chemical properties.

4 How many electrons do atoms of the following elements have in their highest energy level (outermost shell)?
   a beryllium, Be   b boron, B   c potassium, K   d helium, He
   e argon, Ar   f radium, Ra   g radon, Rn   h iodine, I

5 Why are the noble gases so unreactive?

Key points

- The atomic (proton) number of an element determines its position in the periodic table.
- The number of electrons in the outermost shell (highest energy level) of an atom determines its chemical properties.
- The group number in the periodic table equals the number of electrons in the outermost shell.
- The noble gases in Group 0 are unreactive because of their very stable electron arrangements.

3.2 Group 1 – the alkali metals

Learning objectives

After this topic, you should know:

- how the Group 1 elements behave

- how the properties of the Group 1 elements change going down the group.

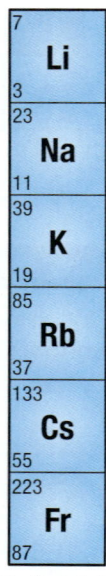

Figure 1 *The alkali metals (Group 1)*

The first group (Group 1) on the left-hand side of the periodic table is called the **alkali metals**. This group consists of the metals lithium (Li), sodium (Na), potassium (K), rubidium (Rb), caesium (Cs), and francium (Fr) (Figure 1). You will probably only see the first three of these as the others are too reactive to use in schools.

Properties of the alkali metals

All the alkali metals are very reactive. They have to be stored in oil (see Figure 2). This stops them reacting with oxygen in the air. Their reactivity increases as you go down the group. So lithium is the least reactive alkali metal and francium is the most reactive.

All the alkali metals have a very low density compared with other metals. In fact lithium, sodium, and potassium all float on water. The alkali metals are also all very soft. They can be cut with a knife. They have a silvery, shiny surface when you first cut them. However, this quickly goes dull as the metal reacts with oxygen in the air. This forms a layer of oxide on the shiny surface.

The properties of this unusual group of metals result from their electronic structure. The atoms of alkali metals all have one electron in their outermost shell (highest energy level). This gives them similar properties. It also makes them very reactive. This is because they need to lose just one electron to get a stable (noble gas) electronic structure.

They react with non-metals, losing their single outer electron. They form a metal ion carrying a 1+ charge, for example, Na^+ and K^+. They always form ionic compounds.

Melting points and boiling points

The Group 1 metals melt and boil at relatively low temperatures for metals. Going down the group, the melting points and boiling points get lower and lower. In fact, caesium turns into a liquid at just 29 °C.

Reaction with water

When you add lithium, sodium, or potassium to water the metal floats on the water, moving around and fizzing. The fizzing happens because the metal reacts with the water to form hydrogen gas. Potassium reacts so vigorously with the water that the hydrogen produced catches fire. It burns with a lilac flame.

The reaction between an alkali metal and water also produces a metal hydroxide, which explains the name 'alkali metals'.

The hydroxides of the alkali metals are all soluble in water. The solution is colourless with a high pH. (**Universal indicator** turns purple.)

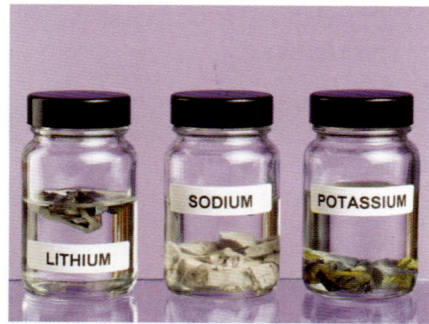

Figure 2 *The alkali metals have to be stored in oil because they are so reactive*

$$\text{sodium} + \text{water} \rightarrow \text{sodium hydroxide} + \text{hydrogen}$$
$$2Na(s) + 2H_2O(l) \rightarrow 2NaOH(aq) + H_2(g)$$

$$\text{potassium} + \text{water} \rightarrow \text{potassium hydroxide} + \text{hydrogen}$$
$$2K(s) + 2H_2O(l) \rightarrow 2KOH(aq) + H_2(g)$$

Demonstration

Reactions of alkali metals with water

The reaction of the alkali metals with water can be demonstrated by dropping a small piece of the metal into a trough of water. This must be done with great care. The reactions are vigorous, releasing a lot of energy. Hydrogen gas is also given off.

■ Describe your observations in detail.

Safety: Wear your safety goggles when watching this demonstration.

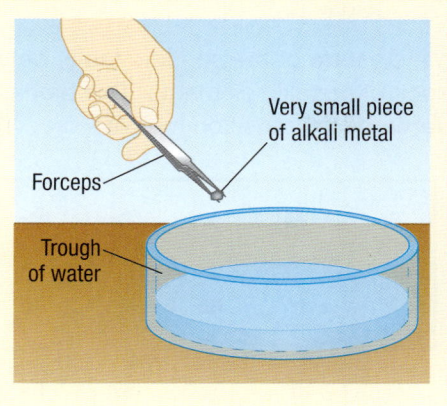

Figure 3 *Lithium, sodium, and potassium reacting with water (the lithium is on the left of the trough, the potassium has burning hydrogen above it, and the sodium is the molten silvery ball on the right)*

Other reactions

The alkali metals also react vigorously with non-metals such as chlorine gas. They produce metal chlorides, which are white solids. Their chlorides all dissolve readily in water to form colourless solutions.

The reactions get more and more vigorous as you go down the group. This is because it becomes easier to lose the single electron in the outer shell to form ions with a 1+ charge.

$$sodium + chlorine \rightarrow sodium\ chloride$$
$$2Na(s) + Cl_2(g) \rightarrow 2NaCl(s)$$

They react in a similar way with fluorine, bromine, and iodine.

All of these ionic compounds of the alkali metals and non-metals are also white and dissolve easily in water. The solutions formed are all colourless.

∞ links

For further information on reactivity within groups, look at 3.5 'Explaining trends'.

Summary questions

1 Why are the alkali metals stored under oil?

2 What is the trend in the melting points of the alkali metals as their atomic number increases?

3 Explain why the alkali metals form ions with a 1+ charge.

4 Caesium is near the bottom of Group 1 in the periodic table.
What do you think would happen if it was dropped into water? Include an explanation of your observations and a balanced symbol equation with state symbols in your answer.

5 Write a balanced symbol equation, including state symbols, for the reaction of caesium (Cs) with:
   a iodine
   b bromine.

Key points

■ The elements in Group 1 of the periodic table are called the alkali metals.

■ Their melting points and boiling points decrease going down the group.

■ The metals all react with water to produce hydrogen and an alkaline solution containing the metal hydroxide.

■ They form 1+ ions in reactions to make ionic compounds. These are generally white and dissolve in water, giving colourless solutions.

■ The reactivity of the alkali metals increases going down the group.

The transition elements

Learning objectives

After this topic, you should know:
- the properties of the transition elements
- how the transition elements compare with the alkali metals.

In the centre of the periodic table, between Groups 2 and 3, there is a large block of metallic elements. This block, shown below, contains the **transition elements** or transition metals.

45 Sc 21	48 Ti 22	51 V 23	52 Cr 24	55 Mn 25	56 Fe 26	59 Co 27	59 Ni 28	63.5 Cu 29	65 Zn 30
89 Y 39	91 Zr 40	93 Nb 41	96 Mo 42	98 Tc 43	101 Ru 44	103 Rh 45	106 Pd 46	108 Ag 47	112 Cd 48
	178 Hf 72	181 Ta 73	184 W 74	186 Re 75	190 Os 76	192 Ir 77	195 Pt 78	197 Au 79	201 Hg 80

Figure 1 *The transition elements. The more common metals are shown in bold type.*

Physical properties

The transition elements have the properties of 'typical' metals. Their metallic bonding and giant structures explain most of their properties.

Transition elements:
- are good conductors of electricity and heat
- are hard and strong
- have high densities
- have high melting points (with the exception of mercury, which is a liquid at room temperature).

The transition elements have very high melting points compared with those of the alkali metals in Group 1 (see Figure 2). They are also harder, stronger, and much denser.

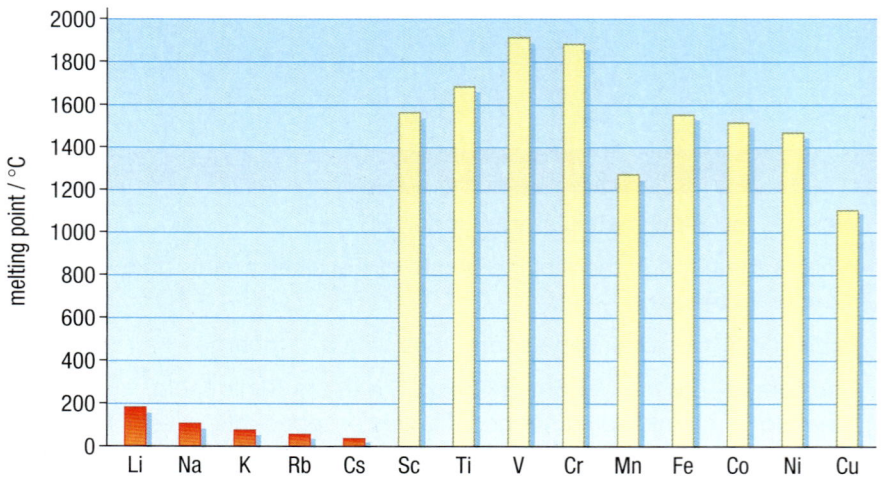

Figure 2 *The melting points of the transition elements are much higher than those of the Group 1 elements*

Chemical properties

The transition elements are much less reactive than the metals in Group 1. This means they do not react as readily with oxygen or water as the alkali metals do. So if they corrode, they do so very slowly. Together with their physical properties, this makes the transition elements very useful as structural materials.

Compounds of transition elements

Many of the transition elements form coloured compounds. These include some very common compounds that are used in the laboratory. For example, copper(II) sulfate is blue (from the copper ions, Cu^{2+}).

Figure 3 *Compounds of transition elements are coloured (as opposed to the mainly white compounds of the alkali metals). The colours of many minerals, rocks and gemstones are due to transition element ions. A reddish-brown colour in a rock is often due to iron ions, Fe^{3+}. The blue colour of sapphires and the green of emeralds are both due to transition element ions in the structures of their crystals.*

Notice that the name of a compound containing a transition element usually includes a Roman number. For example, you will have seen copper(II) sulfate or iron(III) oxide.

This is because transition elements can form more than one ion. For example, iron may exist as Fe^{2+} or Fe^{3+}. Copper can form Cu^+ and Cu^{2+}, and chromium Cr^{2+} and Cr^{3+}. Compounds of these ions are different colours. For example, iron(II) ions (Fe^{2+}) give compounds a green colour, but iron(III) ions (Fe^{3+}) give a reddish-brown colour.

Transition elements and their compounds are also very important in the chemical industry as catalysts. For example, nickel is used as a catalyst in the manufacture of margarine.

Demonstration

The colours of vanadium ions

Your teacher can show you the range of colours that different ions of vanadium can have.

Find out the names and formulae of the ions responsible for each colour.

Study tip

The charge on the metal ion is given in the name of many transition metal compounds, for example, copper(II) sulfate contains Cu^{2+} ions.

Key points

- Compared with the alkali metals, transition elements have much higher melting points and densities. They are also stronger and harder, but are much less reactive.

- The transition elements do not react vigorously with oxygen or water.

- A transition element can form ions with different charges, in compounds that are often coloured.

- Transition elements and their compounds are important industrial catalysts.

Summary questions

1 **a** List the properties of a typical transition element.
   b How is mercury unlike a typical transition metal?

2 Write down the name of the following compounds of transition elements:
   a $FeCl_2$ **b** Cr_2O_3

3 Copper (Cu) can form ions that carry a 1+ or a 2+ charge. Write down the name and formula of each compound copper can form with oxygen.

4 Find out the name and formula of the transition metal compound used as a catalyst in the manufacture of sulfuric acid.

3.4

Group 7 – the halogens

Learning objectives

After this topic, you should know:

- how the Group 7 elements behave

- how the properties of the Group 7 elements change going down the group.

Figure 1 *The Group 7 elements*

Properties of the halogens

The Group 7 elements are called the **halogens**. They are a group of poisonous non-metals that have coloured vapours. They have fairly typical properties of non-metals.

- They have low melting points and boiling points. Their melting points and boiling points increase going down the group (see Figure 2).
- They are also poor conductors of heat and electricity.

As elements, the halogens all exist as molecules made up of pairs of atoms. The atoms in each pair are joined to each other by a covalent bond.

	Melting point (°C)	Boiling point (°C)
F——F (F_2)	−220	−188
Cl——Cl (Cl_2)	−101	−35
Br——Br (Br_2)	−7	59
I——I (I_2)	114	184

Figure 2 *The halogens all form molecules made up of a pair of atoms, joined by a covalent bond. This type of molecule is called a 'diatomic molecule'.*

Reactions of the halogens

The electronic structure of the halogens determines the way they react with other elements. They all have seven electrons in their outermost shell (highest energy level). So they need to gain just one more electron to achieve the stable electronic structure of a noble gas. This means that the halogens take part in both ionic and covalent bonding.

	How the halogens react with hydrogen
$F_2(g) + H_2(g) \rightarrow 2HF(g)$	Explosive even at −200 °C and in the dark
$Cl_2(g) + H_2(g) \rightarrow 2HCl(g)$	Explosive in sunlight, but slow in the dark
$Br_2(g) + H_2(g) \rightarrow 2HBr(g)$	300 °C + platinum catalyst
$I_2(g) + H_2(g) \rightleftharpoons 2HI(g)$	300 °C + platinum catalyst (very slow, reversible)

Look at the table of reactions of the halogens with hydrogen gas. It shows a general trend that you find in Group 7 – **the elements get less reactive going down the group**.

The halogens also all react with metals. They gain a single electron to give them a stable arrangement of electrons. They form ions with a 1− charge, for example, F^-, Cl^-, Br^-. Examples of their ionic compounds include sodium chloride, NaCl, and iron(III) bromide, $FeBr_3$.

Study tip

In Group 7, reactivity *decreases* as you go down the group. However, in Group 1, reactivity *increases* going down the group.

Look at the diagram of the ions in calcium chloride, $CaCl_2$, below:

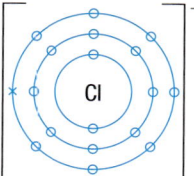 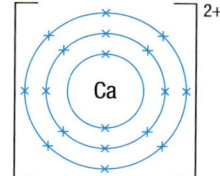 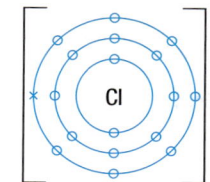

When a halogen reacts with another non-metal, the atoms of the halogen **share electrons** with the atoms of the other element. This gives the atoms of both elements the stable electronic structure of one of the noble gases. Therefore, the compounds of halogens with non-metals contain covalent bonds.

Look at the **dot and cross diagram** of hydrogen chloride, HCl, below, showing only the outer shell electrons:

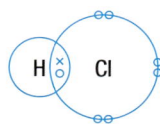

(Note that the circles need not be shown in dot and cross diagrams.)

Displacement reactions between halogens

You can use a more reactive halogen to displace a less reactive halogen from solutions of its salts.

Bromine displaces iodine from solution because it is more reactive than iodine. Chlorine will displace both iodine and bromine.

For example, chlorine will displace bromine:

chlorine + potassium bromide → potassium chloride + bromine
$$Cl_2(aq) \ + \ \ \ \ \ 2KBr(aq) \ \ \ \ \ \to \ \ \ \ \ 2KCl(aq) \ \ \ \ + \ Br_2(aq)$$

Obviously fluorine, the most reactive of the halogens, would displace all of the others. However, it reacts so violently with water that you cannot carry out reactions of fluorine in aqueous solutions.

Summary questions

1 In the Group 7 elements, what is the trend going down the group in:
   a their melting points?   b their reactivity?

2 a Using the data in the table in Figure 2, give the state of each Group 7 element at 20 °C.
   b What is the general name given to the Group 7 elements?

3 a Show the ions in lithium fluoride, LiF, in a diagram showing all the electrons.
   b Draw a dot and cross diagram, showing only the outer shell electrons, of a molecule of hydrogen iodide.

4 a Write a word equation for the reaction of lithium with bromine.
   b Write a word equation for the reaction of bromine water with potassium iodide solution.

5 Write a balanced symbol equation, including state symbols, for the reaction of:
   a sodium metal with iodine vapour
   b chlorine water with sodium iodide solution

??? Did you know …?

Many early chemists were badly hurt or even killed as they tried to make pure fluorine. They called it 'the gas of Lucifer'. It was finally produced by the French chemist Henri Moissan, who died aged just 55. His life was almost certainly shortened by his work with fluorine.

Practical

Displacement reactions

Add bromine water to potassium iodide solution in a test tube. Then try some other combinations of solutions of halogens and potassium halides.

■ Record your results in a table.
■ Explain your observations.

Safety: Wear chemical splash-proof eye protection. Chlorine and bromine are toxic.

Key points

■ **The halogens all form ions with a single negative charge in their ionic compounds with metals.**

■ **The halogens form covalent compounds by sharing electrons with other non-metals.**

■ **A more reactive halogen can displace a less reactive halogen from a solution of one of its salts.**

■ **The reactivity of the halogens decreases going down the group.**

3.5 Explaining trends

Learning objectives

After this topic, you should know:

- the trends in reactivity in Group 1 and Group 7

- how electronic structure can explain trends in reactivity in these groups.

As you saw in Topic 3.2, the Group 1 elements get more reactive going down the group:

Li
Na
K | getting **more** reactive
Rb
Cs ↓

The opposite trend is observed in the Group 7 elements (see Topic 3.4):

F
Cl
Br | getting **less** reactive
I
At ↓

These trends can be explained by looking at the electronic structures and how the atoms tend to lose or gain electrons in their reactions.

Reactivity within groups

As you go down a group the number of shells occupied by electrons increases, by one extra electron shell per period, and the atoms get bigger.

This has two effects:

- larger atoms lose electrons more easily going down a group
- larger atoms gain electrons less easily going down a group.

This happens because the outer electrons are further away from the attractive force of the nucleus. Not only that, the inner shells 'screen' or 'shield' the outer electrons from the positive charge in the nucleus. You can see this effect with the alkali metals and the halogens. Remember that the atoms of alkali metals tend to lose electrons when they form chemical bonds. On the other hand, the atoms of the halogens tend to gain electrons.

Explaining the trend in Group 1

Reactivity increases going down Group 1 because the atoms get bigger so the single electron in the outermost shell (highest energy level) is attracted less strongly to the positive nucleus. The electrostatic attraction gets weaker because the distance between the outer electron and the nucleus increases.

Also the outer electron experiences a shielding effect from inner shells of electrons. This reduces the attraction between the oppositely charged outer electron and the nucleus.

The size of the positive charge on the nucleus does become larger as you go down a group, as more protons are added. This suggests that the attraction for the outer electron should get stronger. However, the greater distance and the shielding effect of inner electrons outweigh the increasing nuclear charge. So the change from Li to Li^+ takes more energy than Na changing to Na^+.

Therefore in Group 1, the outer electron gets easier to remove going down the group and the elements get more and more reactive.

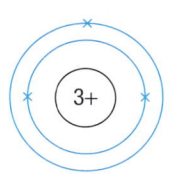

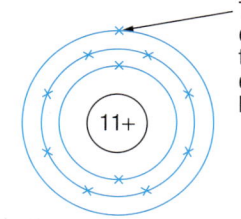

This electron is easier to remove than the outer electron in a lithium atom

Li atom

Na atom

Figure 1 *Sodium's outer electron is further from the nuclear charge and is shielded by more inner shell electrons than lithium's outer electron*

Explaining the trend in Group 7

Reactivity decreases going down Group 7. To explain this, consider the same factors you looked at with the alkali metals:

- the size of the atom
- the shielding effect of inner electrons
- the nuclear charge.

When Group 7 elements react, their atoms gain an electron into their outermost shell (highest energy level). Going down the group, the outermost shell gets further away from the attractive force of the nucleus so it is harder to attract and gain an incoming electron.

The outer shell will also be shielded by more inner electrons, again reducing the attraction of the nucleus for an incoming electron.

The effect of the increased nuclear charge going down the group (which helps atoms gain an incoming electron) is outweighed by the effect of increased distance and shielding by more inner electrons. So iodine is less reactive than fluorine. The attraction for the incoming electron when F changes to F^- is much greater than when I changes to I^-.

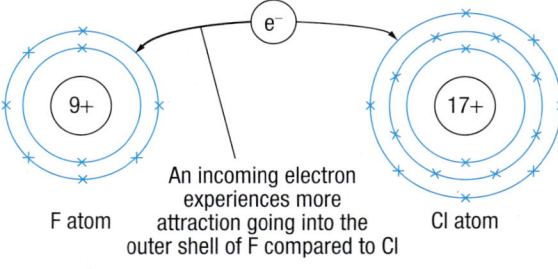

F atom

An incoming electron experiences more attraction going into the outer shell of F compared to Cl

Cl atom

Figure 2 *F forms F– more readily than Cl forms Cl–*

Key points

- You can explain trends in reactivity as you go down a group in terms of the attraction between electrons in the outermost shell and the nucleus.

- This electrostatic attraction depends on:
  - the distance between the outermost electrons and the nucleus
  - the number of occupied inner shells (energy levels) of electrons, which provide a shielding effect.

- In deciding how easy it is for atoms to lose or gain electrons from their outermost shell, these two factors outweigh the increased nuclear charge due to extra protons in the nucleus, going down a group.

Summary questions

1 Which one of these atoms will have electrons in its outermost shell (highest energy level) which experience the greatest attractive force between themselves and the nucleus?

   Helium   Lithium   Fluorine   Potassium   Argon

   Explain your answer.

2 Explain why potassium is more reactive than lithium.

3 Explain why fluorine is more reactive than bromine.

4 Predict the difference in reactivity, giving reasons, seen in:
   a Group 2, between magnesium and calcium
   b Group 6, between oxygen and sulfur.

Chapter summary questions

1 a Where in the periodic table will you find:
  i the halogens?
  ii the noble gases?
  iii the alkali metals?
  iv the transition elements?

b In which groups would you find the four elements described in **i**, **ii**, **iii**, and **iv** below?
  i This is a dense metal with a high melting point. It reacts only very slowly with water but will react when heated with steam.
  ii This is a metal that can be cut with a knife and is stored under oil. It reacts violently with water and forms ions with a 1+ charge.
  iii This is a very unreactive, monatomic gas.
  iv This toxic gas is the most reactive of the all the non-metallic elements. It forms ions with a 1− charge and will also form covalent compounds.

2 Astatine (At) is a halogen whose atomic number is 85. It lies at the bottom of Group 7.

a How many electrons occupy its outermost shell (highest energy level)? How did you work out the answer?

b Predict the state of astatine at 20 °C.

c For the compound **sodium astatide**, predict:
  i its type of bonding
  ii its colour
  iii its chemical formula
  iv the balanced symbol equation for its formation from its elements
  v whether or not you would see signs of a reaction if a solution of sodium astatide was mixed with chlorine water. Explain how you arrived at your answer.

3 Rubidium (Rb) is in Group 1 of the periodic table, lying directly beneath potassium.

a Predict the physical properties of rubidium, to include:
  i its hardness
  ii its electrical conductivity
  iii its melting point.

b i What will be the charge on a rubidium ion?
  ii Copy and complete the table below:

Rubidium compound	Chemical formula
Rubidium iodide	
Rubidium fluoride	
Rubidium hydroxide	

iii What is the colour of the rubidium compounds in the table in part **ii**? Comment on their solubility in water.

c Write down word equations and balanced symbol equations, including state symbols, for the following reactions:
  i rubidium and water
  ii rubidium and chlorine.

d Will the reactions of rubidium in part **c** be more or less vigorous than the same reactions using potassium? Explain your answer by comparing the electronic structures of the two elements.

4 a Draw dot and cross diagrams to show the bonding in the following substances. (Draw only the outer shell electrons.)

  i hydrogen bromide
  ii iodine

b Draw a diagram showing all the electrons in lithium chloride.

5 Copper is a typical transition element. Sodium is a Group 1 element.

a i How would you be able to tell which compound was which if you were given unlabelled samples of copper(II) sulfate and sodium sulfate?
  ii Why does copper in copper(II) sulfate have a Roman numeral after its name but sodium in sodium sulfate does not?

b How do the physical properties of copper differ from those of sodium?

c How does the reactivity of copper compare with that of sodium? Use a test with water to illustrate your answer.

Practice questions

1 The metals in Group 1 of the periodic table are chemically very reactive but have physical properties very different from most other metals.

  a State **two** physical properties whose values are much lower for Group 1 metals than for most other metals. (2)

  b A small piece of lithium is added to a large trough of water. A similar piece of potassium is added to a different trough of water. A reaction occurs in both troughs.

    i State **two** observations that could be made in the lithium reaction. (2)

    ii State **two** observations that could be made in the potassium reaction that would **not** be made in the lithium reaction. (2)

    iii A piece of litmus paper was dipped in the solution formed in the lithium reaction. State and explain the final colour of the litmus. (2)

    iv Write a symbol equation for the reaction between potassium and water. (2)

  c Lithium and potassium also react with chlorine. Potassium reacts more vigorously than lithium. The electronic structures of the metals are

    Li 2,1   K 2,8,8,1

    Use these electronic structures to explain why potassium reacts more vigorously than lithium. (3)

2 The elements in the centre of the periodic table are called transition elements and they have several properties in common with each other.

  a One property is the formation of ions with different charges.

    State the formula of the metal ion in $FeCl_2$ and in $FeCl_3$. (2)

  b Another property is the formation of coloured compounds.

    State the colour of $Fe(OH)_2$ and of $Fe(OH)_3$. (2)

  c The name of the compound manganese(IV) oxide contains the Roman numeral with a value of 4.

    i What does this numeral indicate about the manganese in this compound? (1)

    ii Manganese(IV) oxide speeds up the decomposition of hydrogen peroxide. What name is used for compounds that speed up reactions but remain chemically unchanged at the end of the reaction? (1)

3 The elements in Group 7 of the periodic table are called halogens. They form salts called halides.

  a Write a symbol equation for the salt formed when calcium reacts with chlorine. (2)

  b When a halogen is added to a solution of a halide a displacement reaction sometimes occurs.

    Copy and complete the following word equations with the names of the products where a reaction occurs. If no reaction occurs, write 'no reaction'.

    i iodine + lithium bromide → .... (1)

    ii chlorine + sodium bromide → .... (1)

    iii bromine + potassium iodide → .... (1)

  c Fluorine is more reactive than chlorine. The electronic structures of these halogens are:

    F 2,7   Cl 2,8,7

    Use these electronic structures to explain why fluorine reacts more vigorously than chlorine. (3)

4.1

Useful metals

Learning objectives

After this topic, you should know:

- what transition metals are and why they are so useful
- why copper is such a useful metal
- why alloys are often more useful than pure metals.

Transition metals

In the centre of the periodic table there is a large block of metallic elements. Most are called transition metals and have similar properties. Like all metals, the transition metals are very good conductors of electricity and heat. They are strong but can also be bent or hammered into useful shapes.

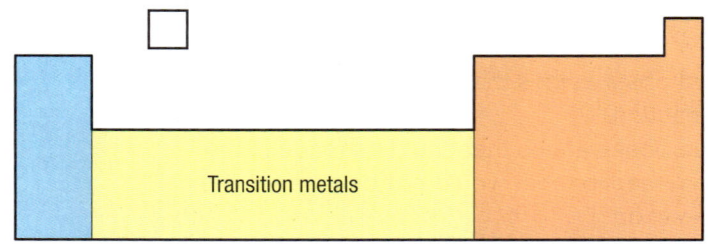

Transition metals

Figure 1 *The transition metals are found in the yellow block of the periodic table shown here*

The properties of the transition metals mean that they can be used in many different ways. You will find them in buildings and in cars, trains, and other types of transport. Their strength makes them useful as building materials. They are used in heating systems and for electrical wiring because heat energy and electricity pass through them easily.

Copper is a very useful transition metal. It can be bent but is still hard enough for plumbers to use as water tanks or pipes and it does not react with water. Copper also conducts electricity and heat very well. So it is ideal where we need:

- pipes that will carry water, or
- wires that will conduct electricity.

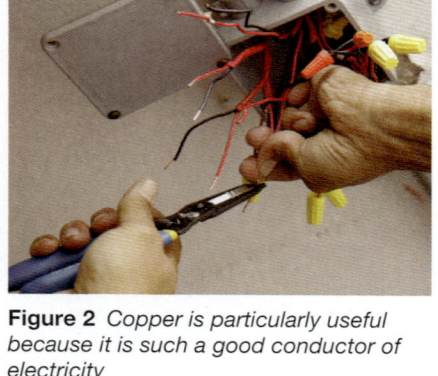

Figure 2 *Copper is particularly useful because it is such a good conductor of electricity*

⚭ links

For more information on the transition metals, look back to 3.3 'The transition elements'.

Figure 3 *Transition metals are used in many different ways because of their useful properties*

Copper alloys

Bronze was probably the first alloy made by humans, about 5500 years ago. It is usually made by mixing copper with tin. It is used to make ship's propellers because of its toughness and resistance to corrosion.

Brass is made by alloying copper with zinc. Brass is much harder than copper but it is workable. It can be hammered into sheets and pressed into intricate shapes. This property is used to make musical instruments, such as trumpets.

Aluminium alloys

Aluminium has a low density for a metal. It can be alloyed with a wide range of other elements. There are over 300 alloys of aluminium available. These alloys have very different properties. Engineers use 'light-weight', but strong, aluminium alloys to build aircraft whilst others can be used as armour-plating on tanks and other military vehicles.

Gold alloys

As with copper and iron, gold and aluminium can be made harder by adding other elements. Gold is usually alloyed with copper when used to make jewellery. Pure gold wears away more easily than its alloy with copper. By varying the proportions of the two metals you also get different shades of 'gold' objects.

Figure 4 *The Statue of Liberty in New York contains over 80 tonnes of copper*

Figure 5 *Alloying with copper makes gold more hardwearing. This is especially important in wedding rings, which many married people wear most of the time.*

??? Did you know ... ?

The purity of gold is often expressed in 'carats', where 24-carat gold is almost pure gold (99.9%). If you divide the carat number by 24, you get the fraction of gold in your jewellery. So an 18-carat gold ring will contain $\frac{3}{4}$ (75%) gold.

Summary questions

1 **a** Write a list of the properties of a typical transition metal.
  b Where in the periodic table will you find the transition metals?
  c Which metal mentioned on this page is **not** a transition metal?

2 **a** Why is copper used so much in plumbing?
  b Why is gold alloyed with copper in many wedding rings?

3 Silver and gold are transition metals that are better conductors of electricity than copper. Why is copper used to make electric cables instead of either of these metals?

4 Why are aluminium alloys used extensively in the aircraft industry?

5 Explain in detail why copper alloys are more suitable for making coins than pure copper metal.

Key points

- The transition metals are found in the central block of elements in the periodic table.

- Transition metals have properties that make them useful for building and making things. For example, copper is used in wiring because of its high electrical conductivity.

- Copper, gold, and aluminium are all alloyed with other metals to make them harder.

4.2 Iron and steels

Learning objectives

After this topic, you should know:

- how iron ore is reduced

- how to explain oxidation and reduction in terms of the addition and removal of oxygen

- the main types and uses of steels.

Iron ore (the rock from which iron is extracted) contains iron combined with oxygen in iron(III) oxide along with rocky impurities. Iron is less reactive than carbon. So, iron can be extracted by using carbon to remove oxygen from the iron(III) oxide in the ore. Iron is extracted in a **blast furnace**. The raw materials fed into the furnace, which is run continuously, are:

- iron ore (called haematite)
- coke (a cheap form of carbon made from coal)
- air
- limestone (used to remove impurities by reacting with them to form slag).

The reaction which happens in the blast furnace to remove the oxygen is:

$$\text{carbon} + \text{oxygen} \rightarrow \text{carbon dioxide}$$
$$C(s) \ + \ O_2(g) \ \rightarrow \ CO_2(g)$$

In this reaction, the carbon has been oxidised. It has had oxygen added to it to form carbon dioxide. One way that chemists define **oxidation** is 'the addition of oxygen'.

Another reaction inside a blast furnace is:

$$\text{carbon dioxide} + \text{carbon} \rightarrow \text{carbon monoxide}$$
$$CO_2(g) \ + \ C(s) \ \rightarrow \ 2CO(g)$$

In this reaction, carbon dioxide is **reduced**, as it loses an oxygen atom to form carbon monoxide. One way that chemists define **reduction** is 'the removal of oxygen'. (Note also that carbon is oxidised in this reaction; forming carbon monoxide). The carbon monoxide then reduces the iron(III) oxide:

$$\text{iron(III) oxide} + \text{carbon monoxide} \rightarrow \text{iron} + \text{carbon dioxide}$$
$$Fe_2O_3(s) \ + \ 3CO(g) \ \rightarrow 2Fe(l) + \ 3CO_2(g)$$

The molten iron is tapped off from the bottom of the furnace.

Iron straight from the blast furnace has limited uses. It contains about 96% iron and contains impurities, mainly carbon. This makes the iron brittle, although it is very hard and cannot be easily compressed. When molten it can be run into moulds and cast into different shapes. This **cast iron** is used to make wood-burning stoves, man-hole covers on roads, and engines. Despite being brittle, cast iron can withstand high forces of compression without breaking. This is why it is used in man-hole covers.

The iron from the blast furnace can be treated to remove some of the carbon.

Removing all the carbon and other impurities from cast iron gives us pure iron. This is relatively soft and easily-shaped. However, it is too soft for most uses. If you want to make iron really useful you have to make sure that it contains tiny amounts of other elements. These include carbon and metals, such as nickel and chromium. A metal that is mixed with other elements is called an alloy.

Figure 1 *The iron which has just come out of a blast furnace contains about 96% iron. The main impurity is carbon.*

Study tip

You need to know how the hardness of steels is related to their carbon content.

Steels

Steels are alloys of iron. By adding elements in carefully controlled amounts, scientists can change the properties of the steels.

Steel is not a single substance. Like all alloys, it is a mixture. There are lots of different types of steel. All of them are alloys of iron with carbon and/or other elements.

Carbon steels

The simplest steels are the **carbon steels**. These are made by removing most of the carbon from cast iron, just leaving small amounts of carbon (from 0.03% to 1.5%). These are the cheapest steels to make. They are used in many products, such as the bodies of cars, knives, machinery, ships, containers, and structural steel for buildings.

Often these carbon steels have small amounts of other elements in them as well. High carbon steel, with a relatively high carbon content, is very strong but brittle. On the other hand, low carbon steel is soft and easily shaped. It is not as strong, but is much less likely to shatter on impact with a hard object.

Mild steel is one type of low carbon steel. It contains less than 0.1% carbon. It is very easily pressed into shapes. This makes it particularly useful in mass production, such as making car bodies.

Alloy steels

Low-alloy steels are more expensive than carbon steels because they contain between 1% and 5% of other metals. Each of these metals produces a steel that is well-suited for a particular use.

Even more expensive are the **high-alloy steels**. These contain a much higher percentage of other metals. The chromium–nickel steels are known as **stainless steels**. They are used to make cooking utensils and cutlery. They are also used to make chemical reaction vessels. This is because they combine hardness and strength with great resistance to corrosion. Unlike most other steels, they do not rust!

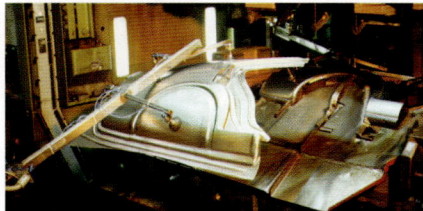
Figure 2 *Low carbon steel called mild steel is easily pressed into shapes*

Figure 3 *The properties of alloy steels make them ideal for use in bridges*

Figure 4 *The resistance of stainless steels to rusting make them ideal for making utensils and cutlery*

 Did you know … ?

Nickel–steel alloys are used to build long-span bridges, bicycle chains and military armour-plating. This is because they are very resistant to stretching forces. Tungsten steel operates well under very hot conditions so it is used to make high-speed tools such as drill bits.

Summary questions

1 **a** Why is iron from a blast furnace very brittle?
   b How does cast iron differ from pure iron in its composition and properties?
   c Give one use of cast iron which relies on its high strength in compression.

2 **a** Name the raw materials used in a blast furnace for the extraction of iron.
   b Name the main reducing agent in the blast furnace and write a balanced symbol equation for its formation. Include state symbols.
   c A small amount of iron(III) oxide is reduced by solid carbon in the blast furnace. Write a word equation and a balanced symbol equation with state symbols for this reaction.
   d Explain why carbon monoxide, as opposed to solid carbon, reduces most of the iron(III) oxide in the blast furnace.

3 Make a table to summarise the main useful properties of low carbon steel, high carbon steel, and chromium–nickel steel.

4 **a** Why are surgical instruments made from steel containing chromium and nickel?
   b Explain which type of steel is used to make:
   i car bodies   **ii** railway tracks.

Key points

- Iron is extracted from iron ore by reducing it using carbon in a blast furnace. The main reducing agent is carbon monoxide gas.

- Pure iron is too soft for it to be very useful.

- Carefully controlled quantities of carbon and other elements are added to iron to make steel alloys with different properties.

- Important examples of steels are:
  – low carbon steels which are easily shaped
  – high carbon steels which are very hard
  – stainless steels which are resistant to corrosion.

4.3 The reactivity series

Learning objectives

After this topic, you should know:

- what the reactivity series is

- how some common metals react with water and with dilute acid.

Practical

Predicting reactions

Calcium is a metal that is more reactive than magnesium, but less reactive than lithium.

- Predict what you will see when a piece of calcium is added to a beaker of water and name any products that might be formed.

Try out the reaction in a large beaker (or trough) of water. Have an inverted test tube full of water ready to collect any gas given off and test the solution left with universal indicator paper. Test if the gas collected is hydrogen, using a lighted spill.

- Evaluate your prediction by comparing your observations with your ideas before trying out the experiment.

- Write a word and balanced symbol equation for the reaction between calcium and water.

Safety: Do not touch calcium metal or the solution formed in the reaction. Wear eye protection. Calcium is highly flammable.

Metals are important in all our lives. For example, in transport metals are used to make bicycles, cars, ships, trains, and aeroplanes. In this chapter you will look at the chemistry involved in getting these metals from their raw materials, **ores**. Ores are rocks from which it is economical to extract the metals that they contain. With new techniques, scientists can now extract metals from rock once thought of as waste.

Most metals in ores are chemically bonded to other elements in compounds. So, how can the metals be extracted from their ores? To answer this you must understand the **reactivity series** of metals. The reactivity series is a list of metals in order of their reactivity, with the most reactive metals at the top and the least reactive ones at the bottom (see Table 1 on the next page).

Metals plus water

You can start putting the metals in order of reactivity by looking at their reactions with water. Most metals do not react vigorously with water. Metals such as copper, which does not react at all with water, can be used to make water pipes. However, there is a great range in reactivity between different metals. For example, you have seen how the alkali metals in Group 1 react with water in Topic 3.2 *Group 1 – the alkali metals*. They react vigorously, giving off hydrogen gas and leaving alkaline solutions. The reactivity increases going down Group 1, so of the metals you observed, the order of reactivity is **1** potassium (most reactive), **2** sodium, and **3** lithium (least reactive).

Magnesium lies somewhere between lithium and copper in the reactivity series. If magnesium is left in a beaker of water, it takes several days to collect enough gas to test with a lighted spill. The resulting 'pop' shows that the gas is hydrogen.

Although magnesium reacts very slowly with water, you can speed up the reaction by heating and reacting the metal with steam.

Demonstration

Magnesium and steam

Watch your teacher demonstrate the reaction below:

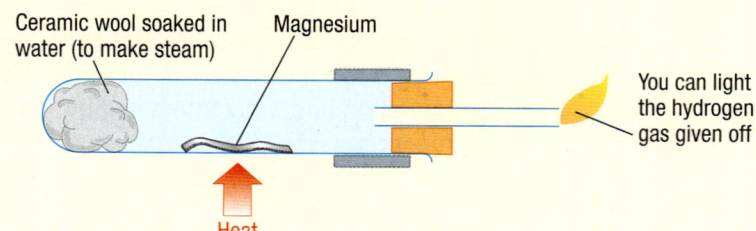

Ceramic wool soaked in water (to make steam)   Magnesium

You can light the hydrogen gas given off

Heat

- Describe what happens as the magnesium reacts with the steam.
- Which gas burns off at the end of the tube?
- The white solid left in the test tube is magnesium oxide (MgO). Write a word equation and a balanced symbol equation, including state symbols, for the reaction of magnesium and steam.

Metals plus dilute acid

You have now seen how a range of metals react with water, and that you can use your observations of these reactions to place the metals into an order of reactivity. However, where the reactions are very slow, the task of ordering the metals is difficult. With these metals, look at their reactions with dilute acid to arrive at an order of reactivity.

Here is a table that summarises the reactions of some important metals with water and dilute acid:

Table 1 *The reactivity series*
Note: Aluminium is protected by a layer of aluminium oxide so will not undergo the reactions above unless the oxide layer is removed. This is why this fairly reactive metal can be used outside, for example in sliding patio doors, without corroding.

Order of reactivity	Reaction with water	Reaction with dilute acid
potassium	fizz, giving off hydrogen, leaving an alkaline solution of metal hydroxide	explode
sodium		
lithium		
calcium		fizz, giving off hydrogen and forming a salt
magnesium	react with steam, giving off hydrogen and forming the metal oxide	
aluminium		
zinc		
iron		
tin	slight reaction with steam	react slowly with warm acid
lead		
copper	no reaction, even with steam	no reaction
silver		
gold		

Summary questions

1 Write the word equation and balanced symbol equation, including state symbols, for the reaction between:
  a lithium and water
  b zinc and steam.

2 A student added a piece of magnesium ribbon to dilute sulfuric acid.
  a List three ways she could tell that a chemical reaction was taking place.
  b Write down the general equation that describes the reaction between a metal and an acid. (See Topic 7.2.)
  c Write the word equation and balanced symbol equation, including state symbols, for the reaction between magnesium and dilute sulfuric acid.

3 Explain these facts in terms of chemical reactivity:
  a Gold, silver, and platinum are used to make jewellery.
  b Potassium, lithium, and sodium are stored in jars of oil.
  c Food cans are plated with tin, but not zinc.

4 Explain why aluminium can be used outdoors, for example, as window frames, even though it is quite high in the reactivity series.

Practical

Metals and acid

You are given coarse-grained filings of the metals copper, aluminium, zinc, iron, and magnesium to put into an order of reactivity according to their reactions with dilute hydrochloric acid.

Plan a test to put the metals in order of reactivity based on your observations. Your plan should include the quantities of reactants you intend to use. (You will have access to a balance and measuring cylinders.)

Let your teacher check your plan before you start your tests.

- Give a brief outline of your method, including how you will make it as fair a test as possible.
- Record your results in a suitable table.
- Put the metals in order of reactivity according to your observations.
- Identify any hazards.

links

For more about the reaction of metals with dilute acid, see 7.2 'Making salts from metals or insoluble bases'.

Key points

- The metals can be placed in order of reactivity by their reactions with water and dilute acid.

- Hydrogen gas is given off if metals react with water or dilute acids. The gas 'pops' with a lighted spill.

4.4 Displacement reactions

Learning objectives

After this topic, you should know:

- how displacement reactions take place

- the position of hydrogen in the reactivity series

- the definitions of reduction and oxidation in terms of electron transfer.

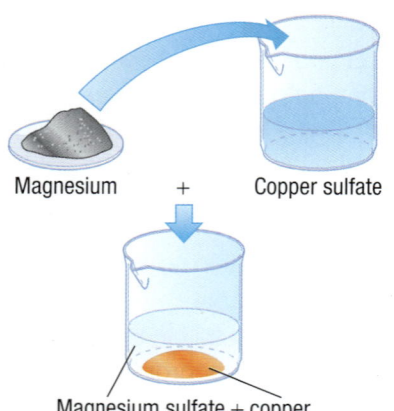

Magnesium + Copper sulfate

Magnesium sulfate + copper

Figure 1 *Magnesium displaces copper from copper(II) sulfate solution*

Practical

Displacing a metal from solution

Set up the test tube as shown:

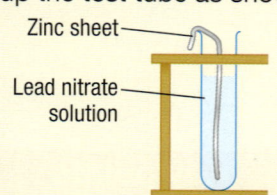

Zinc sheet

Lead nitrate solution

Leave the tube until you see metal crystals on the surface of the zinc sheet.

- What do the crystals look like?
- Explain what happens.

Safety: Lead compounds are toxic. Wear chemical splash-proof eye protection.

In Topic 4.3 *The reactivity series*, you have seen how to use the reactions of metals with water and dilute acid to get an order of reactivity. You can also judge reactivity by putting the metals 'into competition' with each other. One metal starts off as the element and the other metal as an ion in a solution of one of its salts. For example, you might have magnesium metal, Mg(s), and copper(II) ions, Cu^{2+}(aq), in a solution of copper(II) sulfate.

A more reactive metal will displace a less reactive metal from an aqueous solution of one of its salts.

In this case, magnesium is more reactive than copper. Therefore the copper ions will be displaced from solution to form copper metal, Cu(s). In this reaction, the magnesium metal forms aqueous magnesium ions, Mg^{2+}(aq), and dissolves into the solution (see Figure 1). This is a **displacement reaction**.

The word equation is:

$$\text{magnesium} + \text{copper(II) sulfate} \rightarrow \text{magnesium sulfate} + \text{copper}$$

The balanced symbol equation is:

$$Mg(s) + CuSO_4(aq) \rightarrow MgSO_4(aq) + Cu(s)$$

The **ionic equation** (showing only the atoms and ions that change in the reaction):

$$Mg(s) + Cu^{2+}(aq) \rightarrow Mg^{2+}(aq) + Cu(s)$$

This shows that magnesium atoms have a greater tendency to form positive ions than copper atoms.

Zinc is more reactive than lead – it is higher up the reactivity series. Therefore, zinc displaces lead from its solution:

$$\text{zinc} + \text{lead nitrate} \rightarrow \text{zinc nitrate} + \text{lead}$$
$$Zn(s) + Pb(NO_3)_2(aq) \rightarrow Zn(NO_3)_2(aq) + Pb(s)$$

You will see the lead metal forming as crystals on the zinc.

Practical

Predicting reactions

Copy and fill in the table below.

Solution \ Metal	Magnesium	Copper	Zinc	Iron
Magnesium sulfate	✗			
Copper sulfate		✗		
Zinc sulfate			✗	
Iron sulfate				✗

Predict which will react (enter a tick in your table), and which won't (enter a cross).

Try out the reactions on a spotting tile.

Safety: Wear eye protection. Some chemicals are harmful.

Hydrogen in the reactivity series

The positions of the non-metals hydrogen and carbon in the reactivity series can be determined using displacement reactions. You can think of the metal plus acid reactions as displacement of hydrogen ions, $H^+(aq)$, from solution. Copper cannot displace the hydrogen from an acid, whereas lead can. So hydrogen is positioned between copper and lead.

Oxidation and reduction

Oxidation can be defined as the chemical addition of oxygen, and **reduction** as the removal of oxygen. A wider definition involves the transfer of electrons rather than oxygen atoms.

Oxidation is the loss of electrons.   **Reduction is the gain of electrons.**

You can apply these definitions to displacement reactions in solution. Take as an example, the displacement of copper by iron. This reaction is used in industry to extract copper metal from copper sulfate:

$$Fe(s) + Cu^{2+}(aq) \rightarrow Fe^{2+}(aq) + Cu(s)$$

The iron added to the copper sulfate solution is cheap scrap iron. Consider what happens to each reactant in the ionic equation above:

$$Fe(s) \rightarrow Fe^{2+}(aq) + 2e^-$$

This is called a **half equation** - you can see more examples on page 66. Iron atoms lose two electrons to form iron(II) ions. This is oxidation (the loss of electrons). We say that iron atoms have been **oxidised**.

These two electrons from iron are gained by the copper(II) ions as they form copper atoms:

$$Cu^{2+}(aq) + 2e^- \rightarrow Cu(s)$$

This is reduction (the gain of electrons). The copper(II) ions have been reduced.

This is why these displacement reactions are also known as redox reactions (**red**uction–**ox**idation).

Using displacement in industry

In industry, copper can be extracted from copper sulfate solution by adding scrap iron. Iron is more reactive than copper, so it can displace copper from its solutions, for example, from copper(II) sulfate solution:

iron + copper(II) sulfate $\rightarrow$ iron(II) sulfate + copper

??? Did you know … ?

The copper(II) sulfate solution used in the extraction of copper from its ores can be obtained by treating low-grade copper ores with bacteria, in a process called bioleaching.

Summary questions

1 Which of the following pairs of metals and solutions will result in a reaction? If a reaction is predicted, write a balanced symbol equation, including state symbols.
   a iron + zinc sulfate   b zinc + copper sulfate
   c magnesium + iron(II) chloride
2 Hydrogen gas is used in the reduction of tungsten oxide, WO_3, to extract tungsten metal.
   a What does this tell you about the position of tungsten in the reactivity series?
   b Write a balanced symbol equation for the reduction of tungsten oxide by hydrogen.
3 a Write the ionic equation, including state symbols, for the reaction between zinc and iron(II) sulfate.
   b Explain in terms of the transfer of electrons which species (atom, molecule, or ion) is oxidised and which is reduced in this reaction.

Key points

- A more reactive metal will displace a less reactive metal from its aqueous solution.

- Hydrogen can be given a position in the reactivity series on the basis of displacement reactions.

- Oxidation is the loss of electrons.

- Reduction is the gain of electrons.

4.5

Metal carbonates

Learning objectives

After this topic, you should know:

- what happens when dilute acid is added to a metal carbonate

- how the test for carbon dioxide gas works

- what happens in the thermal decomposition of metal carbonates

- some uses of limestone.

Buildings and statues made of limestone suffer badly from damage by acid rain. You might have noticed statues where the fine features have been lost. Limestone, which is quarried from the ground, is mostly calcium carbonate. This reacts with acid, giving off carbon dioxide gas in the reaction.

Metal carbonates react with acids to give a salt, water, and carbon dioxide. For calcium carbonate the reaction with hydrochloric acid is:

calcium carbonate + hydrochloric acid → calcium chloride + water + carbon dioxide

The balanced symbol equation, including state symbols, is:

$$CaCO_3(s) + 2HCl(aq) \rightarrow CaCl_2(aq) + H_2O(l) + CO_2(g)$$

Powdered limestone is used to raise the pH of acidic soils or lakes affected by acid rain, making use of the reaction between calcium carbonate and acid.

Testing for carbon dioxide

You can use a simple test to verify that the gas given off is carbon dioxide. Carbon dioxide turns **limewater** solution cloudy. The test works as follows:

- Limewater is a solution of calcium hydroxide. It is alkaline.
- Carbon dioxide is a weakly acidic gas so it reacts with the alkaline limewater.
- In this reaction tiny solid particles of insoluble calcium carbonate are formed as a precipitate.
- The reaction is:

calcium hydroxide + carbon dioxide → calcium carbonate + water
(limewater)   (an insoluble
                                                    white precipitate)

$$Ca(OH)_2(aq) + CO_2(g) \rightarrow CaCO_3(s) + H_2O(l)$$

- This precipitate of calcium carbonate makes the limewater turn cloudy. This is because light can no longer pass through the solution with the particles of white solid calcium carbonate suspended in it.

Figure 1 *Limestone, made up mainly of calcium carbonate, is attacked and damaged by acids*

 Did you know … ?

Sculptures from the Parthenon (a temple), built by the ancient Greeks in Athens, have had to be removed and replaced by copies to avoid any more damage from acid pollution from vehicle exhausts.

Practical

Acid plus metal carbonates

Set up the apparatus as shown.

Try the test with some other carbonates, such as those of magnesium, copper, zinc, and sodium.

Record your observations.

- What conclusion can you draw?

Dilute hydrochloric acid

Calcium carbonate

Limewater

Safety: Wear eye protection. Some of the chemicals are harmful.

Thermal decomposition of carbonates

The calcium carbonate in limestone decomposes when it is heated. The reaction produces calcium oxide and carbon dioxide:

$$CaCO_3(s) \rightarrow CaO(s) + CO_2(g)$$
$$\text{(lime or quicklime)}$$

This reaction takes place in industry in a lime kiln. The calcium oxide made can be reacted with water to make a cheap alkali, calcium hydroxide. The **thermal decomposition** reaction shown above is also important when limestone is used in making **cement** and glass, as well as in the extraction of iron in a blast furnace.

So do other metal carbonates also decompose when heated?

Figure 2 *The thermal decomposition of copper(II) carbonate*

Practical

Investigating carbonates

You can investigate the thermal decomposition of carbonates by heating samples in a Bunsen flame. You will have samples of the carbonates listed below.

Powdered carbonate samples: sodium carbonate, potassium carbonate, calcium carbonate, magnesium carbonate, zinc carbonate, copper carbonate

- What observations might tell you if a sample decomposes when you heat it?
- How could you test any gas given off?

Safety: It is important to remove the delivery tube from the limewater before you stop heating the carbonate. If you don't, the cold limewater will be 'sucked back' into the hot boiling tube causing it to smash.
You must wear eye protection when doing this practical.
Some chemicals will be harmful.

Investigations like this show that many metal carbonates decompose when they are heated in a Bunsen flame. They form the metal oxide and carbon dioxide – just as calcium carbonate does. Sodium and potassium carbonate do not decompose at the temperature of the Bunsen flame. They need a higher temperature.

Magnesium carbonate decomposes like this:

$$MgCO_3(s) \rightarrow MgO(s) + CO_2(g)$$

Summary questions

1 Give a general word equation for:

   a the reaction of a metal carbonate plus an acid,

   b the thermal decomposition of a metal carbonate.

2 Write a word equation plus a balanced symbol equation, including state symbols, for the reaction of sodium carbonate powder with dilute hydrochloric acid.

3 Copper(II) carbonate, $CuCO_3$, undergoes thermal decomposition. Write the balanced symbol equation, including state symbols, for this reaction.

4 Write the balanced symbol equation for the reaction of copper carbonate with dilute nitric acid.

5 Explain in detail why limewater turns cloudy in the test for carbon dioxide.

Key points

- Carbonates react with dilute acid to form a salt, water, and carbon dioxide.

- Limewater turns cloudy in the test for carbon dioxide gas. A precipitate of insoluble calcium carbonate causes the cloudiness.

- Metal carbonates decompose on heating to form the metal oxide and carbon dioxide.

- Limestone, containing mainly calcium carbonate ($CaCO_3$), is quarried and can be used as a building material, or powdered and used to control acidity in the soil. It can also be used in the manufacture of cement, glass, and iron, as well as producing calcium oxide (lime) when heated.

Chapter summary questions

1 Imagine that a new metal, given the symbol X, has been discovered. It lies between calcium and magnesium in the reactivity series.

a Describe its reaction when heated in steam and give a word equation and a balanced symbol equation, including state symbols. (Metal X forms 2+ ions.)

b Describe the reaction of X with dilute sulfuric acid and give word and symbol equations.

c Why can't you be sure how X will react with cold water?

d X is added to a solution of copper(II) nitrate.
   i Explain what you would expect to see happen, including a balanced symbol equation with state symbols.
   ii Write an ionic equation showing what happens in the change.
   iii Explain which species is oxidised and which is reduced, using ionic half equations.

e Another new metal, Y, does not react with water or steam, but there is a slight reaction with warm dilute acid.
   i Where would you place metal Y in the reactivity series?
   ii Explain what you would expect to happen if metal Y was added to magnesium sulfate solution.

2 Decide which of the pairs below will react. If they do react, complete the word equations and balanced symbol equations, including state symbols:

a zinc + copper(II) oxide

b iron + zinc nitrate

c iron + magnesium oxide

d magnesium + copper(II) sulfate

3 The reaction between aluminium powder and iron(III) oxide is used in the rail industry.

a i Write a word equation and a balanced symbol equation for the reaction that takes place.
   ii What is the name for this type of reaction?

b Compare the reaction described above with the reaction you would expect to see between powdered aluminium and copper(II) oxide.

c Predict what would happen if you heated a mixture of aluminium oxide and iron.

d Explain why the uses of aluminium metal are surprising given its position in the reactivity series.

4 Choose from this list to name the metal described in **a** to **e**.

chromium

gold

aluminium

copper

magnesium

a A precious metal, often alloyed with copper to make it more hard-wearing.

b A metal that is alloyed with nickel and iron in stainless steel.

c A low density metal used to make alloys for the aircraft industry.

d A metal that is extracted from its solution using scrap iron.

e A metal that reacts very slowly with cold water but reacts vigorously when heated in steam, forming a white basic oxide.

Practice questions

1 One way to compare the reactivities of metals is to add them to acids. Some students added samples of five metals to separate tubes containing dilute hydrochloric acid. The teacher provided the metal labelled X. The amount of metal was the same in each experiment.

a To make the comparison fair, some other factors should be the same in each experiment.

  i State **two** properties/factors applying to the acid that should be the same. (2)

  ii State **one** other factor applying to the metal that should be the same. (1)

b The students recorded their observations in a table.

Metal	Observation
copper	nothing happened
iron	very few bubbles
magnesium	very large numbers of bubbles
X	very few bubbles
zinc	quite a lot of bubbles

  i Which metal in the table is the most reactive? (1)

  ii Why is it not possible to deduce that iron is the least reactive metal in the table? (1)

c The teacher suggested that a better method to compare the reactivities of the metals is to see how long it takes to collect a certain volume of the gas formed.

  Name **two** pieces of apparatus needed in the suggested method. (2)

2 The reactivities of metals can be compared by using displacement reactions.

a A more reactive metal will displace a less reactive metal from its salt in solution. An equation for a displacement reaction is:

  zinc + copper(II) sulfate → zinc sulfate + copper

  Write a symbol equation for this reaction. (2)

b The equation for another displacement reaction is:

  $Fe + CuCl_2 \rightarrow FeCl_2 + Cu$

  Write the simplest ionic equation for this reaction. (1)

c The reaction in **b** is a redox reaction.

  i State the meaning of the term 'redox'. (1)

  ii State and explain, in terms of redox and electrons, what happens to the iron in this reaction. (2)

5.1

Electrolysis

Learning objectives

After this topic, you should know:

- what happens in electrolysis
- the type of substances that can be electrolysed
- the products of electrolysis.

Figure 1 *The first person to explain electrolysis was Michael Faraday. He worked on this and many other problems in science nearly 200 years ago.*

The word **electrolysis** means 'breaking down using electricity'. In electrolysis, an electric current is used to break down an ionic substance. The substance that is broken down by electrolysis is called the **electrolyte**.

To set up an electrical circuit for electrolysis, there are two electrodes which dip into the electrolyte. The electrodes are conducting rods. One of these is connected to the positive terminal of a power supply. This positive electrode is called the **anode**. The other electrode is connected to the negative terminal and is called the **cathode**.

The electrodes are often made of an unreactive (or **inert**) substance. This is often graphite or sometimes platinum. This is so the electrodes do not react with the electrolyte or the products made in electrolysis.

During electrolysis, positively charged ions move to the cathode (negative electrode). At the same time, the negative ions move to the anode (positive electrode).

When the ions reach the electrodes they lose their charge and become elements. Gases may be given off or metals deposited at the electrodes. This depends on the compound used and whether it is molten or dissolved in water.

Demonstration

The electrolysis of molten lead bromide

- This demonstration needs a fume cupboard because bromine is toxic and corrosive.
- When does the bulb light up?
- What is observed at each electrode?

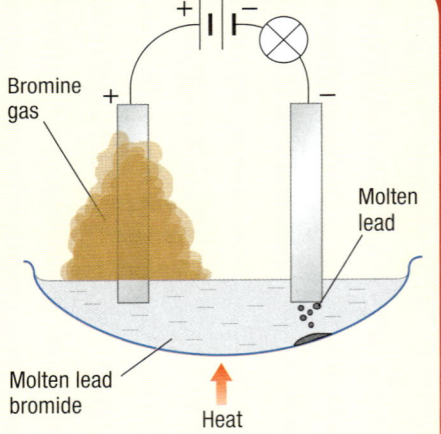

Figure 2 *Passing electricity through molten lead bromide. It forms molten lead and brown bromine gas as the electrolyte is broken down by the electricity.*

Safety: Warn asthmatics. Wear chemical splashproof eye protection. Lead bromide and bromine are toxic.

Figure 2 above shows how electricity breaks down lead bromide into lead and bromine. The reaction that is happening is:

$$\text{lead bromide} \rightarrow \text{lead} + \text{bromine}$$
$$PbBr_2(l) \rightarrow Pb(l) + Br_2(g)$$

Lead bromide is an ionic substance. Ionic substances do not conduct electricity when they are solid as their ions are in fixed positions in their giant lattice. But once they are melted, the ions are free to move around within the liquid and carry their charge towards the electrodes.

The positive lead ions (Pb^{2+}) move towards the cathode (negative electrode). At the same time, the negatively charged bromide ions (Br^-) move towards the anode (positive electrode).

Notice the state symbols in the equation. They tell us that the lead bromide and the lead are molten at the temperature in the dish. The bromine is given off as a gas.

Electrolysis of solutions

Many ionic substances have very high melting points so it takes a lot of energy to melt them and free the ions to move to electrodes in electrolysis. However, some ionic substances dissolve in water and when this happens the ions also become free to move around.

When electrolysing ionic compounds in solution, and not as molten compounds, it is more difficult to predict what will be formed. This is because water also forms ions so the products at each electrode are not always exactly what you would expect (see Topic 5.2 *Changes at the electrodes*). Only metals of very low reactivity, below hydrogen in the reactivity series, are deposited from their aqueous solutions in electrolysis.

When you electrolyse an aqueous solution of copper(II) bromide, copper ions (Cu^{2+}) move to the cathode (negative electrode). The bromide ions (Br^-) move to the anode (positive electrode). Copper(II) bromide is split into its elements at the electrodes (see Figure 3):

$$copper(\text{II}) \text{ bromide} \rightarrow copper + bromine$$
$$CuBr_2(aq) \rightarrow Cu(s) + Br_2(aq)$$

The state symbols in the equation tell us that the copper bromide is dissolved in water, the copper is formed as a solid, and the bromine formed dissolves in the water.

Covalent compounds cannot usually be electrolysed unless they react (or ionise) in water to form ions. For example, acids in water always contain $H^+(aq)$ ions plus negatively charged aqueous ions.

🔗 links
Remind yourself of the relative reactivity of metals and hydrogen in 4.3 'The reactivity series' and 4.4 'Displacement reactions'.

🔗 links
For more information about the effect of water in electrolysis, see 5.2 'Changes at the electrodes' and 5.4 'Electrolysis of brine'.

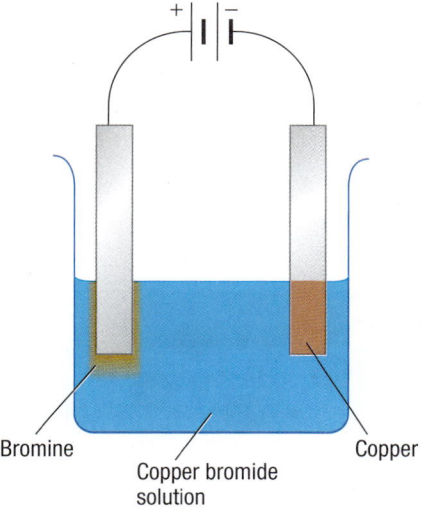

Figure 3 *If we dissolve copper bromide in water, we can decompose it by electrolysis. Copper metal is formed at the cathode (negative electrode). Brown bromine appears in solution around the anode (positive electrode).*

Summary questions

1   **a**   What is electrolysis?
    b   What is the name of the substance broken down by electrolysis?
    c   What type of bonding is present in compounds that can be electrolysed?

2   Predict the products formed at the cathode and anode when the following compounds are melted and then electrolysed:
    a   zinc iodide   **d**   sodium oxide
    b   lithium bromide   **e**   potassium chloride
    c   iron(III) fluoride

3   Write a balanced symbol equation, including state symbols, for the electrolytic decomposition of molten sodium chloride.

4   **a**   Which of the following solutions would deposit a metal at the cathode during electrolysis?

       $KBr(aq)$, $CaCl_2(aq)$, $FeBr_2(aq)$, $CuCl_2(aq)$, $Na_2SO_4(aq)$, $AgNO_3(aq)$

    b   How did you decide on your answer to part **a**?

5   Solid ionic substances do not conduct electricity. Using words and diagrams, explain why they conduct electricity when molten or in aqueous solution, but not when solid.

Key points

- Electrolysis breaks down a substance using electricity.

- Ionic compounds can only be electrolysed when they are molten or dissolved in water. This is because their ions are then free to move and carry their charge to the electrodes.

- In electrolysis, positive ions move to the cathode (negative electrode) whilst negative ions move to the anode (positive electrode).

5.2 Changes at the electrodes

Learning objectives

After this topic, you should know:

- what happens to the ions during electrolysis
- how to represent what happens at each electrode
- how water affects the products of electrolysis.

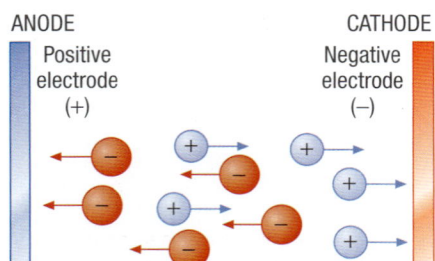

ANODE   CATHODE

Positive electrode (+)   Negative electrode (−)

Figure 1 *An ion always moves towards the oppositely charged electrode*

🔗 links

For more information on reduction and oxidation in terms of electron transfer, look back to 4.4 'Displacement reactions'.

Study tip

Oxidation and reduction reactions do not have to involve oxygen. More generally they involve the transfer of electrons. Remember **OILRIG** – **O**xidation **I**s **L**oss (of electrons), **R**eduction **I**s **G**ain (of electrons).

During electrolysis, mobile ions move towards the electrodes. The direction they move in depends on their charge. As you saw in Topic 5.1, positive ions move towards the cathode (negative electrode). Negative ions move towards the anode (positive electrode).

When ions reach an electrode, they either lose or gain electrons, depending on their charge.

Negatively charged ions *lose* electrons to become neutral atoms. Positively charged ions *gain* electrons to become neutral atoms.

The easiest way to think about this is to look at an example.

Think back to the electrolysis of molten lead bromide. The lead ions (Pb^{2+}) move towards the negative electrode. When they get there, each ion gains two electrons to become a neutral lead atom.

Gaining electrons is called **reduction**. You say that the lead ions are **reduced**. 'Reduction' is simply another way of saying 'gaining electrons'.

The negatively charged bromide ions (Br^-) move towards the positive electrode. Once there, each ion loses its one extra electron to become a neutral bromine atom. Two bromine atoms then form a covalent bond to make a bromine molecule, Br_2.

Losing electrons is called **oxidation**. You say that the bromide ions are **oxidised**. 'Oxidation' is another way of saying 'losing electrons'.

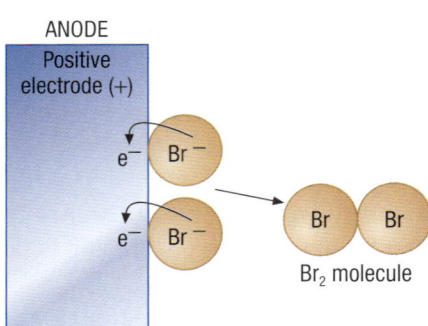

Half equations

You can represent what is happening at each electrode using **half equations**.

At the cathode (negative electrode) you get reduction of a positive ion:

$$Pb^{2+} + 2e^- \rightarrow Pb$$

At the anode (positive electrode) you get oxidation of a negative ion:

$$2Br^- \rightarrow Br_2 + 2e^-$$

Sometimes half equations at the anode can show the electrons being removed from negative ions, like this:

$$2Br^- - 2e^- \rightarrow Br_2$$

You can write the half equation for negative ions either way. They both show the same oxidation of the negatively charged ions.

The effect of water

In aqueous solutions, electrolysis is more complex because of the ions formed by water as it ionises:

$$H_2O(l) \rightleftharpoons H^+(aq) + OH^-(aq)$$
$$\text{hydrogen ions} \quad \text{hydroxide ions}$$

There is a rule for working out what will happen. Remember that if two elements can be produced at an electrode, the less reactive element will usually be formed. In solutions, you will usually have positively charged metal ions and H^+ ions (from water) attracted to the cathode (negative electrode).

Look at Figure 2. It shows what happens in the electrolysis of a solution of a potassium compound. Hydrogen is less reactive than potassium, so hydrogen is produced at the cathode rather than potassium.

At cathode (−):

$$2H^+(aq) + 2e^- \rightarrow H_2(g)$$

So what happens at the anode in the electrolysis of aqueous solutions?

Hydroxide ions, $OH^-(aq)$, from water are often discharged. When hydroxide ions are discharged, you see oxygen gas given off at the anode (positive electrode).

At anode (+):

$$4OH^-(aq) \rightarrow 2H_2O(l) + O_2(g) + 4e^-$$

This happens unless the solution contains a reasonably high concentration of a halide ion, that is, Group 7 ions, such as $Cl^-(aq)$. In this case, the halide ion is discharged and the halogen is formed:

$$2Cl^-(aq) \rightarrow Cl_2(g) + 2e^-$$

So the 'order of discharge' (starting with the easiest) at the anode is:

halide ion > hydroxide > all other negatively charged ions.

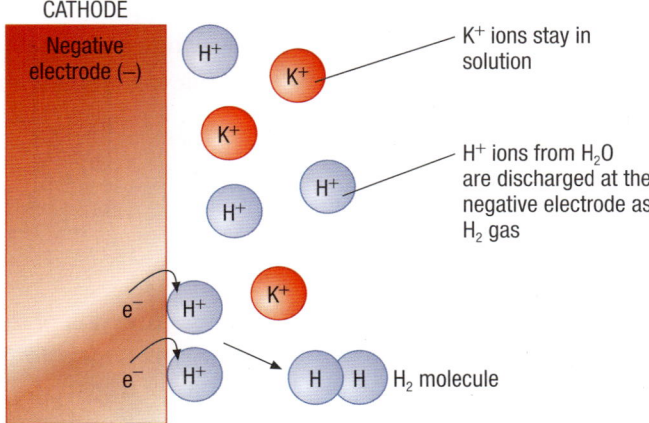

Figure 2 *Here is the cathode in the electrolysis of a solution of a potassium compound. Hydrogen is less reactive than potassium, so hydrogen gas is given off at the electrode.*

Summary questions

1 a i How do negatively charged ions become neutral atoms in electrolysis?
   ii Is this reduction or oxidation of the negatively charged ion?
  b i How do positively charged ions become neutral atoms in electrolysis?
   ii Is this reduction or oxidation of the positively charged ion?

2 Predict what is formed at each electrode in the electrolysis of:
  a molten lithium oxide
  b copper chloride solution
  c sodium sulfate solution.

3 Copy and balance the following half equations where necessary:
  a $Cl^- \rightarrow Cl_2 + e^-$
  b $Br^- \rightarrow Br_2 + e^-$
  c $Mg^{2+} + e^- \rightarrow Mg$
  d $Al^{3+} + e^- \rightarrow Al$
  e $K^+ + e^- \rightarrow K$
  f $H^+ + e^- \rightarrow H_2$
  g $O^{2-} \rightarrow O_2 + e^-$
  h $OH^- \rightarrow O_2 + H_2O + e^-$

Key points

- In electrolysis, the ions move towards the oppositely charged electrodes.

- At the electrodes, negative ions are oxidised (at the anode) whilst positive ions are reduced (at the cathode).

- When electrolysis happens in aqueous solution, the less reactive element, either hydrogen or the metal, is usually produced at the cathode. At the anode, you often get oxygen gas given off from discharged hydroxide ions produced from water.

5.3 The extraction of aluminium

Learning objectives

After this topic, you should know:

- how aluminium is obtained from aluminium oxide
- why cryolite is used in the process
- what happens at each electrode in the process.

You already know that aluminium is a very important metal. The uses of the metal or its alloys include:

- pans
- overhead power cables
- aeroplanes
- cooking foil
- drink cans
- window and patio door frames
- bicycle frames and car bodies.

Figure 1 *Aluminium alloys have a low density but are very strong*

Extracting aluminium from its ore

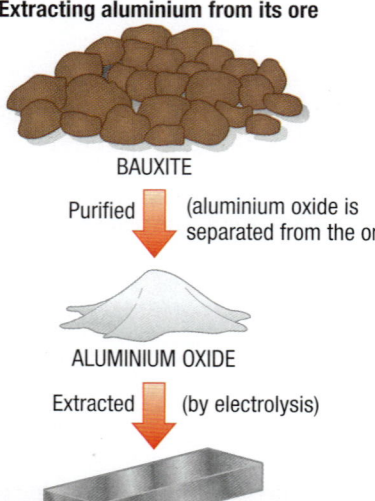

BAUXITE

Purified (aluminium oxide is separated from the ore)

ALUMINIUM OXIDE

Extracted (by electrolysis)

ALUMINIUM METAL

Figure 2 *Extracting aluminium from its ore. This process requires a lot of energy. The energy is needed for melting and electrolysing the oxide.*

Aluminium is quite a reactive metal. It is less reactive than magnesium but more reactive than zinc or iron. Carbon is not reactive enough to use in its extraction so you must use electrolysis. The compound electrolysed is aluminium oxide, Al_2O_3.

Aluminium oxide is extracted from the ore called bauxite. The ore is mined by open-cast mining. Bauxite contains mainly aluminium oxide. However, it is mixed with other rocky impurities. So the first step is to separate aluminium oxide from the ore. The impurities contain a lot of iron(III) oxide. This colours the waste solution from the separation process rusty brown. The brown wastewater has to be stored in large lagoons.

Electrolysis of aluminium oxide

To electrolyse the aluminium oxide it must first be melted. This enables the ions to move to the electrodes.

Unfortunately aluminium oxide has a very high melting point of 2050 °C. However, chemists have found a way of saving at least some energy. This is done by mixing the aluminium oxide with molten cryolite. Cryolite is another ionic compound. The molten mixture can be electrolysed at about 850 °C. The electrical energy transferred to the electrolysis cells keeps the mixture molten (see Figure 3).

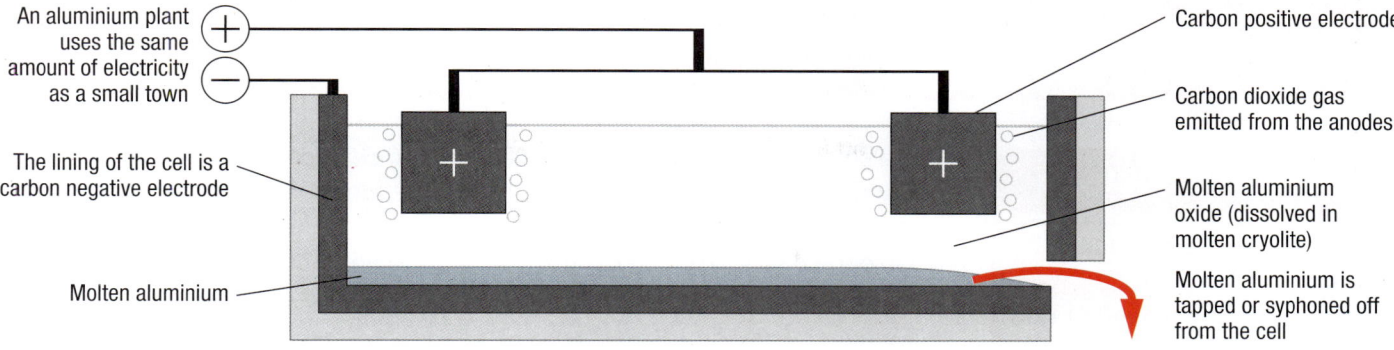

An aluminium plant uses the same amount of electricity as a small town

The lining of the cell is a carbon negative electrode

Molten aluminium

Carbon positive electrode

Carbon dioxide gas emitted from the anodes

Molten aluminium oxide (dissolved in molten cryolite)

Molten aluminium is tapped or syphoned off from the cell

Figure 3 *A cell used in the extraction of aluminium by electrolysis*

The overall reaction in the electrolysis cell is:

aluminium oxide $\xrightarrow{\text{electrolysis}}$ aluminium + oxygen

$$2Al_2O_3(l) \xrightarrow{} 4Al(l) + 3O_2(g)$$

At the cathodes (negative electrodes):

Each aluminium ion (Al^{3+}) gains three electrons. The ions turn into aluminium atoms. We say that the Al^{3+} ions are reduced (as they gain electrons) to form Al atoms:

$$Al^{3+}(l) + 3e^- \rightarrow Al(l)$$

The aluminium metal formed is molten at the temperature of the cell and collects at the bottom. It is siphoned or tapped off.

At the anodes (positive electrodes):

Each oxide ion (O^{2-}) loses two electrons. The ions turn into oxygen atoms. We say that the O^{2-} ions are oxidised (as they lose electrons) to form oxygen atoms. These bond in pairs to form molecules of oxygen gas (O_2):

$$2O^{2-}(l) \rightarrow O_2(g) + 4e^-$$

The oxygen reacts with the hot carbon anodes, making carbon dioxide gas:

$$C(s) + O_2(g) \rightarrow CO_2(g)$$

So the carbon anodes gradually burn away. They need to be replaced in the cells regularly.

 Did you know …?

Recycling one aluminium can can save enough energy to run a TV for about 3 hours.

Summary questions

1 a Explain why aluminium oxide must be molten for electrolysis to take place.
  b Why is aluminium oxide dissolved in molten cryolite in the extraction of aluminium?

2 Why are the carbon anodes replaced regularly in the industrial electrolysis of aluminium oxide?

3 a Write half equations for the changes at each electrode in the electrolysis of molten aluminium oxide.
  b Explain which ions are oxidised and which ions are reduced in the electrolysis of molten aluminium oxide.

4 a Explain why the extraction of aluminium requires so much energy.
  b Explain why aluminium metal was only discovered in the early 1800s despite it being the most common metallic element in the Earth's crust.

Key points

- Aluminium oxide, from the ore bauxite, is electrolysed in the manufacture of aluminium metal.

- The aluminium oxide is mixed with molten cryolite to lower its melting point.

- Aluminium forms at the cathode (negative electrode) and oxygen forms at the anode (positive electrode).

- The carbon anodes are replaced regularly as they gradually burn away.

5.4 Electrolysis of brine

The electrolysis of **brine** (concentrated sodium chloride solution) is a very important industrial process. When an electric current is passed through brine you get three products:

- Chlorine gas is produced at the anode (positive electrode).
- Hydrogen gas is produced at the cathode (negative electrode).
- Sodium hydroxide solution is also formed.

You can summarise the electrolysis of brine as:

$$\text{sodium chloride solution} \xrightarrow{\text{electrolysis}} \text{hydrogen gas} + \text{chlorine gas} + \text{sodium hydroxide solution}$$

At the anode (+):

The negative chloride ions (Cl^-) are attracted to the positive electrode. When they get there, they each lose one electron. The chloride ions are oxidised, as they lose electrons. The chlorine atoms bond together in pairs and are given off as chlorine gas (Cl_2). The half equation at the anode is:

$$2Cl^-(aq) \rightarrow Cl_2(g) + 2e^-$$

[This can also be written as: $2Cl^-(aq) - 2e^- \rightarrow Cl_2(g)$]

At the cathode (–):

There are H^+ ions in brine, formed when water breaks down:

$$H_2O(l) \rightleftharpoons H^+(aq) + OH^-(aq)$$

These positive hydrogen ions are attracted to the negative electrode. The sodium ions (Na^+) are also attracted to the same electrode. But remember in Topic 5.2, you saw what happens when two different ions are attracted to an electrode. It is the less reactive element that gets discharged. In this case, hydrogen ions are discharged and sodium ions stay in solution.

When the H^+ ions get to the negative electrode, they each gain one electron. The hydrogen ions are reduced, as they each gain an electron. The hydrogen atoms formed bond together in pairs and are given off as hydrogen gas (H_2). The half equation at the cathode is:

$$2H^+(aq) + 2e^- \rightarrow H_2(g)$$

The remaining solution:

You can test the solution around the cathode (negative electrode) with indicator. It shows that the solution is alkaline. This is because you can think of brine as containing aqueous ions of Na^+ and Cl^- (from salt) and H^+ and OH^- (from water). The $Cl^-(aq)$ and $H^+(aq)$ ions are removed during electrolysis. So this leaves a solution containing $Na^+(aq)$ and $OH^-(aq)$ ions, that is, a solution of sodium hydroxide, $NaOH(aq)$.

Look at the way brine is electrolysed in industry in Figure 1.

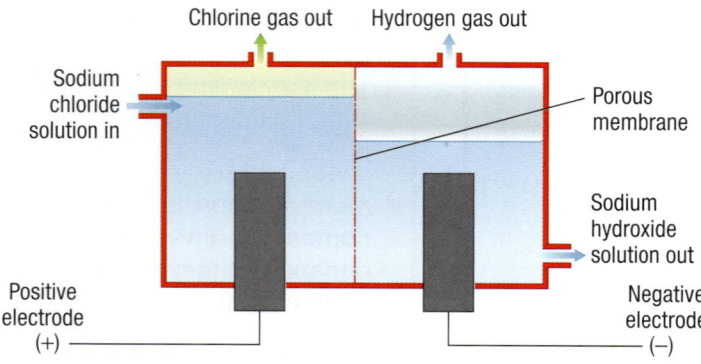

Figure 1 In industry, brine can be electrolysed in a cell in which the two electrodes are separated by a porous membrane. This is called a diaphragm cell.

Practical

Electrolysing brine in the lab

Turn off the electricity once the tubes are nearly full of gas to avoid inhaling chlorine gas (toxic).

■ How can you positively test for the gases collected?

Test the solution near the cathode (negative electrode) with universal indicator solution.

■ What does the indicator tell us?

Safety: Wear eye protection. Do not smell the gas.

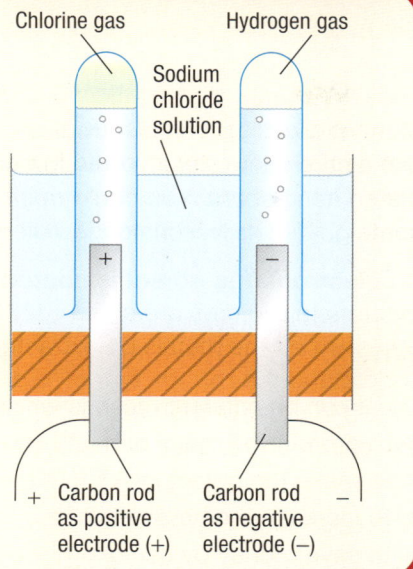

Chlorine gas   Hydrogen gas

Sodium chloride solution

+   −

+ Carbon rod as positive electrode (+)   Carbon rod as negative electrode (−) −

links

For information about what happens when two ions are attracted to an electrode, look back to 5.2 'Changes at the electrodes'.

Using chlorine

You can react chlorine with the sodium hydroxide produced in the electrolysis of brine. This makes a solution of bleach. Bleach is very good at killing bacteria.

Chlorine is also important in making many other disinfectants, as well as plastics such as PVC.

Using hydrogen

The hydrogen made by electrolysing brine is particularly pure. This makes it very useful in the food industry. Margarine is made by reacting hydrogen with vegetable oils.

Using sodium hydroxide

The sodium hydroxide from the electrolysis of brine is an important alkali in industry. It is used to make soap and paper, as well as bleach (see above).

Figure 2 *The chlorine made when we electrolyse brine is used to kill bacteria in drinking water, and also in swimming pools*

Summary questions

1 a What are the **three** products made when brine is electrolysed?
  b What are they used for?

2 For the electrolysis of brine, write half equations, including state symbols, for the reactions:
  a at the anode
  b at the cathode.

3 It is also possible to electrolyse *molten* sodium chloride.
  a Compare the products formed with those from the electrolysis of sodium chloride solution.
  b Explain any differences.

Key points

■ When brine is electrolysed, three products are formed – chlorine gas, hydrogen gas, and sodium hydroxide solution (an alkali).

■ Chlorine is used to make bleach, which kills bacteria, and to make plastics.

■ Hydrogen is used to make margarine.

■ Sodium hydroxide is used to make bleach, paper, and soap.

5.5 Electroplating

Learning objectives

After this topic, you should know:

■ why objects are electroplated

■ how to electroplate a metal object.

Figure 1 *Chromium-plated objects look very shiny and attractive. The chromium layer does not corrode away so it protects the steel beneath from rusting.*

Most of us will use an electroplated object at some time each day. You might use a chromium-plated kettle to boil water or ride a bicycle with chromium-plated handlebars. You could open a tin-plated steel can for a meal or put on some gold- or silver-plated jewellery.

An electroplated object is coated with a thin layer of metal by electrolysis.

Why are objects electroplated?

There can be different reasons why scientists electroplate objects. These include:

■ to protect the metal beneath from corroding

■ to make the object look more attractive

■ to increase the hardness of a surface and its resistance to scratching

■ to save money by using a thin layer of a precious metal instead of the pure expensive metal. This also helps people who are allergic to nickel – a metal often used to make cheap jewellery.

Electroplating saves money by making cheaper jewellery. However, using electroplating to protect large metal surfaces against rusting and damage makes things more expensive. In the long term, though, this can still make economic sense for some objects because they don't have to be replaced so often.

Electroplating a metal object

You can try to zinc-plate some copper foil in the experiment below.

Figure 2 *So-called 'tin' cans actually contain very little tin. The layer on the steel can be only a few thousandths of a millimetre thick! The tin keeps air and water away from the iron in steel and stops it rusting – at least until the tin gets scratched! Tin is quite a soft metal, unlike chromium.*

Practical

Zinc-plating copper metal

Your teacher will melt some wax in a metal tray. Using tongs you can dip in a cleaned piece of copper foil.

Let the wax set. Then scratch a simple design in the wax. You want the design to be plated with zinc so get this area as free from wax as possible.

Set up the apparatus as shown in the diagram. The zinc-plating solution is made by adding $30\,cm^3$ of $0.4\,mol/dm^3$ sodium hydroxide solution to $5\,cm^3$ of $0.1\,mol/dm^3$ zinc sulfate solution and stirring. Using a small current for a long time will give best results.

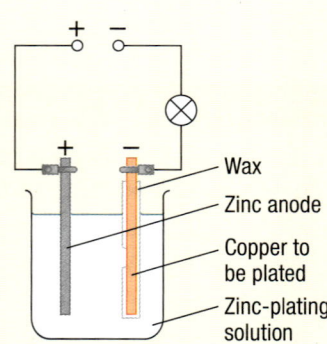

When you have finished, rinse the copper foil in water, dry, then scrape off the rest of the wax.

■ What happens at the negative electrode?

Safety: Wear gloves and eye protection. Sodium hydroxide solution is an irritant.

Explaining the zinc-plating experiment

The metal object to be plated (the copper foil in this case) is used as the cathode (negative electrode). The anode (positive electrode) is made from the plating metal (in this case zinc). The electrolysis takes place in a solution containing ions of the plating metal. In the previous experiment you used zinc ions in the solution.

At the anode (+) made of the plating metal:

Zinc atoms in the electrode are oxidised. They lose two electrons each and form zinc (Zn^{2+}) ions which leave the anode and go into the solution. The half equation is:

$$Zn(s) \rightarrow Zn^{2+}(aq) + 2e^-$$

At the cathode (–) to be plated:

Zinc ions (Zn^{2+}) from the solution are reduced. They gain two electrons and form zinc atoms which are deposited on the copper cathode. The half equation is:

$$Zn^{2+}(aq) + 2e^- \rightarrow Zn(s)$$

Copper plating

Here are the half equations at each electrode if an object is electroplated by copper:

At the anode (positive electrode): $Cu(s) \rightarrow Cu^{2+}(aq) + 2e^-$

At the cathode (negative electrode) to be plated: $Cu^{2+}(aq) + 2e^- \rightarrow Cu(s)$

The electrolyte used contains copper(II) ions, such as a solution of copper(II) sulfate.

Did you know ... ?

It is not only metal objects that can be electroplated. Scientists can now electroplate plastic objects as well. The object to be plated is first coated in a 'paint' containing tiny particles of graphite. Once dry, the object has a layer of graphite that will conduct electricity. It can then be electroplated.

Summary questions

1 Name **four** metals that are used to electroplate objects.

2 List the reasons why objects get electroplated.

3 Look at the experiment described in the Practical box on the previous page.
   a Where are zinc atoms oxidised?
   b What is formed when zinc atoms are oxidised?

4 Draw a simple electrical circuit that could be used to copper-plate a steel ball.

5 In making 'chrome' objects, chromium metal is used to electroplate a steel object. The steel is first electroplated with nickel because chromium does not stick well on steel.
   a Give the half equation at the cathode for the nickel-plating process. Include state symbols in your answer. (The formula of a nickel ion is Ni^{2+}.)
   b Give the half equation at the cathode for the chromium-plating process, including state symbols. (The formula of a chromium ion is Cr^{3+}.)

Key points

- Objects can be electroplated to improve their appearance, to protect their surface, and to use smaller amounts of precious metals.

- The object to be electroplated is made the cathode (negative electrode) in an electrolysis cell. The plating metal is made the anode (positive electrode). The electrolyte contains ions of the plating metal.

5.6

Electrolysing copper sulfate solution

Learning objectives

After this topic, you should know:

- what happens at the electrodes in the electrolysis of copper(II) sulfate solution using inert and active electrodes

- the half equations at the electrodes in the electrolysis of copper(II) sulfate solution

- how to investigate the electrolysis of copper(II) sulfate solution.

You have now seen how electrolysis can be used to extract highly reactive metals, like aluminium. The other uses of electrolysis, such as electroplating, depend on electrodes which react – active electrodes.

In some of the experiments in this chapter, you have used carbon electrodes. These are called 'inert' electrodes. They take no part in the electrolysis. The hot carbon electrodes just carry electrons to and from the electrolyte. (Note that carbon electrodes can react with oxygen produced at an anode, forming carbon dioxide gas, as in the extraction of aluminium – see Topic 5.3 *The extraction of aluminium*.)

However, some metal electrodes do take part in electrolysis. These are called active electrodes. Let's look at copper as an example of an active electrode. You will electrolyse copper sulfate solution.

Required practical

Investigate the electrolysis of copper sulfate solution with graphite and copper electrodes

Make sure your copper electrodes are clean and shiny.

Use a pencil to mark one electrode + and the other −.

Weigh them with an accurate balance.

Set up the apparatus as shown in Figure 1.

- What do you observe?

After 10 minutes, rinse the electrodes with distilled water.

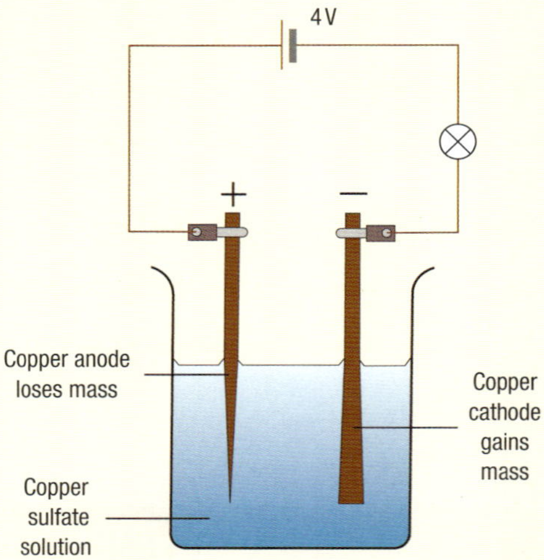

Dry them by dipping them in propanone and letting it evaporate.

Re-weigh the electrodes when they are dry.

- What do you find?
- Why would it be a good idea to repeat your experiment?

Now carry out the electrolysis of copper sulfate solution, but this time using carbon electrodes.

- What difference do you notice at the anode (positive electrode)?

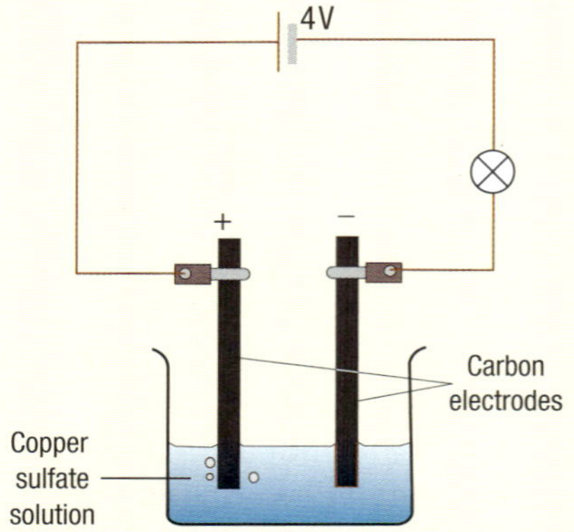

Safety: Wear eye protection. Propanone is flammable.

Figure 1 *The electrolysis of copper(II) sulfate solution with copper electrodes*

Figure 2 *The electrolysis of copper(II) sulfate solution with carbon electrodes*

When copper sulfate solution is electrolysed using copper electrodes, the mass of both electrodes changes. Remember that metals are always formed at the cathode (negative electrode). So not surprisingly, the cathode gets heavier.

The strange thing is that the anode loses mass. The copper anode is an active electrode. You will find that:

loss in mass of active electrode = **gain in mass of active electrode**
at the anode   **at the cathode**

Comparing inert and active electrodes

Let's compare the half equations at each electrode in the electrolysis of copper sulfate solution using carbon and copper electrodes:

Table 1 *Comparing the reactions at the cathode and anode in the electrolysis of copper(II) sulfate solution using different electrodes*

	Carbon electrodes	Copper electrodes
At the cathode (−)	$Cu^{2+}(aq) + 2e^- \longrightarrow Cu(s)$	$Cu^{2+}(aq) + 2e^- \longrightarrow Cu(s)$
	the reactions are the same	
At the anode (+)	$4OH^-(aq) \longrightarrow 2H_2O(l) + O_2(g) + 4e^-$ Oxygen gas is given off.	$Cu(s) \longrightarrow Cu^{2+}(aq) + 2e^-$ No oxygen is seen. Copper atoms on the anode lose 2 electrons and enter the solution as Cu^{2+} ions.

The copper anode is an active electrode. As copper atoms form on the cathode, copper atoms are lost from the anode. As copper ions leave the solution at the cathode, they are replaced in the solution at the anode.

Copper ions, $Cu^{2+}(aq)$, make copper sulfate solution blue.

- What do you think would happen to the colour of the copper sulfate solution using carbon electrodes?
- Why won't the blue colour fade using copper electrodes?

Did you know …?

Copper electrodes are used in a solution of copper(II) ions to purify copper. The impurities that collect as sludge at the bottom of the electrolysis cells contain the precious metals silver and gold.

Summary questions

1 Draw a table to compare the observations made at each electrode when copper(II) sulfate is electrolysed using copper electrodes and using carbon electrodes.

2 Explain why the copper electrodes in the practical in this topic are called 'active' electrodes whereas the carbon electrodes are called 'inert' electrodes.

3 a Write the half equations for the changes at each electrode in the electrolysis of copper(II) sulfate using copper electrodes.
  b Explain in terms of oxidation and reduction the changes that occur at each electrode in part **a**.

Key points

- The electrolysis of copper(II) sulfate solution using copper electrodes results in the copper anode (+) losing the same mass as the copper cathode (−) gains.

- In the electrolysis of copper (II) sulfate solution using carbon electrodes, copper is still deposited at the cathode (−). However, at the anode (+) oxygen gas is given off from hydroxide ions that come from the water in the copper(II) sulfate solution.

Chapter summary questions

1 Select **A** or **B** to describe correctly the anode and cathode in electrolysis for statements **a–f**.

> **A** Anode (+)   **B** Cathode (–)

a Positive ions move towards this.

b Oxidation happens here.

c This is connected to the negative terminal of the power supply.

d This is connected to the positive terminal of the power supply.

e Reduction happens here.

f Negative ions move towards this.

2 a Which of the following ions would move towards the anode (+) and which towards the cathode (–) during electrolysis?

> *potassium iodide calcium fluoride oxide magnesium aluminium bromide*

b Write a half equation for the discharge of:
  i magnesium ions
  ii bromide ions.

3 The diagram shows the electrolysis of sodium chloride solution in the laboratory.

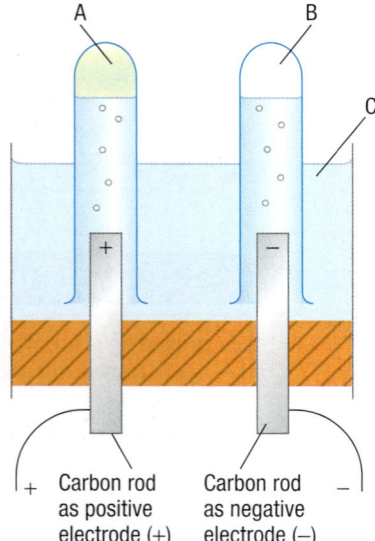

a Identify the products **A**, **B**, and **C** on the diagram.

b Give two uses for substance **A**.

c What would be the pH of the solution around the cathode?

d How would you carry out a positive test on product **B**?

e Write the half equations, including state symbols, for the changes at the anode and cathode.

4 Water can be slightly acidified and broken down into hydrogen and oxygen using electrolysis. The word equation for this reaction is:

> water → hydrogen + oxygen

a Write a balanced symbol equation for this reaction, including state symbols.

b In water, a small percentage of molecules ionise (split up). Write a balanced symbol equation, including state symbols, for the ionisation of water molecules.

c Write half equations to show what happens at the positive and negative electrodes in the electrolysis of water.

d Where does the energy needed to split water into hydrogen and oxygen come from during electrolysis?

5 Copy and complete the following half equations:

a $Li^+ \rightarrow Li$

b $Sr^{2+} \rightarrow Sr$

c $F^- \rightarrow F_2$

d $O^{2-} \rightarrow O_2$

6 Electrolysis can be used to produce a thin layer of metal on the surface of a metal object.

Using words and diagrams, describe how you would cover a small piece of iron with tin using an electrolyte of tin(II) nitrate. Make sure that you write down the half equations that describe the changes at the anode and cathode, and include the words 'oxidation' and 'reduction'.

7 A student half-filled two small beakers with copper(II) sulfate solution and labelled them **A** and **B**. He electrolysed solution **A** using copper electrodes, and solution **B** using carbon electrodes.

a Describe what would happen at each electrode in each of the solutions.

b Write the half equations at each electrode in:
  i beaker **A**
  ii beaker **B**.

c If the electrolysis cells were left running for long enough, eventually the current would stop in both circuits.

Explain why the current stops in each of the two beakers.

Practice questions

1 The diagram shows apparatus for the electrolysis of lead bromide.

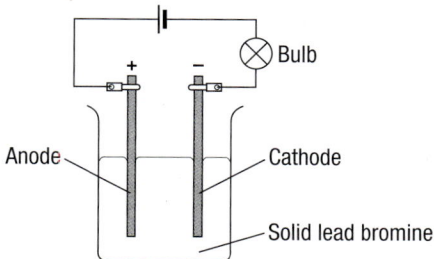

When the lead bromide is heated, it melts and observations can be recorded.

a A brown gas is seen at the anode. Name this gas and write a symbol equation for its formation. (3)

b Silvery globules are seen at the cathode. Name the substance in the globules and write a symbol equation for its formation. (3)

c The bulb does not light until the lead bromide is molten. What happens in the connecting wires when the bulb is lit? (1)

d Why does no electrolysis occur until the lead bromide has melted? (1)

e Why is the reaction at the cathode described as reduction? (1)

2 A student used the following apparatus to electrolyse dilute sulfuric acid.

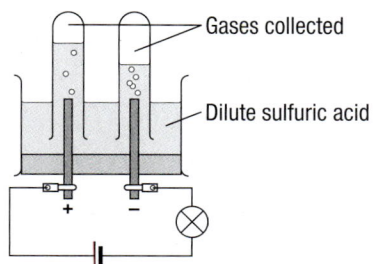

a What is the formula of the only positive ion in dilute sulfuric acid? (1)

b Write the equation for the reaction of the OH⁻ ion that occurs at the anode. (2)

c Which ion in dilute sulfuric acid is attracted to an electrode but is not discharged? (1)

d Predict what happens to the concentration of the sulfuric acid as electrolysis occurs. (1)

e Why is the volume of hydrogen gas collected approximately double the volume of the oxygen gas? (1)

3 The electrolysis of sodium chloride solution is used to manufacture hydrogen and chlorine.

a Why is the H^+ ion discharged in preference to the Na^+ ion at the cathode? (1)

b Why is the Cl^- ion discharged in preference to the OH^- ion at the anode? (1)

c Name the solution formed in this process. (1)

d i State **one** use for the hydrogen formed. (1)
ii State **one** use for the chlorine formed. (1)

4 Objects can be electroplated. For example, a copper spoon can be plated with nickel.

a i What should be used as the anode and cathode in this process? (1)
ii Suggest the name of a solution that could be used as the electrolyte. (1)
iii Write the half equation for the reaction at the anode. (1)

b Suggest **one** physical property of an object that would change when it is electroplated. (1)

c Objects made of plastic can be electroplated. Why are they coated with graphite before they are electroplated? (1)

5 Aluminium is made by the electrolysis of aluminium oxide.

Aluminium oxide is an ionic compound containing aluminium ions (Al^{3+}) and oxide ions (O^{2-}).

The diagram below shows the apparatus used to electrolyse aluminium oxide.

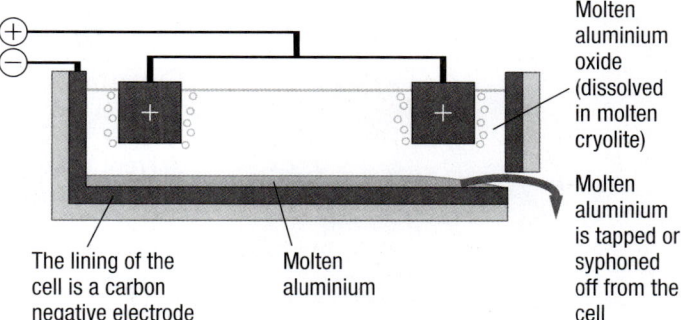

Molten aluminium oxide (dissolved in molten cryolite)

Molten aluminium is tapped or syphoned off from the cell

The lining of the cell is a carbon negative electrode

Molten aluminium

Use information in the diagram and your knowledge and understanding of this extraction to explain, as fully as you can, how aluminium and carbon dioxide are formed in this process. (6)

Separating mixtures

Learning objectives

After this topic, you should know:

- what a mixture is

- how to separate the components in a range of mixtures by:
  - filtration
  - distillation
  - crystallisation.

When analytical chemists are given an unknown sample to identify, it is often a mixture of different substances.

A mixture is made up of two or more elements or compounds which are not chemically combined together.

Mixtures are different to chemical compounds. Look at the table below:

Compounds	Mixtures
Compounds have a fixed composition (the ratio of elements present is always the same in any particular compound).	Mixtures have no fixed composition (the proportions vary depending on the amount of each substance mixed together).
Need chemical reactions to separate the elements in a compound.	The different elements or compounds in a mixture can be separated again more easily (by physical means using the differences in properties of each substance in the mixture).
There are chemical bonds between atoms of the different elements in the compound.	There are no chemical bonds between atoms of the different substances in a mixture.

Before the substances in a mixture are identified, they are separated from each other. As it states in the table, scientists can use physical means to achieve the separation. The techniques available include:

- filtration
- crystallisation
- distillation
- chromatography (see Topic 6.2).

These techniques all rely on differences in the physical properties of the substances in the mixture, such as solubilities or boiling points.

Filtration

This technique is used to separate substances that are insoluble in a particular solvent from those that are soluble. For example, you have probably previously tried to separate a mixture of sand and salt (sodium chloride) in science lessons (see Figure 1).

The sand collected on the filter paper can be washed with distilled water (to remove any salt left in solution on it) and dried in a warm oven to obtain the pure sand.

Crystallisation

To obtain a sample of pure salt from the sand/salt mixture, following filtration, you would need to get the sodium chloride from the solution in water (called the filtrate). The water is evaporated from the sodium chloride solution by heating in an evaporating dish on a water bath. The water bath is a gentler way of heating than heating the evaporating dish directly on a tripod and gauze.

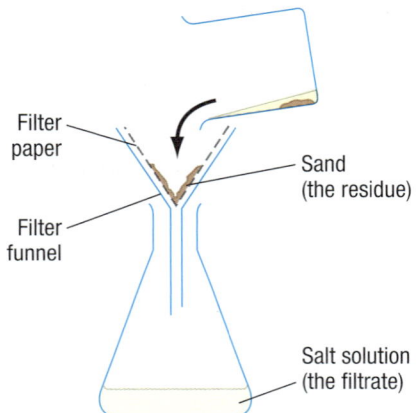

Figure 1 *Filtering a mixture of sand, salt, and water in the lab*

Filter paper
Filter funnel
Sand (the residue)
Salt solution (the filtrate)

Figure 2 *Filtering in the home – some people like filter coffee made from ground-up coffee beans. The solid bits that are insoluble in water get left on filter paper as a residue.*

Heating should be stopped when the solution is at the point of crystallisation, that is, when small crystals first appear around the edge of the solution or when crystals appear in a drop of solution extracted from the dish with a glass rod. Look at Figure 3. The rest of the water is then left to evaporate off the saturated solution at room temperature to get a good sample of crystals. A flat-bottomed crystallisation dish or Petri dish can be used for this final step to give a large surface area for the water to evaporate from.

Distillation

Crystallisation separates a soluble solid from a solvent but sometimes people need to collect the solvent itself instead of just letting it evaporate off into the air. For example, some countries with a lack of freshwater sources purify seawater to obtain usable water. Distillation allows us to do this.

In distillation a solution is heated and boiled to evaporate the solvent. The vapour given off is then cooled and condensed back into a liquid for collection (see Figure 4). The dissolved solids remain in the heated flask.

Distillation can also be used to separate mixtures of miscible liquids, such as ethanol and water. Miscible refers to liquids that dissolve in each other – they do not form separate layers as in immiscible liquids. The liquids will have different boiling points, so this property can be used to distil off and collect the liquid with the lowest boiling point first.

However, it is difficult to get pure liquids from mixtures of liquids by simple distillation, as vapour is given off from liquids before they actually reach their boiling point. This is especially the case where the liquids in the mixture have similar boiling points. So to aid separation you can add a fractionating column to the apparatus in Figure 4. This is usually a tall glass column filled with glass beads, fitted vertically on top of the flask being heated. The vapours must pass over and between the glass beads before they reach the condenser. The substance with the higher boiling point will condense more readily on the glass beads and drop back down into the flask beneath.

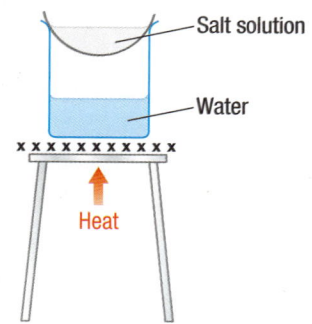

Figure 3 *Crystallising sodium chloride from its solution in water*

∞ links

To see crystallisation in action, see 7.2 'Making salts from metals or insoluble bases' and 7.3 'Making salts by neutralisation and precipitation'.

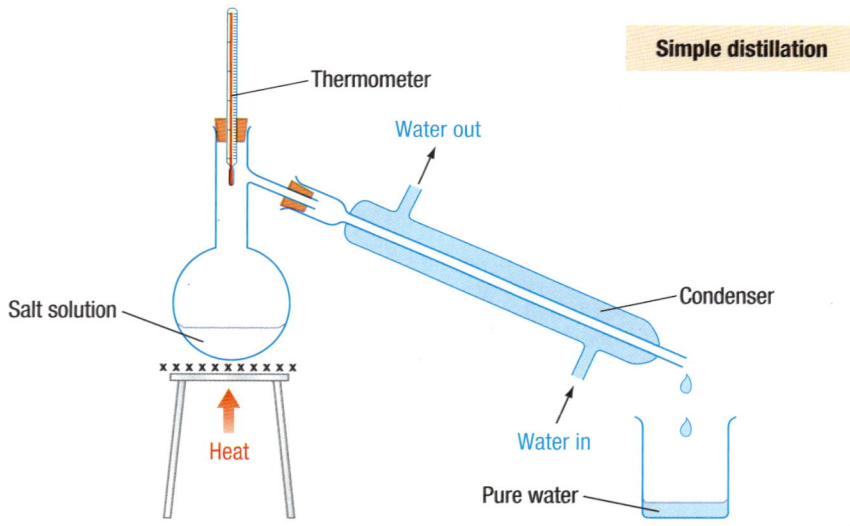

Figure 4 *Distilling pure water from a salt solution*

Summary questions

1 Define what a mixture is.

2 A mixture has 'no fixed composition'. What does this mean?

3 a Draw the apparatus you could use to separate ethanol and water.
   b Explain why you would be able to collect a more pure sample of ethanol using this apparatus than by using simple distillation.

4 Sulfur is soluble in the flammable liquid xylene but not in water. Sodium nitrate is soluble in water but not in xylene.
   Describe **two** ways to separate a mixture of sulfur powder and sodium nitrate to collect pure samples of each solid.

Key points

- A mixture is made up of two or more elements or compounds which are not chemically combined together.

- Mixtures can be separated by physical means, such as filtration, crystallisation, and distillation.

6.2 Paper chromatography

Learning objectives

After this topic, you should know:

- how chromatography can be used to distinguish pure substances from impure substances
- how to interpret chromatograms
- how to determine R_f values from chromatograms.

Scientists have many instruments that they can use to identify unknown compounds. Many of these are more sensitive, automated versions of the techniques, such as paper chromatography, that you use in school labs. For example, **chromatography** can be used to separate and identify mixtures of amino acids. The amino acids are colourless, but appear as purple spots on the paper when sprayed with a locating agent and dried, as shown in Figure 1.

You will have tried paper chromatography before, and probably used it to separate dyes in inks or food colourings (see Figure 2).

Chromatography always involves a mobile phase and a stationary phase. The mobile phase moves through the stationary phase, carrying with it the components of the mixture under investigation. Each component in the mixture will have a different attraction for the mobile phase and the stationary phase. A substance with stronger forces of attraction between itself and the mobile phase, than between itself and the stationary phase, will be carried a greater distance in a given time. A substance with a stronger force of attraction to the stationary phase will not travel as far in the same time.

So in paper chromatography the mobile phase is the solvent chosen, and the stationary phase is the paper. In Figure 1, amino acid X from the mixture M has the strongest attraction to the solvent, and amino acid Z has the strongest attraction to the paper.

Given an unknown organic solution, chromatography can usually tell you if it is a single compound or a mixture. If the unknown sample is a mixture of compounds, there will probably be more than one spot formed on the chromatogram. On the other hand, a single spot indicates the possibility of a pure substance.

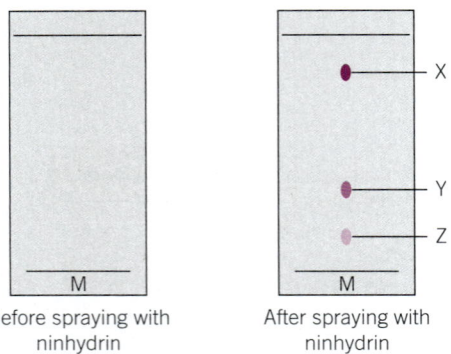

Figure 1 A chromatogram produced by a mixture of amino acids. The spots are made by different amino acids forming coloured compounds with the locating agent (ninhydrin).

Identifying unknown substances using chromatography

Once the compounds in a mixture have been separated using chromatography, they can be identified. You can compare spots on the chromatogram with others obtained from known substances:

Figure 2 Black ink can be separated out into its different colours on a chromatogram

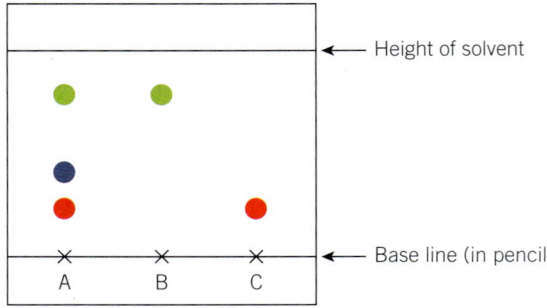

Figure 3 This chromatogram shows that A is a mixture of three substances, B and C plus one other unknown substance

Following the chromatography in Figure 3, mixture A still has one substance left unknown. A scientist making the chromatogram often does not know which pure compounds to include in their experiment to make a positive identification. It is also not practical to store actual chromatograms or their images, even on a computer. To make valid comparisons, every variable that affects a chromatogram would need to be exactly the same in all the chromatograms.

It is far more effective to measure data taken from any chromatogram of the unknown sample, then match it against a database. So the data is presented as an **R_f (retention factor)** value. This is a numeric value between 0 and 1, calculated by dividing the distance a spot travels up the paper (measured to the centre of the spot) by the distance the solvent front travels:

$$R_f = \frac{\textbf{distance moved by substance}}{\textbf{distance moved by solvent}}$$

As the number generated in the calculation is a ratio, it does not matter how long you run your chromatography experiment or what quantities you use. For comparisons against an R_f database to be valid, you just have to ensure that the solvent and the temperature used are the same as those quoted in the database or databook. Look at Figure 4 to see how to get the measurements to calculate R_f values.

Practical

Finding R_f values

Using a capillary tube, pencil, pipette, water, boiling tube, and a narrow strip of chromatography paper, find out the R_f values of the different dyes in the mixture of food colourings provided.

Present your evidence clearly. Include your dried chromatogram, calculations, and an evaluation.

Summary questions

1 Describe in detail how you can positively identify the dyes mixed to make a food colouring from its chromatogram.

2 What is the R_f value of this substance X?

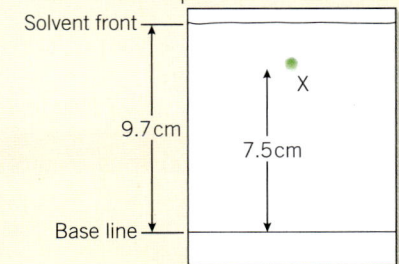

3 The R_f values of two substances, Y and Z, were taken from a chromatogram run in 50% water and 50% ethanol solvent at 20 °C. The R_f value of Y was 0.54 and of Z was 0.79.

What can you deduce about Y and Z from these values, and how could you use them to identify Y and Z?

4 In order to positively identify a compound from a chromatogram, explain why the solvent and temperature must be the same as those used to generate the R_f values in a database.

Worked example

Find the R_f value of compounds A and B using the chromatogram below.

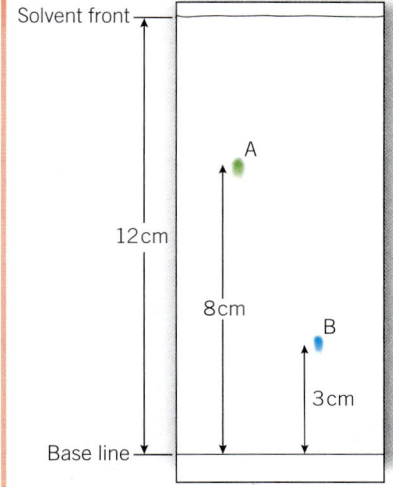

Figure 4 *The R_f value of an unknown substance, in a particular solvent at a given temperature, can be compared with values in a database to identify the substance*

Solution

The R_f value of A $= \dfrac{8}{12} = 0.67$ (2 s.f.)

The R_f value of B $= \dfrac{3}{12} = 0.25$

Key points

- Scientists can analyse unknown substances in solution by using paper chromatography.

- R_f values can be measured and matched against databases to identify specific substances.

- $R_f = \dfrac{\textbf{distance moved by substance}}{\textbf{distance moved by solvent}}$

6.3 Testing for gases

Learning objectives

After this topic, you should know:

■ the tests and the positive results for the gases:
  – hydrogen
  – oxygen
  – carbon dioxide
  – chlorine.

Many of the reactions you will study in chemistry give off gases as a product. So chemists have devised quick and easy tests to identify different gases.

Test for hydrogen

The reaction between zinc and dilute acid is a convenient way to make some hydrogen gas to test:

$$\text{zinc} + \text{sulfuric acid} \rightarrow \text{zinc sulfate} + \text{hydrogen}$$
$$Zn(s) + H_2SO_2(aq) \rightarrow ZnSO_4(aq) + H_2(g)$$

If you want the gas to be produced more quickly, a few crystals of copper(II) sulfate can be added, or magnesium can be used instead of zinc.

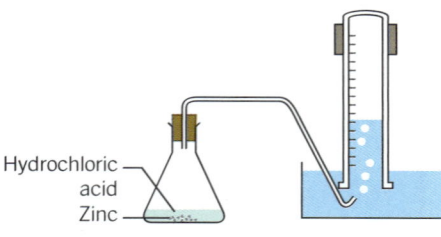

Hydrochloric acid
Zinc

Figure 1 *Collecting hydrogen over water*

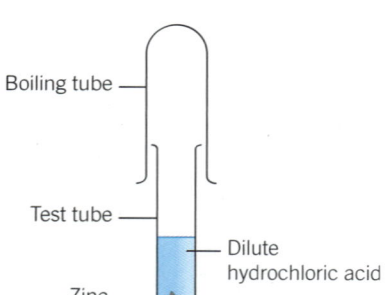

Boiling tube

Test tube

Dilute hydrochloric acid

Zinc

Figure 2 *Collecting hydrogen by upward delivery (downward displacement of air)*

⚭ links

To learn how you can use the gases given off in reactions to monitor rates of reaction, see 9.1 'How fast?'.
To learn about the reaction of metals with dilute acids, see 4.3 'The reactivity series'.

Study tip

When asked how to identify a given gas, always give the test **and** its result.

Practical

Testing for hydrogen gas

Collect a test tube of hydrogen gas, using either of the sets of apparatus shown in Figures 1 and 2.

■ Record your observations when you hold a lighted splint at the open end of the test tube of hydrogen gas.

■ Explain your observations.

■ What do the methods of collecting hydrogen gas tell you about its properties?

Safety: Wear eye protection. Hydrogen gas is flammable.

Positive test for hydrogen: a lighted splint 'pops'.

Test for oxygen

A convenient way to make some oxygen gas to test is the decomposition of hydrogen peroxide solution, with a little manganese(IV) oxide added as a catalyst.

$$\overset{\text{manganese(IV) oxide}}{\text{hydrogen peroxide} \rightarrow \text{water} + \text{oxygen}}$$
$$2H_2O_2(aq) \rightarrow 2H_2O(l) + O_2(g)$$

Practical

Testing for oxygen gas

Put $15\,cm^3$ hydrogen peroxide solution in a small conical flask.
Add a small amount of manganese(IV) oxide from the end of a spatula.
Insert a glowing splint (made by blowing out a lighted splint) in the mouth of the flask.

■ Record and explain your observations.

Safety: Wear eye protection.

Positive test for oxygen gas: a glowing splint relights.

Test for carbon dioxide

You can make carbon dioxide gas to test by reacting marble chips (calcium carbonate) and dilute hydrochloric acid:

calcium carbonate	+	hydrochloric acid	→	calcium chloride	+	water	+	carbon dioxide
$CaCO_3(s)$	+	$2HCl(aq)$	→	$CaCl_2(aq)$	+	$H_2O(g)$	+	$CO_2(g)$

Positive test for carbon dioxide: limewater (calcium hydroxide solution) turns cloudy white.

Read more about this test in Topic 4.5.

Test for ammonia

Ammonia is the only common alkaline gas so it will make **damp red litmus paper turn blue**. It also has a characteristic sharp, choking smell.

Ammonia gas also forms a white smoke of ammonium chloride when hydrogen chloride gas, from concentrated hydrochloric acid, is held near it, for example, holding the stopper from a bottle of concentrated hydrochloric acid near the ammonia gas:

$$NH_3(g) + HCl(g) \rightarrow NH_4Cl(s)$$

You first met this reaction in Topic 1.2.

Test for chlorine

Chlorine is a toxic gas, so care must be taken when working with this gas. Chlorine has a characteristic sharp, choking smell. Your teacher will show you the test for chlorine gas.

Positive test for chlorine gas: damp red litmus paper turns blue and then white (as it gets bleached).

Practical

Testing for chlorine gas

Your teacher will carefully add concentrated hydrochloric acid (corrosive) to a spatula of moistened potassium manganate(VII) crystals in a boiling tube held in a rack inside a fume cupboard.

A piece of damp blue litmus paper can be held in the mouth of the boiling tube.

■ Record and explain your observations.

⚭ links

You might be asked to test the chlorine gas given off during the electrolysis of a chloride solution, as in 5.4 'Electrolysis of brine'.

Summary questions

1 Explain why hydrogen gas 'pops' when a lighted splint is applied.

2 **a** Write a word equation for the reaction of magnesium carbonate plus dilute sulfuric acid, which produces carbon dioxide gas.
   b Carbon dioxide gas is denser than air. Suggest a way to collect the gas from the reaction in part **a** and then test it with limewater.

3 During the electrolysis of a chloride solution, a student predicted that a mixture of chlorine and oxygen gases would be given off from the anode. Describe how you could test this prediction.

4 **a i** Write a balanced symbol equation to describe the reaction between magnesium and dilute hydrochloric acid, to produce hydrogen gas.
     ii Write an ionic equation for the reaction in part **i**.
   b In the reaction in part **a**, which species is reduced and which is oxidised? Explain your answer.

5 Plan an investigation to see how the proportions of hydrogen and air in a test tube affect the loudness of the 'pop' you hear when a lighted splint is held at the mouth of the tube.

Key points

■ Hydrogen gas burns rapidly with a 'pop' when you apply a lighted splint.

■ Oxygen gas relights a glowing splint.

■ Carbon dioxide gas turns limewater milky (cloudy).

■ Chlorine gas bleaches damp blue litmus paper white.

■ Ammonia gas has a characteristic sharp, choking smell. It also makes damp red litmus paper turn blue and forms a white smoke of ammonium chloride with hydrogen chloride gas (from concentrated hydrochloric acid).

6.4 Tests for positive ions

Learning objectives

After this topic, you should know:

- how to identify different positive ions by:
  - flame tests
  - precipitation reactions.

Scientists working in environmental monitoring, industry, medicine, and forensic science need to analyse and identify substances. To identify unknown substances, there are a variety of different chemical tests.

Flame tests

Some metal ions produce flames with a characteristic colour. To carry out a flame test you do the following:

- Put a small amount of the compound to be tested in a nichrome wire loop. (The wire loop should be dipped in concentrated hydrochloric acid and heated to clean it first. Then it should be dipped in the acid again before dipping in the metal compound.)
- Then hold the loop in the roaring blue flame of a Bunsen burner.
- Use the colour of the Bunsen flame to identify the metal ion in the compound.

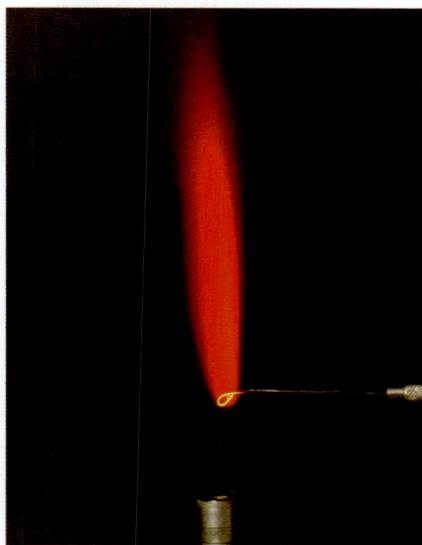

Figure 1 *A flame test can identify most Group 1 or Group 2 metals in a compound*

Metal ion	Flame colour
Lithium (Li$^+$)	crimson
Sodium (Na$^+$)	yellow
Potassium (K$^+$)	lilac
Calcium (Ca^{2+})	red
Barium (Ba^{2+})	green

Reactions with sodium hydroxide

The reactions with sodium hydroxide solution can also help us identify some positive ions. Aluminium ions, calcium ions, and magnesium ions all form *white precipitates* with sodium hydroxide solution. So, if a white precipitate forms you know an unknown compound contains either Al^{3+}, Ca^{2+}, or Mg^{2+} ions.

For example, the ionic equation for the reaction with aluminium ions is:

$$Al^{3+}(aq) + 3OH^-(aq) \rightarrow Al(OH)_3(s)$$

If you add more and more sodium hydroxide then the precipitate formed with aluminium ions dissolves. However, the white precipitate formed with calcium or magnesium ions will not dissolve. Calcium and magnesium ions can be distinguished by a flame test. Calcium ions give a brick-red flame but magnesium ions produce no colour at all.

Some metal ions form *coloured precipitates* with sodium hydroxide.

If you add sodium hydroxide solution to a substance containing:

- copper(II) ions, a blue precipitate is formed
- iron(II) ions, a green precipitate is formed
- iron(III) ions, a brown precipitate is formed.

For example, adding sodium hydroxide solution to iron(II) ions gives the ionic equation:

Figure 2 *This distinctive precipitate that forms when you add sodium hydroxide solution tells us that Cu^{2+} ions are present*

$$Fe^{2+}(aq) + 2OH^-(aq) \rightarrow Fe(OH)_2(s)$$

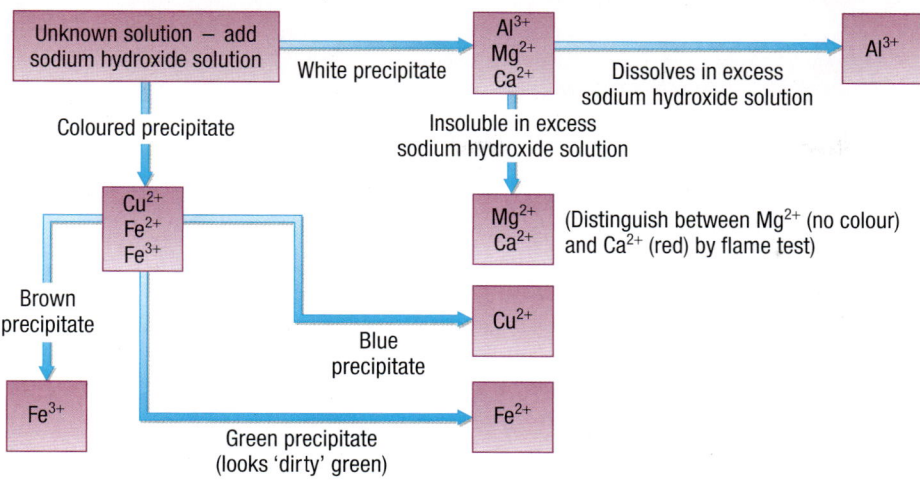

Figure 3 *Sodium hydroxide solution provides a very useful test for many positive ions*

Required practical

Identifying positive ions

Try to identify the metal ions in some unknown compounds provided for you.

Safety: Wear chemical splashproof eye protection.

Summary questions

1 a What colour precipitate does dilute sodium hydroxide produce with aluminium, calcium, and magnesium ions?

  b How can you distinguish between these three metal ions?

  c Write an ionic equation, including state symbols, for the precipitation of magnesium ions from solution by aqueous hydroxide ions.

2 a Draw a flow chart to describe how to carry out a flame test.

  b If the flame test was carried out on barium chloride, describe what you would see.

3 Copy the table below and complete the missing cells **a** to **j**:

Add sodium hydroxide solution	Flame test	Metal ion
nothing observed	lilac	**f**
white precipitate	red	**g**
a	**d**	Fe^{3+}
white precipitate which dissolves in excess sodium hydroxide solution	nothing observed	**h**
green precipitate which slowly turns reddish-brown	nothing observed	**i**
b	**e**	Na^+
c	crimson	**j**

Key points

- Most Group 1 and Group 2 metal ions can be identified in their compounds using flame tests.

- Sodium hydroxide solution can be used to identify different metal ions, depending on the precipitate that is formed.

6.5 Tests for negative ions

Learning objectives

After this topic, you should know:

■ how to identify different negative ions, in particular:
  - carbonates
  - halides
  - sulfates.

You can also do chemical tests to identify some negative ions in ionic compounds.

Carbonates

If you add a dilute acid to a carbonate, it fizzes and produces carbon dioxide gas. This is a good test to see if an unknown substance is a carbonate.

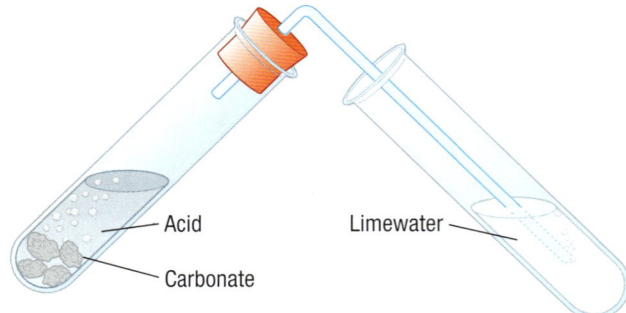

Acid

Limewater

Carbonate

Figure 1 *The test for a carbonate*

🔗 links

For more information on carbonates, look back to 4.5 'Metal carbonates'.

If the carbonate is sodium carbonate, the reaction you get is:

$$Na_2CO_3(s) + 2HCl(aq) \rightarrow 2NaCl(aq) + H_2O(l) + CO_2(g)$$

You can represent the reaction by just showing the ions that change in the reaction in an ionic equation:

$$\underset{\text{acid}}{2H^+(aq)} + \underset{\text{carbonate ion}}{CO_3{}^{2-}(s)} \rightarrow CO_2(g) + H_2O(l)$$

In limewater, the carbon dioxide reacts with calcium hydroxide. It forms a white precipitate of calcium carbonate which turns the limewater cloudy white.

Halide ions (chlorides, bromides, and iodides)

A very simple test shows whether chloride, bromide, or iodide ions are present in a compound. First you add dilute nitric acid, and then add silver nitrate solution. If a precipitate forms, there are halide ions present.

(You add the nitric acid to dissolve the compound and to remove any carbonate ions as they would also form a precipitate with the silver ions, and so interfere with the test.)

Study tip

In the test for a halide ion: add dilute **nitric** acid before the silver **nitrate**. Do not add any other acid as they will produce precipitates with silver nitrate solution.

The colour of the precipitate tells us which halide ion is present:
■ Chloride ions (Cl⁻) give a white precipitate of silver chloride.
■ Bromide ions (Br⁻) give a cream precipitate of silver bromide.
■ Iodide ions (I⁻) give a yellow precipitate of silver iodide.

If the unknown halide was sodium chloride, the precipitation reaction would be:

$$NaCl(aq) + AgNO_3(aq) \rightarrow NaNO_3(aq) + AgCl(s)$$

Here is the ionic equation, where X⁻ is the halide ion:

$$Ag^+(aq) + X^-(aq) \rightarrow AgX(s)$$

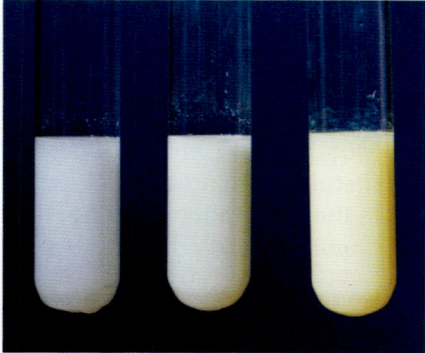

Figure 2 *One simple test with silver nitrate solution can tell us if an unknown substance contains chloride, bromide, or iodide ions*

Did you know … ?

Silver compounds are sensitive to light. If a silver halide precipitate is left for a few minutes in bright sunlight it slowly darkens as silver metal is formed.

Sulfates

You can test for sulfate ions by adding dilute hydrochloric acid, followed by barium chloride solution. (Add the dilute hydrochloric acid first to remove carbonate ions that would form a precipitate with the barium ions.) A white precipitate tells us sulfate ions are present. The white precipitate is the insoluble salt barium sulfate, $BaSO_4$.

If the unknown compound was potassium sulfate, then the equation for the precipitation reaction would be:

$$K_2SO_4(aq) + BaCl_2(aq) \rightarrow 2KCl(aq) + BaSO_4(s)$$

Here is the ionic equation for this reaction:

$$Ba^{2+}(aq) + SO_4^{2-}(aq) \rightarrow BaSO_4(s)$$

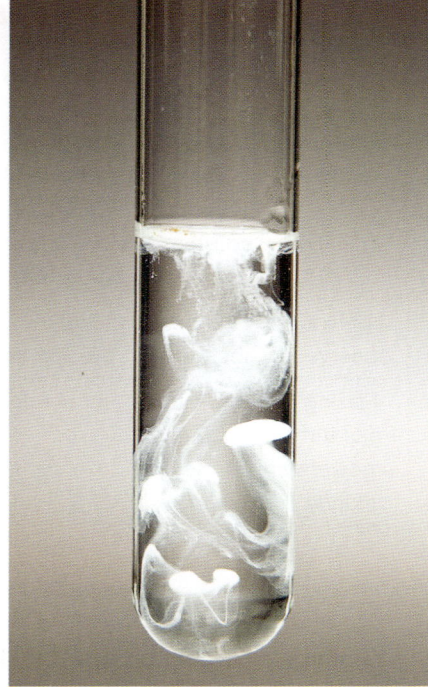

Figure 3 *The white precipitate of barium sulfate forming*

Practical

Identifying unknown ionic compounds

Now you know the tests for some positive and negative ions, you can try to identify some unknown compounds.

Safety: Wear chemical splashproof eye protection.

Study tip

In the test for the sulfate ion: add **hydrochloric** acid before the barium **chloride**. Do not add sulfuric acid – it contains sulfate ions!

Summary questions

1 List the reagents you would need to assemble in a lab to test a sample of an unknown substance for carbonate, halide, and sulfate ions.

2 An unknown compound is a white solid which dissolves in water to produce a colourless solution. When this solution is acidified with nitric acid and then silver nitrate is added, a yellow precipitate is produced. A flame test on the unknown compound produces a lilac flame. Deduce the name of the unknown compound and give your reasoning.

3 Write a word equation and a balanced symbol equation, including state symbols, for the following reactions:
   a magnesium chloride solution + silver nitrate solution
   b potassium carbonate powder + hydrochloric acid
   c aluminium sulfate solution + barium chloride solution

4 Write ionic equations, including state symbols, to summarise the reactions that help identify:
   a bromide ions
   b sulfate ions
   c carbonate ions.

5 In the test for halide ions, explain why nitric acid (rather than other acids) is added to a sample before you add silver nitrate solution.

Key points

- You identify carbonates by adding dilute acid, which produces carbon dioxide gas. The gas turns limewater cloudy.

- You identify halides by adding nitric acid, then silver nitrate solution. This produces a precipitate of silver halide (chloride = white, bromide = cream, iodide = yellow).

- You identify sulfates by adding hydrochloric acid, then barium chloride solution. This produces a white precipitate of barium sulfate.

Chapter summary questions

1 What would be entered in **a** to **f** to complete the table?

Add dilute acid	Add sodium hydroxide solution	Add dilute nitric acid followed by silver nitrate solution	Flame test	Substance
Nothing observed	Nothing observed	Yellow precipitate	Yellow	e
Fizzing – gas turns limewater cloudy	Pale blue precipitate formed	Not needed	Nothing observed	f
a	b	c	d	Calcium chloride

2 An unknown compound gave the following positive tests when analysed:

- The Bunsen flame turned crimson in a flame test.
- When dissolved in dilute hydrochloric acid and barium chloride solution was added, a white precipitate was formed.

Name the unknown compound.

3 The label on a jar of white crystals has been damaged by water and cannot be read. A science technician thinks it is probably a jar of potassium bromide.

a Describe how the technician could positively identify the crystals as potassium bromide.

b Give the formula of potassium bromide.

c What type of bonding will be found in potassium bromide?

4 a What are the differences between a mixture and a compound?

b Name the technique you would use to separate and collect:
  i hydrated copper(II) sulfate, $CuSO_4 \cdot 5H_2O$, from its aqueous solution
  ii a precipitate of lead iodide from the solution formed when aqueous solutions of lead nitrate and sodium iodide are mixed
  iii water from a solution of potassium chloride
  iv ethanol from a mixture of water and ethanol.

c Write a balanced symbol equation, including state symbols, for the reaction in part **b ii** above.

5 Chemists find that quick and simple tests are useful for identifying the gaseous products of chemical reactions.

a Name and give the chemical formula of the gas identified by:
  i its property as a bleaching agent
  ii its potentially explosive reaction with oxygen
  iii its precipitation reaction with an alkaline solution containing calcium ions
  iv the white smoke formed when it reacts with hydrogen chloride gas
  v a combustion reaction that re-ignites wood.

b Sometimes the gaseous products of a reaction are collected before testing takes place.

Using the gases identified in part **a**:
  i name three gases that can be collected over water
  ii explain why the other two gases from part **a** cannot be collected using this method.

c Gases can also be collected by upward delivery into an up-turned test tube. This method relies on the low density of the gas compared with air. The density of a gas depends on its relative formula mass, M_r (See Topic 8.3).
  i Calculate the M_r values of the five gases identified in part **a**.
  ii Use this information to suggest which two gases could be collected effectively using upward delivery.

6 a You can use paper chromatography to identify solutes using R_f values from a database. Look at the chromatogram below and calculate the R_f value of the unknown substances X and Y.

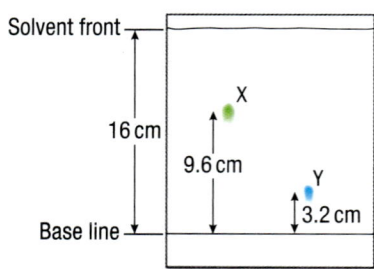

b The solvent used to obtain the chromatogram above was ethanol.

Explain the different R_f values of substance X and substance Y.

c In a database, apart from the solvent used, what other variable is quoted to ensure that R_f values are standardised?

Practice questions

1 Potash alum is a compound with the formula $KAl(SO_4)_2$

A student dissolved some potash alum in water and did some tests on the solution.

a Some of the solution was sprayed into a bunsen flame.
   i State the colour of the flame. (1)
   ii Write the formula of the ion responsible for this colour. (1)

b The student added a small amount of some sodium hydroxide solution to some potash alum solution.
   i State the observation that could be made. (1)
   ii Write the name of the compound responsible for this observation. (1)

c Another student added some sodium hydroxide solution to some potash alum solution and recorded his observation as a colourless solution.

   Explain why he did not make the same observation as the student in part **b**. (2)

d Describe how the student should do a test to show that the potash alum contains sulfate ions. (3)

2 A solution contains dissolved sodium carbonate, iron(II) chloride, and iron(III) chloride.

a After some dilute nitric acid is added, bubbles are seen.
   i Name the gas that forms. (1)
   ii Describe how you could test for this gas. (2)

b Next, some silver nitrate solution is added. A white precipitate forms.
   i Name the white precipitate. (1)
   ii Write the simplest ionic equation for the formation of the white precipitate. (2)

c A student adds some sodium hydroxide solution to test for positive ions.

   Why will this test not clearly identify the three positive ions in the solution? (3)

3 Many methods of separation depend on a property that is different for each substance in a mixture.

For each of the mixtures in this question, suggest a suitable method of obtaining the first substance from the mixture and state the property that the method depends on.

a Water from a mixture with sodium chloride (2)

b Calcium carbonate from a mixture with sodium chloride solution (2)

c Blue ink from a mixture with red ink (2)

4 A scientist used chromatography to identify the artificial colours in a flavoured drink.

The result of the experiment is shown in the diagram.

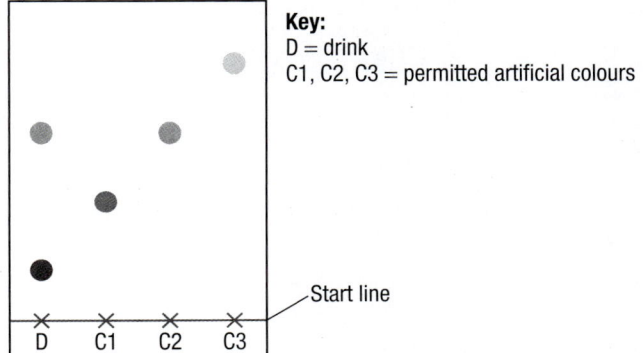

Key:
D = drink
C1, C2, C3 = permitted artificial colours

Describe how to do the experiment that gave this result. (6)

Acids and alkalis

After this topic, you should know:

- why solutions are acidic or alkaline
- what bases and alkalis are
- how to measure acidity.

Figure 1 *Acids and bases are all around us, in many of the things you buy at the shops, in our schools and factories – and in our bodies too*

Acids and bases are an important part of our understanding of chemistry. They play an important part inside us and all other living things.

What are acids and bases?

When you dissolve a substance in water you make an aqueous solution. Whether the solution formed is acidic, alkaline, or neutral depends on which substance has been dissolved.

- Soluble hydroxides are called **alkalis**. Their solutions are alkaline. An example is sodium hydroxide solution.
- **Bases**, which include alkalis, are substances that can neutralise acids. Metal oxides and metal hydroxides are bases. Examples include iron oxide and copper hydroxide.
- **Acids** include citric acid, sulfuric acid, and ethanoic acid. All acids taste very sour, although many acids are far too dangerous to put in your mouth. Ethanoic acid (in vinegar) and citric acid (in citrus fruit and fizzy drinks) are acids that are edible.
- Pure water is **neutral**.

One acid used in science labs is hydrochloric acid. This is formed when the gas hydrogen chloride (HCl) dissolves in water:

$$HCl(g) \xrightarrow{\text{water}} H^+(aq) + Cl^-(aq)$$

All acids form H^+ ions when added to water. It is these $H^+(aq)$ ions that make a solution acidic.

Because alkalis are bases that dissolve in water and form solutions, they are the bases that are often used in experiments. Sodium hydroxide solution is often found in school labs. You get sodium hydroxide solution when you dissolve solid sodium hydroxide in water:

$$NaOH(s) \xrightarrow{\text{water}} Na^+(aq) + OH^-(aq)$$

All alkalis form hydroxide ions (OH^-) when added to water. It is these aqueous hydroxide ions, $OH^-(aq)$, that make a solution alkaline.

Measuring acidity or alkalinity

Indicators are substances which change colour when added to acids and alkalis. Litmus paper is a well-known indicator, but there are many more.

Scientists use the **pH scale** to show how acidic or alkaline a solution is. The scale runs from 0 (most acidic) to 14 (most alkaline). You can use universal indicator (UI) to find the pH of a solution. It is a special indicator made from a number of dyes. It turns a range of colours as the pH changes. Anything in the middle of the pH scale (pH 7) is neutral, that is, neither acidic nor alkaline.

You can use the mathematical symbols '>' (read as 'is greater than') and '<' ('is less than') when interpreting pH values.

You can say:

pH < 7 indicates an acidic solution – that is, pH values less than 7 are acidic.

pH > 7 indicates an alkaline solution – that is, pH values greater than 7 are alkaline.

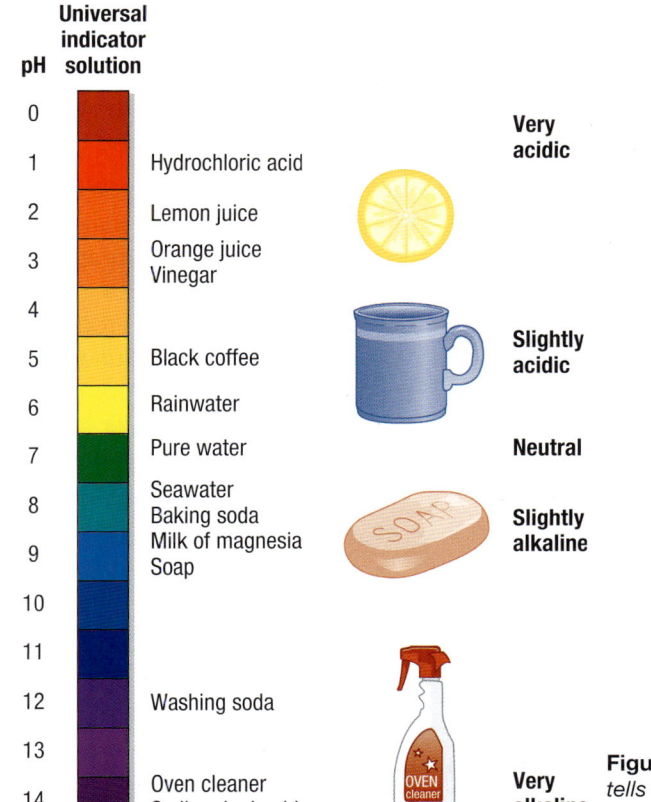

Universal indicator solution

pH		
0		**Very acidic**
1	Hydrochloric acid	
2	Lemon juice	
3	Orange juice / Vinegar	
4		
5	Black coffee	**Slightly acidic**
6	Rainwater	
7	Pure water	**Neutral**
8	Seawater / Baking soda	**Slightly alkaline**
9	Milk of magnesia / Soap	
10		
11		
12	Washing soda	
13		**Very alkaline**
14	Oven cleaner / Sodium hydroxide	

Figure 2 *The pH scale tells us how acidic or alkaline a solution is*

Practical

Which is the most alkaline cleaning product?

Compare the alkalinity of various cleaning products.

You can test washing-up liquids, shampoos, soaps, hand-washing liquids, washing powders/liquids, and dishwasher powders/tablets.

You could use a pH sensor and data logger to collect your data.

■ What are the advantages of using a pH sensor instead of universal indicator solution or paper?

Safety: Wear chemical splash-proof eye protection and disposable gloves.

Summary questions

1 a What distinguishes alkalis from other bases?
   b What do all alkalis have in common?
   c Show the change that happens when potassium hydroxide (KOH) dissolves in water using an equation, including state symbols.

2 a What ions do all acids produce in aqueous solution?
   b Show the change that happens when hydroiodic acid (HI) dissolves in water, including state symbols.

3 How could you use universal indicator paper as a way of distinguishing between distilled water, sodium hydroxide solution, and ethanoic acid solution?

4 Which would be the more accurate way of finding the pH of a solution, using universal indicator paper or a pH sensor and datalogger? Why?

Key points

■ Acids are substances which produce $H^+(aq)$ ions when added to water.

■ Bases are substances that will neutralise acids.

■ An alkali is a soluble hydroxide. Alkalis produce $OH^-(aq)$ ions when added to water.

■ You can use the pH scale to show how acidic or alkaline a solution is.

7.2 Making soluble salts from metals or insoluble bases

Learning objectives

After this topic, you should know:

- the products made when acids react with metals
- the products made when acids react with bases
- how to prepare crystals of salts from their solutions.

Acids + metals

A **salt** is a compound formed when the hydrogen in an acid is wholly or partially replaced by a metal or ammonium ions. One way to make salts is to react acids directly with metals. This is only possible if the metal is more reactive than hydrogen. If it is, then the metal will react with an acid to form a salt plus hydrogen gas:

$$\text{acid} + \text{metal} \rightarrow \text{a salt} + \text{hydrogen}$$
$$2HCl(aq) + Mg(s) \rightarrow MgCl_2(aq) + H_2(g)$$
$$\text{hydrochloric acid} + \text{magnesium} \rightarrow \text{magnesium chloride} + \text{hydrogen}$$

Note: If the metal is very reactive, the reaction with acid is too violent to carry out safely. So metals such as sodium or potassium are never added to acid.

Crystals of a salt (magnesium chloride in the equation above) can be obtained from the solution. Some of the water is evaporated off from the solution until the point of crystallisation. At this point, the solution becomes saturated and crystals will appear at the edge of a salt solution being heated in an evaporating dish. This point can also be tested by dipping a glass rod into the hot salt solution, removing it and seeing if crystals form on the solution left on the rod as it cools down. The salt solution should then be left at room temperature for the remaining water to evaporate off slowly in order to prepare the best samples of salt crystals. Wear eye protection in case the solution 'spits'.

⃝⃝ links

For more information on neutralisation reactions, see 7.3 'Making salts by neutralisation and precipitation'.

Acid + insoluble base

When you react an acid with a base, a salt, and water is formed.

The general equation which describes this **neutralisation** reaction is:

$$\text{acid} + \text{base} \rightarrow \text{a salt} + \text{water}$$

The salt that you make depends on the metal or the base that you use, as well as the acid. So bases that contain sodium ions will always make sodium salts. Those that contain potassium ions will always make potassium salts.

In terms of the acid used:

- the salts formed when neutralising hydrochloric acid (HCl) are always **chlorides** (containing Cl^- ions)
- sulfuric acid (H_2SO_4) makes salts which are **sulfates** (containing SO_4^{2-} ions)
- nitric acid (HNO_3) always makes **nitrates** (containing NO_3^- ions).

The oxide of a transition metal, such as iron(III) oxide, is an example of a base that can be used to make a salt in this way:

$$\text{acid} + \text{base} \rightarrow \text{a salt} + \text{water}$$
$$\mathbf{6HCl(aq)} + \mathbf{Fe_2O_3(s)} \rightarrow \mathbf{2FeCl_3(aq)} + \mathbf{3H_2O(l)}$$
$$\text{hydrochloric acid} + \text{solid iron(III) oxide} \rightarrow \text{iron(III) chloride solution} + \text{water}$$

Practical

Making a copper salt

You can make copper sulfate crystals from copper(II) oxide (an insoluble base) and sulfuric acid. The equation for the reaction is:

acid   +   base   →   a salt   + water

$H_2SO_4(aq)$ + $CuO(s)$ → $CuSO_4(aq)$ + $H_2O(l)$

sulfuric acid + copper(II) oxide → copper(II) sulfate + water

1   Add insoluble copper oxide to sulfuric acid and stir until no more reacts. Warm gently on a tripod and gauze (do not boil).

Warm gently

2   The solution turns blue as the reaction occurs, showing that copper sulfate is being formed. **Excess** black copper oxide can be seen.

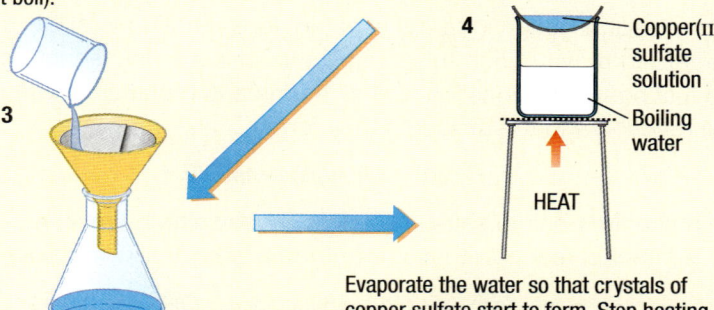

3   When the reaction is complete, filter the solution to remove excess copper oxide

4   Copper(II) sulfate solution — Boiling water — HEAT

Evaporate the water so that crystals of copper sulfate start to form. Stop heating when you see the first crystals appear at the edge of the solution. Then leave for the rest of the water to evaporate off slowly. This will give you larger crystals.

■ What does the copper sulfate look like? Draw a diagram if necessary.

Safety: Wear eye protection. Chemicals in this practical are harmful.

Summary questions

1 Write the general word equation for the reaction between an acid and:
   a a base   **b** a metal.

2 Describe in detail how you could prepare a sample of copper sulfate crystals from its solution.

3 **a** Why can't copper sulfate be prepared by adding copper metal to dilute sulfuric acid?
   b Why is potassium chloride never prepared by reacting potassium metal and dilute hydrochloric acid together?

4 Write balanced symbol equations, including state symbols, for the following reactions:
   a magnesium + hydrochloric acid
   b lithium oxide (in excess) + sulfuric acid

Key points

■ A salt is a compound formed when the hydrogen in an acid is wholly or partially replaced by metal or ammonium ions.

■ When you react an acid with a base a neutralisation reaction occurs.

■ The reaction between an acid and a base produces a salt and water.

■ Salts can also be made by reacting a suitable metal with an acid. This reaction produces hydrogen gas as well as a salt. A sample of the salt made can then be crystallised out of solution by evaporating off the water.

7.3 Making salts by neutralisation or precipitation

Learning objectives

After this topic, you should know:

- the ionic equation to describe neutralisation
- how to prepare salts from an acid and an alkali
- how to prepare insoluble salts.

There are two other important ways of making salts from solutions.

- You can react an acid and an alkali together to form a soluble salt.
- You can make an *insoluble* salt by reacting solutions of two soluble salts together.

Acid + alkali

When an acid reacts with an alkali, a neutralisation reaction takes place.

Hydrochloric acid reacting with sodium hydroxide solution is an example:

acid	+	alkali	→	a salt	+ water
$HCl(aq)$	+	$NaOH(aq)$	→	$NaCl(aq)$	+ $H_2O(l)$

hydrochloric acid + sodium hydroxide solution → sodium chloride + water

When you react an acid with an alkali you need to be able to tell when the acid and alkali have completely reacted. It is not obvious by just observing the reaction – so you need to use an indicator to help.

Think about neutralisation in terms of $H^+(aq)$ ions reacting with $OH^-(aq)$ ions. They react to form water and this can be shown in an ionic equation. An ionic equation just shows the ions, as well as atoms and molecules, that change in a chemical reaction and the products they form:

$$H^+(aq) + OH^-(aq) \rightarrow H_2O(l)$$

You can make ammonium salts, as well as metal salts, by reacting an acid with an alkali. Ammonia reacts with water to form a weakly alkaline solution:

$$NH_3(aq) + H_2O(l) \rightleftharpoons NH_4^+(aq) + OH^-(aq)$$

Ammonia solution reacts with an acid (e.g., nitric acid):

acid	+	ammonia solution	→ an ammonium salt +	water
$HNO_3(aq)$	+	$NH_4^+(aq) + OH^-(aq)$	→ $NH_4NO_3(aq)$	+ $H_2O(l)$

nitric acid + ammonia solution → ammonium nitrate + water

Ammonium nitrate contains a high proportion of nitrogen, and it is very soluble in water. This makes it ideal as a source of nitrogen for plants to take up through their roots. It replaces the nitrogen taken up from the soil by plants as they grow (see Figure 1).

Ammonium salts are made by adding ammonia solution to an acid until there is a small excess of ammonia. You can detect the excess ammonia by using universal indicator paper which will turn blue. You then crystallise the ammonium salt from its solution. The excess ammonia evaporates off.

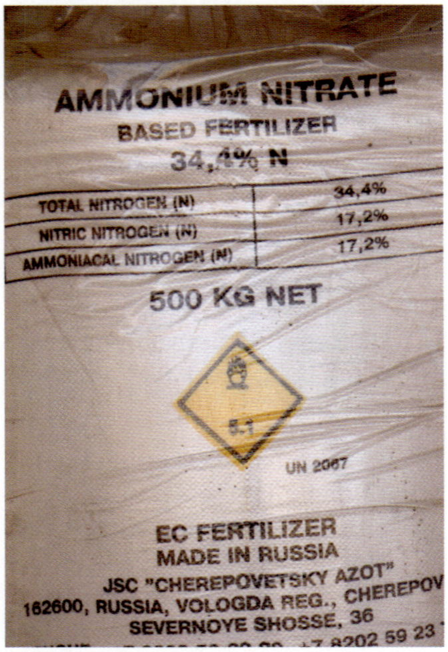

Making insoluble salts

Scientists can sometimes make salts by combining two solutions that contain different soluble salts. When the soluble salts react to make an insoluble salt, we call the reaction a precipitation reaction. This is because the insoluble solid formed in the reaction mixture is called a **precipitate**.

$Pb(NO_3)_2(aq)$ +	$2KI(aq)$	→	$PbI_2(s)$ +	$2KNO_3(aq)$
lead nitrate	+ potassium iodide	→	lead iodide	+ potassium nitrate
solution	solution		precipitate	solution

Figure 1 *Ammonium nitrate (NH_4NO_3) made from ammonia and nitric acid is used as a fertiliser*

Each of the reactant solutions contains one of the ions that make up the insoluble salt. In this case, they are lead ions in lead nitrate and iodide ions in potassium iodide. Lead iodide forms a yellow precipitate that you can filter off from the solution.

Using precipitation

Precipitation reactions are used to remove pollutants from the wastewater produced by factories. The effluent must be treated before it is discharged into rivers and the sea.

Precipitation is used in the removal of metal ions from industrial wastewater. By raising the pH of the water, engineers can make insoluble metal hydroxides precipitate out. This produces a sludge which can easily be removed from the solution.

The cleaned-up water can then be discharged safely into a river or the sea.

Precipitation can be also used to remove unwanted ions from drinking water.

Practical

Making an insoluble salt

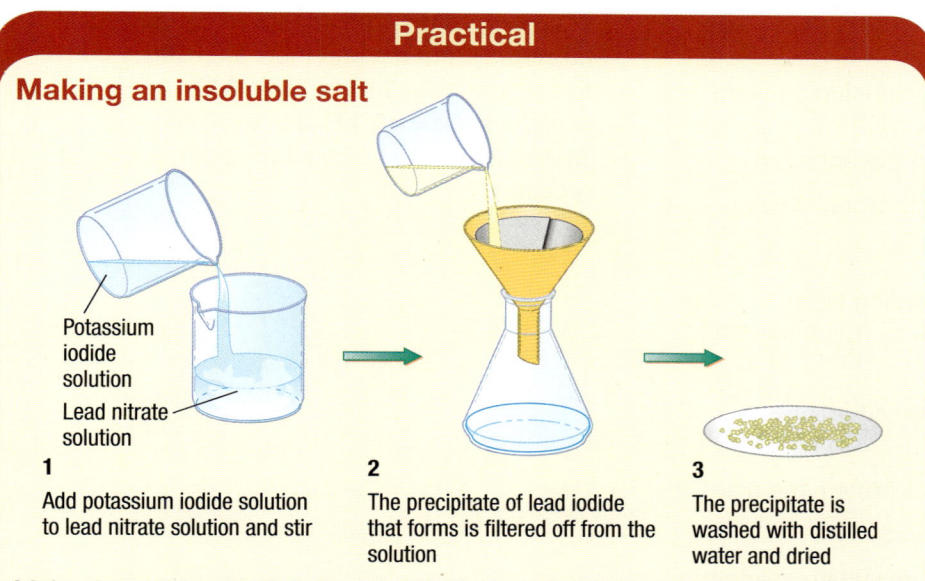

Potassium iodide solution

Lead nitrate solution

1 Add potassium iodide solution to lead nitrate solution and stir

2 The precipitate of lead iodide that forms is filtered off from the solution

3 The precipitate is washed with distilled water and dried

Make the salt lead iodide from lead nitrate solution and potassium iodide solution. The equation for the reaction is shown at the bottom of the previous page.

■ Why is the precipitate of lead iodide washed with distilled water?

Safety: Wear chemical splashproof eye protection. Lead compounds are toxic.

Figure 2 *Water treatment plants use chemical treatments, such as adding lime, to precipitate out metal compounds which can then be removed by filtering*

Summary questions

1 a Write down a general equation for the neutralisation reaction between an acid and an alkali.
 b Write an ionic equation to show what happens in this neutralisation reaction.

2 a Name the salt made when ammonia solution neutralises dilute nitric acid.
 b Why is this salt important?
 c Write a balanced symbol equation for the reaction to make this salt.

3 Write a balanced symbol equation, including state symbols, and a method to show how to make lithium chloride (a soluble salt – look ahead to Topic 8.6 for help if necessary).

4 Lead chloride, $PbCl_2$, is an insoluble salt.
 a Suggest a method to prepare a sample of lead chloride. (Hint: all nitrates are soluble in water.)
 b Write a balanced symbol equation, including state symbols, for the reaction used in part **a**.
 c What type of bonding will be present in the lead chloride?

5 A chemical factory produces aqueous chromium(III) ions as a by-product. Suggest how the chemical company can remove the chromium ions from the wastewater before it is discharged from the factory.

Key points

■ An indicator is needed when a soluble salt is prepared by reacting an alkali with an acid.

■ Insoluble salts can be made by reacting two solutions to produce a precipitate.

■ Precipitation is an important way of removing some metal ions from industrial wastewater.

Chapter summary questions

1 Nickel(II) sulfate crystals can be made from an insoluble oxide base and sulfuric acid.

a i Name the insoluble base that can be used to make nickel(II) sulfate.

ii Write a balanced symbol equation, including state symbols, to show the reaction.

iii What type of reaction is shown in the equation?

b Describe how you could obtain crystals of nickel(II) sulfate from the reaction in part **a ii**.

2 Write balanced symbol equations, including state symbols, to describe the reactions below. (Each reaction forms a salt.)

a Lithium hydroxide solution (in excess) and dilute sulfuric acid

b Iron(III) oxide (an insoluble base) and dilute nitric acid

c Zinc metal and dilute hydrochloric acid

d Barium nitrate solution and potassium sulfate solution (this reaction produces an insoluble salt)

3 a What is the chemical formula of calcium carbonate – the main compound in limestone?

b Calcium carbonate reacts with dilute hydrochloric acid, giving off a gas.

i Write a balanced symbol equation, including state symbols, for this reaction.

ii Draw a diagram of the apparatus you would use to give a positive test for the gas given off.

iii Write a word equation and a balanced symbol equation for the change that happens in a positive test for the gas in part **ii**.

c This is a lime kiln for roasting limestone:

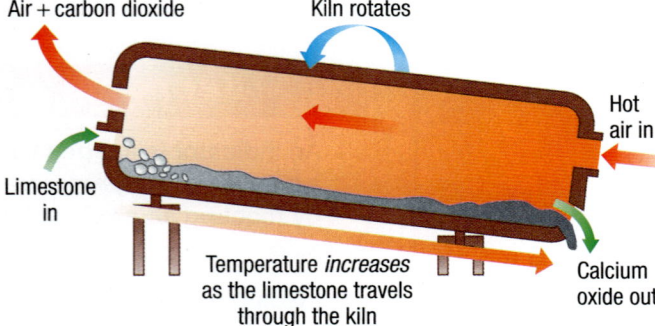

Air + carbon dioxide   Kiln rotates

Hot air in

Limestone in

Temperature *increases* as the limestone travels through the kiln

Calcium oxide out

i Write a word equation and a balanced symbol equation, including state symbols, for the reaction in the lime kiln.

ii What do you call the type of reaction that takes place in a lime kiln?

iii What mass of the gas would be produced if 12.5 tonnes of limestone was roasted in a lime kiln, assuming all the limestone was composed of calcium carbonate and there was 100% conversion to calcium oxide? (See Topic 8.2).
(A_r values: Ca = 40, O = 16, C = 12)

iv State **three** uses of limestone not mentioned in this question.

Practice questions

1 Three solutions are made by dissolving equal amounts of compounds in water. The pH values of these solutions are 2, 7, and 12.

Solution 1 contains hydrochloric acid.
Solution 2 contains sodium chloride.
Solution 3 contains sodium hydroxide.

a Write the appropriate pH value for each solution shown above. (1)

b Name **one** indicator that could be used to identify each solution and explain why it is better than using litmus. (2)

c Solution 4 has a pH value of 9. Solution 5 has a pH value of 11.

Which solution contains more H^+ ions? (1)

2 One way to prepare soluble salts involves adding a metal or a metal compound to a dilute acid.

a Copy and complete the following word equations that show the reactions used to make three salts.
i magnesium + hydrochloric acid →
ii zinc oxide + nitric acid →
iii copper(II) carbonate + sulfuric acid → (6)

b What would you observe in reactions **i** and **iii** but not in reaction **ii**? (1)

c Why should an excess of metal or metal compound be used in each case? (1)

d After each reaction is complete, filtration is used.
i State the name of the substance in solution when reaction **i** is filtered. (1)
ii What does the residue in reaction **ii** contain? (1)
iii What is the colour of the filtrate in reaction **iii**? (1)

e The method used for all the reactions above produces a dilute solution of the salt. Describe how crystallisation is used to obtain a pure dry sample of the salt produced in reaction **iii**. (4)

3 Insoluble salts are made by the precipitation method.

a The word equation for a reaction used to make an insoluble salt is:

magnesium chloride + sodium fluoride → magnesium fluoride + sodium chloride

Write a symbol equation, including state symbols, for this reaction. (3)

b i Name **two** compounds that could be used to make the insoluble salt lead sulfate. (2)
ii When lead sulfate is made the reaction mixture is filtered. Lead sulfate is in the residue.
Why should the residue be washed with distilled water before it is dried? (1)

8.1

Chemical equations

Learning objectives

After this topic, you should know:

- what happens to the atoms in a chemical reaction

- how the mass of reactants compares with the mass of products in a chemical reaction

- how to write balanced symbol equations, including state symbols, to represent reactions.

Chemical equations show the **reactants** (the substances you start with) and the **products** (the new substances made) in a reaction.

The test for hydrogen gas can be represented using a **word equation**:

$$\text{hydrogen} + \text{oxygen} \rightarrow \text{water}$$
$$\text{(reactants)} \qquad \text{(product)}$$

In chemical reactions the atoms get rearranged. Now you can investigate what happens to the mass of reactants compared with the mass of products in a reaction.

Using **symbol equations** helps us to see how much of each substance is involved in a reaction.

For example, calcium carbonate decomposes (breaks down) on heating. You can show the reaction using a symbol equation like this:

$$CaCO_3 \rightarrow CaO + CO_2$$

This equation is **balanced** – there is the same number of each type of atom on both sides of the equation. This is very important, because atoms cannot be created or destroyed in a chemical reaction. This also means that:

The total mass of the products formed in a reaction is equal to the total mass of the reactants.

You can check if an equation is balanced by counting the number of each type of atom on either side of the equation. If the numbers are equal, then the equation is balanced.

You can also add **state symbols** to a balanced symbol equation to give extra information. The state symbols used are:

- (s) for solids
- (l) for liquids
- (g) for gases
- (aq) for substances dissolved in water, that is, **aqueous solutions**.

So the balanced symbol equation, including state symbols, for the decomposition of calcium carbonate is:

$$CaCO_3(s) \rightarrow CaO(s) + CO_2(g)$$

⚲ links

For more information on the decomposition of calcium carbonate, see 4.5 'Metal carbonates'.

Practical

Investigating the mass of reactants and products

You are given solutions of lead nitrate (toxic) and potassium iodide.

Wearing chemical splashproof eye protection, add a small volume of each solution together in a test tube.
- What do you see happen?

The formula of lead nitrate is $Pb(NO_3)_2$ and of potassium iodide is KI.

The precipitate (a solid formed in the reaction between two solutions) is lead iodide, PbI_2 (which is toxic).
- Predict a word equation and a balanced symbol equation, including state symbols, for the reaction.
- How do you think that the mass of the reactants compares with the mass of the products?

Now plan an experiment to test your answer to this question. Your teacher must check your plan before you start the practical work. Wash your hands after the experiment.

Making an equation balance

In the case of hydrogen reacting with oxygen it is not so easy to balance the equation. Firstly, we write the formula of each reactant and product:

$$H_2 + O_2 \rightarrow H_2O$$

Count the atoms on either side of the equation:

Reactants   **Products**
2 H atoms, 2 O atoms   2 H atoms, 1 O atom

So you need another oxygen atom on the product side of the equation. You cannot simply change the formula of H_2O to H_2O_2. (H_2O_2 – hydrogen peroxide – is a bleaching agent which is certainly not suitable to drink!) But you can have **2 water molecules** in the reaction – this is shown in a symbol equation as:

$$H_2 + O_2 \rightarrow \mathbf{2}H_2O$$

Again, count the atoms on either side of the equation:

Reactants   **Products**
2 H atoms, 2 O atoms   4 H atoms, 2 O atoms

Although the oxygen atoms are balanced, you now need two more hydrogen atoms on the reactant side. You do this by putting 2 in front of H_2:

$$\mathbf{2}H_2 + O_2 \rightarrow 2H_2O$$

Now you have:

Reactants   **Products**
4 H atoms, 2 O atoms   4 H atoms, 2 O atoms

… and the equation is balanced.

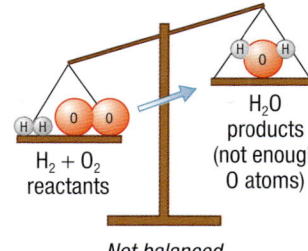

H$_2$ + O$_2$ reactants → H$_2$O products (not enough O atoms)

Not balanced

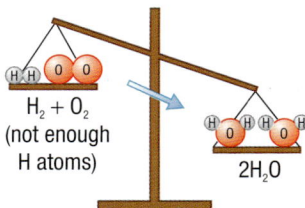

H$_2$ + O$_2$ (not enough H atoms)   2H$_2$O

Still not balanced!

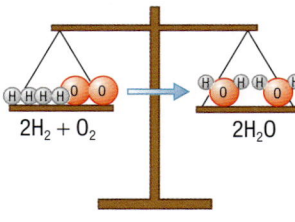

2H$_2$ + O$_2$ 2H$_2$O

Balanced at last!

Look at the chemical equation on the previous page. You can work out the mass of $CaCO_3$, CaO, or CO_2 given the masses of the other two compounds.

Because the total mass of the products formed in a reaction is equal to the total mass of the reactants, you can write:

$$CaCO_3 \rightarrow CaO + CO_2$$
$$\text{Mass} = \quad a \quad = \quad b \quad + \quad c$$

So if the mass of CaO formed is 2.8 g (*b* above) and the mass of CO_2 is 2.2 g (*c* above), then the mass of $CaCO_3$ (*a* above) that you start with must be 2.8 + 2.2 (*b*+*c*) which equals **5.0 g**.

Rearranging the equation for *a*, *b*, and *c* you get $a - c = b$.

So if the reaction started with 100 tonnes of $CaCO_3$ (*a*) and it gave off 44 tonnes of CO_2 (*c*), then the mass of CaO (*b*) made is 100 − 44 ($a - c$) = **56 tonnes**.

Summary questions

1 **a** Why must all symbol equations be balanced?
  b Copy this equation and balance it: $H_2 + Cl_2 \rightarrow HCl$

2 **a** A mass of 33.6 g of magnesium carbonate ($MgCO_3$) completely decomposed when it was heated. It made 16.0 g of magnesium oxide (MgO). What mass of carbon dioxide (CO_2) was produced in this reaction?
  b Write a word equation and a balanced symbol equation, including state symbols, to show the reaction in part **a**.

3 Copy these symbol equations and balance them:
  a $KNO_3 \rightarrow KNO_2 + O_2$ **b** $Li + O_2 \rightarrow Li_2O$ **c** $Fe + O_2 \rightarrow Fe_2O_3$

4 Sodium metal, Na, reacts with water to form a solution of sodium hydroxide, NaOH, and gives off hydrogen gas, H_2.
  Write a balanced symbol equation, including state symbols, for this reaction.

Key points

- No new atoms are ever created or destroyed in a chemical reaction:

  the total mass of reactants = the total mass of products

- There is the same number of each type of atom on each side of a balanced symbol equation.

- You can include state symbols to give extra information in balanced symbol equations. These are (s) for solids, (l) for liquids, (g) for gases, and (aq) for aqueous solutions.

8.2 Relative masses and moles

Learning objectives

After this topic, you should know:

- how electrons are involved in bonding

- what is meant by the relative atomic mass of an element

- how to calculate the relative formula mass of a compound

- how to calculate the number of moles, given the mass (or the mass, given the number of moles) of substance.

🔗 links

For more information about ions and bonding look back to 2.1 'Atoms into ions', 2.2 'Ionic bonding', and 2.5 'Giant ionic structures'.

🔗 links

For more information about atoms sharing electrons and bonding look back to 2.3 'Covalent bonding', 2.6 'Simple molecules', and 2.7 'Giant covalent structures'.

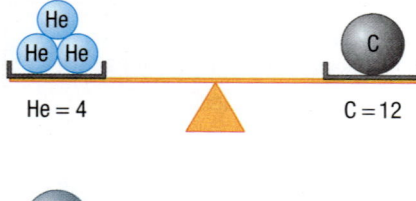

Figure 1 *The relative atomic mass of $^{12}_{6}C$ atom is 12. Compared with this, the A_r of helium is 4 and the A_r of magnesium is 24.*

When different types of atom combine chemically, a compound is formed.

Sometimes atoms react together by **transferring** electrons to form chemical bonds. This happens when metals react with non-metals. If the reacting atoms are all non-metallic, then the atoms **share** electrons to form chemical bonds.

Forming ions

As you will recall from Chapter 2, when a metal bonds with a non-metal, the metal atom gives one or more electrons to the non-metal atom. Both atoms become charged particles called **ions**.

- Metal atoms form positively charged ions. Their atoms lose electrons, for example, Li, which contains three protons and three electrons, forms Li^+, which contains three protons and two electrons.
- Non-metal atoms form negatively charged ions. Their atoms gain electrons, for example, F, which contains nine protons and nine electrons, forms F^-, which contains nine protons and ten electrons.

The ions formed attain the stable electronic structure of the nearest noble gas to them in the periodic table. So the electronic structure of Li^+ is 2 (the same as a helium atom) and F^- is 2,8 (the same as a neon atom).

Within these compounds made from metals and non-metals, there are strong electrostatic attractions between the positive and negative ions, forming the chemical bonds (called **ionic bonds**).

Forming molecules

Non-metal atoms bond to each other in a different way. The outermost shells of their atoms overlap and they share electrons. Each pair of shared electrons forms a chemical bond between the atoms. Molecules are formed, such as hydrogen sulfide, H_2S, and methane, CH_4.

Relative atomic masses

The mass of a single atom is so tiny that it would not be practical to use it in experiments or calculations. So instead of working with the actual masses of atoms, use the *relative* masses of atoms of different elements. We call these **relative atomic masses (A_r)**.

Scientists use an atom of carbon-12 ($^{12}_{6}C$) as a standard atom. They give this a 'mass' of exactly 12 units, because it has 6 protons and 6 neutrons. They then compare the masses of atoms of all the other elements with this standard carbon atom. Relative atomic mass A_r is the average mass of naturally occurring atoms of an element on a scale where ^{12}C has a mass of exactly 12 units.

The A_r takes into account the proportions of any isotopes of the element found naturally (this is why chlorine has a relative atomic mass of 35.5, although you could never have half a proton or neutron in an atom).

Relative formula masses

You can use the A_r of the various elements to work out the **relative formula mass (M_r)** of compounds. This is true whether the compounds are made up of molecules (in which case, the term 'relative molecular mass' can be used), or ions. A simple example is a substance such as sodium chloride.

The A_r of sodium is 23 and the A_r of chlorine is 35.5. So the relative formula mass of sodium chloride (NaCl) is:

$$23 + 35.5 = \mathbf{58.5}$$
$$A_r{:}\,Na \quad A_r{:}\,Cl \quad M_r{:}\,NaCl$$

You can use the same approach with relatively complicated molecules such as sulfuric acid, H_2SO_4. Hydrogen has a A_r of 1, the A_r of sulfur is 32, and the A_r of oxygen is 16. This means that the M_r of sulfuric acid is:

$$(1 \times 2) + 32 + (16 \times 4) = 2 + 32 + 64 = \mathbf{98}$$

Moles

Saying or writing 'relative atomic mass in grams' or 'relative formula mass in grams' is rather clumsy. So chemists have a shorthand word to describe this amount of substance, a **mole**.

They say that the relative atomic mass in grams of carbon (i.e., 12 g of carbon) is a mole of carbon atoms. **One mole is simply the relative atomic mass or relative formula mass of any substance expressed in grams**. A mole of any substance always contains the same number of atoms, molecules, or ions. This huge quantity, called the **Avogadro constant**, is 6.02×10^{23} per mole. So chemists prefer to use the mole when describing relative numbers of particles in a certain mass of substance. They use the equation:

$$\text{number of moles} = \frac{\text{mass (g)}}{A_r} \text{ or } \frac{\text{mass (g)}}{M_r}$$

to tell them the number of moles in a given mass of a substance. So if you have 120 g of carbon you have $\frac{120}{12} = 10$ moles of C atoms, or if you have 9.8 g of sulfuric acid you have $\frac{9.8}{98} = 0.1$ moles of H_2SO_4 molecules.

Re-arranging the equation lets us work out the mass of a certain number of moles of substance:

$$\text{mass (g)} = \text{number of moles} \times A_r$$
$$\text{mass (g)} = \text{number of moles} \times M_r$$

?? Did you know ... ?

If you had as many soft drink cans as there are atoms in a mole, they would cover the surface of the Earth to a depth of 200 miles!

Study tip

You don't have to remember the number 6.02×10^{23}. The relative atomic masses of elements are always provided. But do practise calculating the mass of one mole of different substances from their formula and the relative atomic masses that you are given, as well as the number of moles in a given mass of substance.

Summary questions

1 a Define the term 'relative atomic mass' of an element.
  b Explain why it is possible to have relative atomic masses which are not whole numbers, for example, A_r of chlorine (Cl) is 35.5.

2 What is the relative formula mass of each of the following?
  a S_8 (A_r value: S = 32)
  b MgF_2 (A_r values: Mg = 24, F = 19)
  c $CuSO_4$ (A_r values: Cu = 63.5, S = 32, O = 16)
  d $C_6H_{12}O_6$ (A_r values: C = 12, H = 1, O = 16)

3 The relative atomic mass (A_r) of helium is 4, and that of sulfur is 32.
  a How many times heavier is a sulfur atom than a helium atom?
  b How many moles of helium atoms are there in:
    i 16 g of helium?   ii 0.02 g of helium?
  c How many moles of sulfur atoms are there in:
    i 9.6 g of sulfur?
    ii 16 tonnes of sulfur (where 1 tonne = 1000 kg)?

4 What is the mass of the following?
  a 0.5 moles of aluminium, Al
  b 50 moles of calcium carbonate, $CaCO_3$
  c 0.05 moles of hydrogen, H_2
  d 0.6 moles of phosphorus, P_4
  (Find the A_r values needed from the periodic table.)

Key points

- Compare the masses of atoms by measuring them relative to atoms of carbon-12.
- Work out the relative formula mass of a compound by adding up the relative atomic masses of the elements in it, in the ratio shown by its formula.
- One mole of any substance is its relative formula mass, in grams.
- Number of moles = $\frac{\text{mass (g)}}{A_r}$ or $\frac{\text{mass (g)}}{M_r}$

8.3

Percentages by mass and empirical formulae

Learning objectives

After this topic, you should know:

- how to calculate the percentage of an element in a compound from its formula

- how to calculate the empirical formula of a compound from its percentage composition.

You can use the relative formula mass of a compound to calculate the percentage mass of each element in it.

Working out the percentage by mass of an element in a compound

Worked example 1

What percentage of the mass of magnesium oxide is actually magnesium?

Solution

You need to know the formula of magnesium oxide: MgO

The A_r of magnesium is 24 and the A_r of oxygen is 16.

Adding these together gives us the relative formula mass (M_r) of MgO, that is, $24 + 16 = 40$

So in 40 g of magnesium oxide, 24 g is actually magnesium.

The fraction of magnesium in MgO is:

$$\frac{\text{mass of magnesium}}{\text{total mass of compound}} = \frac{24}{40}$$

so the percentage of magnesium in the compound is:

$$\frac{24}{40} \times 100\% = \mathbf{60\%}$$

Worked example 2

A pure white powder is found at the scene of a crime. It could be strychnine, a deadly poison with the formula $C_{21}H_{22}N_2O_2$, but is it?

When a chemist analyses the powder, she finds that 83% of its mass is carbon. What is the percentage mass of carbon in strychnine? Is this the same as the white powder?

Solution

Given the A_r values: C = 12, H = 1, N = 14, O = 16, the relative formula mass (M_r) of strychnine is:

$$(12 \times 21) + (1 \times 22) + (14 \times 2) + (16 \times 2) = 252 + 22 + 28 + 32 = 334$$

The percentage mass of carbon in strychnine is therefore:

$$\frac{252}{334} \times 100 = \mathbf{75.4\%} \text{ (to 3 s.f.)}$$

This is **not** the same as the percentage mass of carbon in the white powder – so the white powder is not strychnine.

To calculate the percentage of an element in a compound:

- Write down the formula of the compound.
- Using the A_r values from your data sheet in the exam, work out the M_r of the compound. Write down the mass of each element making up the compound as you work it out.
- Write the mass of the element you are investigating as a fraction of the M_r.
- Find the percentage by multiplying your fraction by 100.

Working out the empirical formula of a compound from the masses or percentage of the elements in the compound

You can find the percentage of each element in a compound by experiments. Then you can work out the simplest ratio of each type of atom in the compound. This simplest (whole-number) ratio is called its **empirical formula**.

This is sometimes the same as the actual number of atoms in one molecule (which is called the **molecular formula**) – but not always. For example, the empirical formula of water is H_2O, which is also its molecular formula. However, hydrogen peroxide has the empirical formula HO, but its molecular formula is H_2O_2.

Worked example 3

A hydrocarbon contains 75% carbon and 25% hydrogen by mass. What is its empirical formula? (A_r values: C = 12, H = 1)

Solution

Imagine you have 100 g of the compound. Then 75 g is carbon and 25 g is hydrogen.

Work out the number of moles by dividing the mass of each element by its relative atomic mass:

For carbon: $\frac{75}{12} = 6.25$ moles of carbon atoms

For hydrogen: $\frac{25}{1} = 25$ moles of hydrogen atoms

So this tells us that 6.25 moles of carbon atoms are combined with 25 moles of hydrogen atoms.

This means that the ratio is 6.25 (C) : 25 (H).

So the simplest whole-number ratio is 1 : 4 (by dividing both numbers by the smallest number in the ratio).

In other words each carbon atom is combined with four hydrogen atoms.

So the empirical formula of the compound is **CH_4**.

To work out the formula from percentage masses:

- Change the percentages given to the masses of each element in 100 g of compound.
- Change the masses to moles of atoms by dividing the masses by the A_r values. This tells you how many moles of each different element are present.
- From this you can work out the ratio of atoms of the different elements in the compound.
- Then the *simplest* whole-number ratio gives you the empirical formula of the compound.

Summary questions

1 What is the percentage mass of hydrogen in hydrogen sulfide, H_2S? (A_r values: H = 1, S = 32)

2 Ammonium sulfate, $(NH_4)_2SO_4$, is used as a fertiliser. What is the percentage mass of nitrogen in it? (A_r values: H = 1, N = 14, O = 16, S = 32)

3 A compound contains 40% sulfur and 60% oxygen by mass. What is its empirical formula? (A_r values: S = 32, O = 16)

4 22.55% of the mass of a sample of phosphorus chloride is phosphorus. What is the empirical formula of phosphorus chloride? (A_r values: P = 31, Cl = 35.5)

5 10.8 g of aluminium react exactly with 9.6 g of oxygen. What is the empirical formula of the compound formed? (A_r values: Al = 27, O = 16)

Key points

- The relative atomic masses of the elements in a compound and its formula can be used to work out its percentage composition.

- You can calculate empirical formulae given the masses or percentage composition of elements present.

8.4 Equations and calculations

Learning objectives

After this topic, you should know:
- what balanced symbol equations tell us about chemical reactions
- how to use balanced symbol equations to calculate masses of reactants and products.

Chemical equations can be very useful. When you want to know how much of each substance is involved in a chemical reaction, you can use the balanced symbol equation.

Think about what happens when hydrogen molecules (H_2) react with chlorine molecules (Cl_2). The reaction makes hydrogen chloride molecules (HCl):

$$H_2 + Cl_2 \rightarrow HCl \text{ (not balanced)}$$

This equation shows the reactants and the product – but it is not balanced.

Here is the balanced equation:

$$H_2 + Cl_2 \rightarrow 2HCl$$

This balanced equation tells us that '1 hydrogen molecule reacts with 1 chlorine molecule to make 2 hydrogen chloride molecules'. But the balanced equation also tells us the number of moles of each substance involved. So our balanced equation also tells us that '1 mole of hydrogen molecules reacts with 1 mole of chlorine molecules to make 2 moles of hydrogen chloride molecules'.

1 hydrogen molecule	1 chlorine molecule	2 hydrogen chloride molecules
H_2 +	Cl_2 $\longrightarrow$	$2HCl$
1 mole of hydrogen molecules	1 mole of chlorine molecules	2 moles of hydrogen chloride molecules

Using balanced equations to work out reacting masses

The balanced equation above is really useful, because you can use it to work out what mass of hydrogen and chlorine react together. You can also calculate how much hydrogen chloride is made.

To do this, you need to know that the A_r for hydrogen is 1 and the A_r for chlorine is 35.5:

A_r of hydrogen = 1   ... so mass of 1 mole of H_2 = $2 \times 1 = 2\,g$
A_r of chlorine = 35.5   ... so mass of 1 mole of Cl_2 = $2 \times 35.5 = 71\,g$
M_r of HCl = $(1 + 35.5) = 36.5$ ... so mass of 1 mole of HCl = $36.5\,g$

Our balanced equation tells us that 1 mole of hydrogen reacts with 1 mole of chlorine to give 2 moles of HCl. So turning this into masses:

$$
\begin{aligned}
1 \text{ mole of hydrogen} &= 1 \times 2\,g &&= \mathbf{2\,g} \\
1 \text{ mole of chlorine} &= 1 \times 71\,g &&= \mathbf{71\,g} \\
2 \text{ moles of HCl} &= 2 \times 36.5\,g &&= \mathbf{73\,g}
\end{aligned}
$$

Calculations

These calculations are important when you want to know the mass of chemicals that react together. For example, sodium hydroxide reacts with chlorine gas to make bleach.

Here is the balanced symbol equation for this reaction:

$$2NaOH + Cl_2 \rightarrow NaOCl + NaCl + H_2O$$

sodium hydroxide   chlorine   bleach   salt   water

This reaction happens when chlorine gas is bubbled through a solution of sodium hydroxide.

If you have a solution containing 100 g of sodium hydroxide, how much chlorine gas is needed to convert it to bleach? Too much, and some chlorine will be wasted. Too little, and not all of the sodium hydroxide will react.

(A_r values: H = 1, O = 16, Na = 23 and Cl = 35.5)

Mass of 1 mole of	
NaOH	Cl_2
= 23 + 16 + 1 = 40	= 35.5 × 2 = 71

The table shows that 1 mole of sodium hydroxide has a mass of 40 g.

So 100 g of sodium hydroxide is $\frac{100}{40}$ = 2.5 moles.

The balanced symbol equation tells us that for every 2 moles of sodium hydroxide you need 1 mole of chlorine to react with it.

So you need $\frac{2.5}{2}$ = 1.25 moles of chlorine.

The table shows that 1 mole of chlorine has a mass of 71 g.

So you will need 1.25 × 71 = **88.75 g** of chlorine to react with 100 g of sodium hydroxide.

Summary questions

1 '2HCl' can have two meanings. What are they?

2 Magnesium burns in oxygen with a bright white flame:

$$2Mg(s) + O_2(g) \rightarrow 2MgO(s)$$

What mass of oxygen will react exactly with 6.0 g of magnesium? (A_r values: O = 16, Mg = 24)

3 a An aqueous solution of hydrogen peroxide, H_2O_2, decomposes to form water and oxygen gas. Write a balanced symbol equation, including state symbols, for this reaction.

b When hydrogen peroxide decomposes, what mass of hydrogen peroxide is needed in solution to produce 1.6 g of oxygen gas? (A_r values: H = 1, O = 16)

4 When a small lump of calcium metal, Ca, is added to water it reacts giving off hydrogen gas. A solution of calcium hydroxide, $Ca(OH)_2$, is also formed in the reaction. (A_r values: Ca = 40, O = 16, H = 1)

a Write a balanced symbol equation, including state symbols, for the reaction.

b Calculate how much calcium metal must be added to an excess of water to produce 3.7 g of calcium hydroxide in solution.

Key points

- Balanced symbol equations tell us the number of moles of substances involved in a chemical reaction.

- Use balanced symbol equations to calculate the masses of reactants and products in a chemical reaction.

8.5 The yield of a chemical reaction

Learning objectives

After this topic, you should know:

- what is meant by the yield of a chemical reaction
- what factors can affect the yield
- how to calculate the percentage yield of a chemical reaction.

Many of the substances that you use every day have to be made from other chemicals. This may involve using complex chemical reactions. Examples include the plastics and composites used in your phones and computers, the ink in your pen or printer, and the artificial fibres in your clothes. All of these are made using chemical reactions.

Imagine a reaction: $A + 2B \rightarrow C$

If 1000 kg of C is needed, you can work out how much A and B are required. You need to know the relative formula masses of A, B, and C, and the balanced symbol equation.

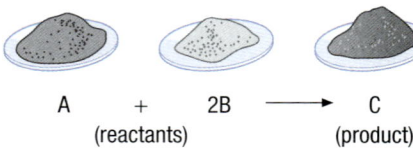

If you carry out the reaction, it is unlikely that you will get as much of C as you worked out. This is because your calculations assumed that all of the reactants A and B would be turned into product C. The amount of product that a chemical reaction produces is called its **yield**.

It is useful to think about reactions in terms of their **percentage yield**. This compares the amount of product that the reaction *really* produces with the maximum amount that it could *possibly* produce, predicted from the balanced symbol equation:

$$\text{Percentage yield} = \frac{\text{amount of product produced}}{\text{maximum amount of product possible}} \times 100\%$$

Calculating percentage yield

Worked example

Limestone is made mainly of calcium carbonate. Crushed lumps of limestone are heated in a rotating lime kiln. The calcium carbonate decomposes to make calcium oxide, and carbon dioxide gas is given off. A company processes 200 tonnes of limestone a day. It collects 98 tonnes of calcium oxide, the useful product. What is the percentage yield of the reaction in the kiln, assuming limestone contains only calcium carbonate? (A_r values: $Ca = 40$, $C = 12$, $O = 16$)

Solution

$$\text{calcium carbonate} \rightarrow \text{calcium oxide} + \text{carbon dioxide}$$
$$CaCO_3(s) \rightarrow CaO(s) + CO_2(g)$$

Work out the relative formula masses of $CaCO_3$ and CaO.

$$M_r \text{ of } CaCO_3 = 40 + 12 + (16 \times 3) = 100$$
$$M_r \text{ of } CaO = 40 + 16 = 56$$

So the balanced symbol equation tells us that:

100 tonnes of $CaCO_3$ could make 56 tonnes of CaO, assuming a 100% yield.

Therefore 200 tonnes of $CaCO_3$ could make a maximum of (56×2) tonnes of CaO = 112 tonnes.

So **percentage yield** $= \dfrac{\text{amount of product produced}}{\text{maximum amount of product possible}} \times 100\%$

$$= \frac{98}{112} \times 100 = \textbf{87.5\%}$$

This yield can be explained as some of the limestone is lost as dust in the crushing process and in the rotating kiln. There will also be some other mineral compounds in the limestone. It is not 100% calcium carbonate as assumed in the calculation.

Very few chemical reactions have a yield of 100% for the following reasons:

- The reaction may be reversible (so as products form they react to re-form the reactants again).
- Some reactants may react to give unexpected or unwanted products in alternative reactions.
- Some of the product may be lost in handling or left behind in the apparatus.
- The reactants may not be pure.
- Some of the desired product may be lost during separation from the reaction mixture.

Sustainable production

Chemical companies use reactions to make products they sell. Ideally, they want to use reactions with high yields (that also happen at a reasonable rate). Making a product more efficiently means making less waste. As much product as possible should be made from the reactants.

Chemical factories (or plants) are designed by chemical engineers. They design a plant to work as safely and economically as possible. It should waste as little energy and raw materials as possible. This helps the company to make money. It is better for the environment too as it conserves our limited resources. It also reduces the pollution from the use of fossil fuels as sources of energy.

⃝⃝ links

For information about using balanced symbol equations to predict reacting masses, look back to 8.4 'Equations and calculations'.

Summary questions

1 Explain generally why it is good for the environment if industry finds ways to make products using high yield reactions and processes that waste as little energy as possible.

2 List the factors that can affect the percentage yield of a reaction.

3 If the percentage yield for a reaction is 100%, 60 g of reactant A would make 80 g of product C. How much of reactant A is needed to make 80 g of product C if the percentage yield of the reaction is only 75%?

4 Ammonia gas (NH_3) is made by heating the gases nitrogen and hydrogen under pressure in the presence of an iron catalyst:

$$N_2(g) + 3H_2(g) \rightarrow 2NH_3(g)$$

If 7.0 g of nitrogen are reacted with excess hydrogen and 1.8 g of ammonia is collected, what is the percentage yield?

5 Sodium hydrogencarbonate, $NaHCO_3$, can be converted into sodium carbonate, Na_2CO_3, by heating. This is a thermal decomposition reaction in which water vapour and carbon dioxide are also products of the reaction. A student started with 16.8 g of sodium hydrogencarbonate and collected 9.20 g of sodium carbonate.

  a Write a balanced symbol equation for the thermal decomposition of sodium hydrogencarbonate.

  b What percentage yield did the student obtain?

Key points

- The yield of a chemical reaction describes how much product is made.

- The percentage yield of a chemical reaction tells us how much product is made compared with the maximum amount that could be made (100%).

- Factors affecting the yield of a chemical reaction include product being left behind in the apparatus, reversible reactions not going to completion, some reactants possibly producing unexpected reactions, and losses in separating the products from the reaction mixture.

8.6 Titrations

Learning objectives

After this topic, you should know:

- how to measure accurately the amount of acid and alkali that react together completely

- how to determine when the reaction is complete.

🔗 **links**

For information about strong and weak acids see 14.3 'Carboxylic acids and esters'.

🔗 **links**

For more information on neutralisation reactions, look back at 7.2 'Making soluble salts from metals or insoluble bases' and 7.3 'Making salts by neutralisation and precipitation'.

Study tip

Use a pipette to measure out a fixed volume of solution. Use a burette to measure the volume of the solution added.

An acid and an alkali (a soluble base) react together and neutralise each other. They form a salt (plus water) in the process.

Suppose you mix a strong acid and a strong alkali. The solution made will be neutral only if you add exactly the right quantities of acid and alkali.

If you start off with more acid than alkali, then the alkali will be neutralised. However, the solution left after the reaction will be acidic, not neutral. This is because some acid will be left over. The acid is said to be *in excess*. If you have more alkali than acid to begin with, then all the acid will be neutralised and the solution left will be alkaline.

You can measure the exact volumes of acid and alkali needed to react with each other using a technique called **titration**. The point at which the acid and alkali have reacted completely is called the **end point** of the reaction. You should judge when the end point is reached by using an acid/base indicator.

Required practical

Carrying out a titration

In this experiment you can carry out a titration. You will find out how much acid is needed to completely react with an alkali.

1 Measure a known volume of alkali into a conical flask using a volumetric **pipette**. Before doing this, you should first wash the pipette with distilled water, and then with some of the alkali.

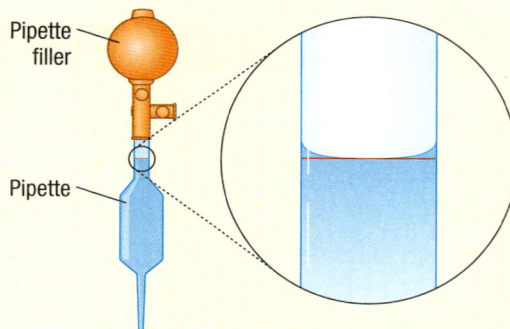

Figure 1 *A volumetric pipette and pipette filler. Fill the pipette until the bottom of the meniscus (curved surface of the solution) coincides with the mark. Allow the liquid to run out of the pipette and touch the tip on the side of the flask to drain out the solution. It is normal for a tiny amount of solution to remain in the pipette.*

2 Now add a few drops of acid/base indicator to the solution in the flask and swirl. Commonly used examples are methyl orange or phenolphthalein.

3 Pour the acid you are going to use into a **burette**. This is a long tube with a tap on one end. The tube has markings on it to enable you to measure volumes accurately (often to the nearest 0.05 cm³). Before doing this, you should first wash the burette with distilled water, and then with some of the acid.

4 Record the reading on the burette. Then open the tap to release a small amount of acid into the flask. Swirl the flask to make sure that the two solutions are mixed.

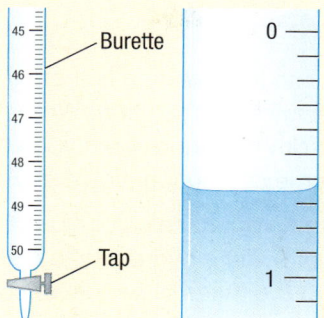

Figure 2 *A burette – use the bottom of the meniscus to read the scale. The reading here is 0.65 cm³.*

5 Keep on repeating step 4 until the indicator in the flask changes colour. This shows when the alkali in the flask has completely reacted with the acid added from the burette. Record the reading on the burette and work out the volume of acid run into the flask. (On your first go at doing this you will probably run too much acid into the flask, so treat this as a rough estimate of how much acid is needed.)

6 Repeat the whole process at least three times. Discard any anomalous results. Then calculate an average (mean) value to give the most accurate results possible. Alternatively, repeat the titration until you get two consecutive identical results. These are called concordant results. In practice, aim for two results within 0.10 cm³ of each other and take their average.

7 Having found out how much acid was required to neutralise the alkali, you can now use your results to calculate the concentration of the alkali.

Safety: Wear eye protection. Chemicals in this practical may be harmful or irritant.

⚬⚬ **links**

For more information on calculations, see 8.7 'Titration calculations'.

Summary questions

1 What word do scientists use for a set of repeat readings with a very narrow range? (Hint: see pages 194–195 in the 'Experimental data handling' section.)

2 a What word is used for the curved surface of a liquid in a measuring cylinder?
  b i Name two measuring instruments used to measure volumes of liquid in a titration.
    ii What is the correct way to read the volume from the level of liquid in one of these measuring instruments.

3 a Describe how to carry out a titration between dilute nitric acid of known concentration and sodium hydroxide solution of unknown concentration.
  b Write a balanced symbol equation, including state symbols, for the reaction in part **a**.

4 a Name two acid/base indicators that can be used in the titration between dilute hydrochloric acid and sodium hydroxide solution.
  b Explain what makes a good indicator for a titration between an acid and an alkali.

Key points

- Titration is used to measure accurately how much acid and alkali react together completely.

- The point at which an acid–alkali reaction is complete is called the end point of the reaction.

- Use an acid/base indicator to show the end point of the reaction between an acid and an alkali.

8.7 Titration calculations

After this topic, you should know:

■ how to calculate concentrations from reacting volumes of solutions

■ how to calculate the amount of acid or alkali needed in a neutralisation reaction.

Calculating concentrations

The concentration of a solute in a solution is the number of moles of solute dissolved in one cubic decimetre of solution. Write these units as **moles per decimetre cubed** or **mol/dm³** for short. So if you know the mass of solute dissolved in a certain volume of solution, you can work out its concentration.

As an example, imagine that you make a solution of sodium hydroxide in water. You dissolve exactly 40 g of sodium hydroxide to make exactly 1 dm³ of solution. You know how to work out the mass of 1 mole of sodium hydroxide (NaOH). Add up the relative atomic masses of sodium, oxygen and hydrogen:

$$23 \, (Na) + 16 \, (O) + 1 \, (H) = 40 \, g = \text{mass of 1 mole of NaOH}$$

Therefore the solution contains 1 mole of sodium hydroxide in 1 dm³ of solution. So the concentration of sodium hydroxide in the solution is 1 mol/dm³. See Worked example 1.

Sometimes you know the concentration of a solution and need to work out the mass of solute in a certain volume. See Worked example 2.

Worked example 1

What if you use 20 g of sodium hydroxide to make 200 cm³ of solution instead of 40 g in 1 dm³? What will be the concentration of the sodium hydroxide in mol/dm³? (Note that 1 dm³ = 1000 cm³.)

Solution

To find the concentration of the solution you must work out how much sodium hydroxide there would be if you had 1000 cm³ (1 dm³) of the solution.

20 g of NaOH are dissolved in 200 cm³ of solution, so

$\dfrac{20}{200}$ g of NaOH would be dissolved in 1 cm³ of solution, and

$\dfrac{20}{200} \times 1000 \, g = \mathbf{100 \, g}$ of NaOH would be dissolved in 1000 cm³ of solution.

The mass of 1 mole of NaOH is 40 g, so

$$100 \, g \text{ of NaOH is } \frac{100}{40} \text{ moles} = \mathbf{2.5 \ moles}$$

2.5 moles of NaOH are dissolved in 1 dm³ of solution. So the concentration of NaOH in the solution is **2.5 mol/dm³**.

Worked example 2

What mass of potassium sulfate, K_2SO_4, is there in 25 cm³ of a 2.0 mol/dm³ solution?

Solution

In 1 dm³ of solution there would be 2 moles of K_2SO_4

The mass of 1 mole of K_2SO_4 is $(2 \times 39) + 32 + (4 \times 16) \, g = 174 \, g$, so in 1000 cm³ of solution there would be

$(174 \times 2) = 348 \, g$ of K_2SO_4, and in 1 cm³ of solution there are $\dfrac{348}{1000}$ g of K_2SO_4

So in 25 cm³ of solution there are $\dfrac{348}{1000} \times 25 \, g$ of $K_2SO_4 = \mathbf{8.7 \, g}$ of K_2SO_4

There is **8.7 g of K_2SO_4** in 25 cm³ of 2.0 mol/dm³ potassium sulfate solution.

Titration calculations

In a titration you always have one solution with a concentration which you know accurately. This is put in the burette. Then you can place the other solution, with an unknown concentration, in a conical flask using a volumetric pipette. This ensures you know the volume of this solution accurately. The result from the titration is used to calculate the number of moles of the substance in the solution in the conical flask. See Worked example 3.

Worked example 3

A student put $25.0 \, cm^3$ of sodium hydroxide solution of unknown concentration into a conical flask using a volumetric pipette. The sodium hydroxide reacted with exactly $20.0 \, cm^3$ of $0.50 \, mol/dm^3$ sulfuric acid added from a burette. What was the concentration of the sodium hydroxide solution in mol/dm^3?

Solution

The equation for this reaction is:

$$2NaOH(aq) + H_2SO_4(aq) \rightarrow Na_2SO_4(aq) + 2H_2O(l)$$

This equation tells us that 2 moles of NaOH reacts with 1 mole of H_2SO_4.

The concentration of the H_2SO_4 is $0.50 \, mol/dm^3$, so

0.50 moles of H_2SO_4 are dissolved in $1000 \, cm^3$ of acid, and

$\dfrac{0.50}{1000}$ moles of H_2SO_4 are dissolved in $1 \, cm^3$ of acid, therefore

$\dfrac{0.50}{1000} \times 20.0$ moles of H_2SO_4 are dissolved in $20.0 \, cm^3$ of acid.

So there are 0.010 moles of H_2SO_4 dissolved in $20.0 \, cm^3$ of acid.

The equation for the reaction tells us that 0.010 moles of H_2SO_4 will react with exactly 2×0.010 moles of NaOH. This means that there must have been 0.020 moles of NaOH in the $25.0 \, cm^3$ of solution in the conical flask. To calculate the concentration of NaOH in the solution in the flask you need to calculate the number of moles of NaOH in $1 \, dm^3$ ($1000 \, cm^3$) of solution.

0.020 moles of NaOH are dissolved in $25.0 \, cm^3$ of solution, so

$\dfrac{0.020}{25}$ moles of NaOH are dissolved in $1 \, cm^3$ of solution, and there will be

$\dfrac{0.020}{25} \times 1000 = 0.80$ moles of NaOH in $1000 \, cm^3$ of solution.

The concentration of the sodium hydroxide solution is $0.80 \, mol/dm^3$.

Key points

- To calculate the concentration of a solution, given the mass of solute in a certain volume:
  1. Calculate the mass (in grams) of solute in $1 \, cm^3$ of solution.
  2. Calculate the mass (in grams) of solute in $1000 \, cm^3$ of solution.
  3. Convert the mass (in grams) to moles.

- To calculate the mass of solute in a certain volume of solution of known concentration:
  1. Calculate the mass (in grams) of the solute there is in $1 \, dm^3$ ($1000 \, cm^3$) of solution.
  2. Calculate the mass (in grams) of solute in $1 \, cm^3$ of solution.
  3. Calculate the mass (in grams) of solute there is in the given volume of the solution.

Summary questions

In a titration, a $12.5 \, cm^3$ sample of nitric acid (HNO_3) reacted exactly with $10.0 \, cm^3$ of $0.40 \, mol/dm^3$ potassium hydroxide solution. Answer questions 1 to 4 below:

1. Write down a balanced symbol equation, including state symbols, for this reaction.

2. Calculate the number of moles of potassium hydroxide used.

3. How many moles of nitric acid react?

4. Calculate the concentration of the nitric acid in mol/dm^3 and in g/dm^3.

8.8 Volumes of gases

After this topic, you should know:

- how to calculate the volume of a gas at room temperature and pressure from its mass and relative formula mass

- how to calculate volumes of gaseous reactants and products from a balanced equation and a given volume of a gaseous reactant or product.

Figure 1 *Crash-test dummies are used to gauge the best volume of gas to cushion the effects of a collision. A sensor detects the sudden deceleration and completes an electrical circuit to ignite the chemicals in the air-bag.*

Having air-bags in cars has saved many lives. When a crash takes place, the chemicals in the bags react together to rapidly give off nitrogen gas, N_2. The reacting chemicals have to release just the right volume of gas to make the air-bag act like a cushion, reducing the damage caused to the occupants of the car.

The designers of air-bags can calculate the volume of gas produced from the balanced equation for the reaction (see the Worked examples below). This tells them how many moles of gas are made, as it has been found that equal numbers of moles of any gas occupy the same volume.

The volume of 1 mole of any gas occupies 24 dm³ (24 000 cm³) at room temperature and pressure, that is, at 20 °C and 1 atmosphere.

24 dm³/mol is known as the molar gas volume. This data is always provided if needed to answer a question. Using the known volume of 1 mole of gas:

$$\text{number of moles of gas} = \frac{\text{volume of gas (dm}^3)}{24\,\text{dm}^3} = \frac{\text{volume of gas (cm}^3)}{24\,000\,\text{cm}^3}$$

Calculating volumes of gaseous reactants or products

You can use a balanced symbol equation to find the numbers of moles of reactants and products in reactions involving gases.

This is straightforward if the reaction involves more than one gas and you are given the volume of one of the gases. The ratio of the numbers of moles in the balanced equation gives you the ratio of the volume of gases involved. This is because the same number of moles of any gas occupies the same volume. For example, as you saw on page 46, hydrogen and chlorine react to make hydrogen chloride:

$$H_2(g) + Cl_2(g) \rightarrow 2HCl(g)$$

Worked example 1

One make of car has an air-bag that is inflated by 70.0 g of nitrogen, N_2, when activated.

What volume would the nitrogen gas occupy at room temperature and pressure?   (A_r of N = 14)

Solution

First of all you have to find out how many moles of nitrogen gas are in 70.0 g of N_2.

You have seen on page 101 that:

$$\text{number of moles} = \frac{\text{mass}}{\text{relative formula mass (in g)}}$$

The relative formula mass of $N_2 = (14 \times 2) = 28$ g

So the number of moles of N_2 gas $= \frac{70.0\,\text{g}}{28\,\text{g}} = 2.5\,\text{mol}$

To find the volume that 2.5 mol of N_2 gas will occupy, you must first rearrange the equation:

$$\text{number of moles of gas} = \frac{\text{volume of gas (dm}^3)}{24\,\text{dm}^3}$$

To get:

$$\text{volume of gas (dm}^3) = \text{number of moles} \times 24\,\text{dm}^3$$

So the volume of nitrogen gas $= 2.5 \times 24\,\text{dm}^3 = \textbf{60 dm}^3$

1 mole of hydrogen gas reacts with 1 mole of chlorine gas to give 2 moles of hydrogen chloride gas. So the ratio of moles of $H_2(g) : Cl_2(g) : HCl(g)$ is 1 : 1 : 2. So if you start with $50\,cm^3$ of $H_2(g)$, it will react with $50\,cm^3$ of $Cl_2(g)$ to give $100\,cm^3$ of $HCl(g)$.

The problems are more complex if you have reactions of solids and/or solutions that produce a gas. However, working logically from the numbers of moles in the balanced equation will allow you to solve them.

Worked example 2

The nitrogen gas produced in an air-bag is formed in two reactions. In the first reaction, a solid called sodium azide, NaN_3, is ignited and decomposes, producing the majority of the gas to fill the air-bag:

$$2NaN_3(s) \rightarrow 2Na(s) + 3N_2(g)$$

What mass of sodium azide would be needed to produce $48.0\,dm^3$ of nitrogen gas at room temperature and pressure? (A_r of N = 14, Na = 23)

Solution

You want to make $48.0\,dm^3$ of nitrogen gas, so convert that to moles using:

$$\text{number of moles of gas} = \frac{\text{volume of gas (dm}^3)}{24\,dm^3} = \frac{48.0\,dm^3}{24\,dm^3} = 2 \text{ moles of } N_2(g)$$

The balanced equation for the reaction shown above tells us that:

2 moles of $NaN_3(s) \rightarrow$ 2 moles of $Na(s)$ + 3 moles of $N_2(g)$

So the ratio you need from the equation is:

2 mol $NaN_3(s)$: 3 mol $N_2(g)$

Dividing this ratio by 3 means that $\frac{2}{3}$ mol $NaN_3(s)$ would give 1 mol $N_2(g)$, and then multiplying by 2 (so you get two moles of N_2 gas) means that $\frac{4}{3}$ mol $NaN_3(s)$ would give 2 mol $N_2(g)$.

Using the equation:

$$\text{mass} = \text{number of moles} \times \text{relative formula mass}$$

you get the mass of $\frac{4}{3}$ moles of $NaN_3(s) = \frac{4}{3} \times [23 + (14 \times 3)]\,g =$ **86.7 g** (answer given to 3 significant figures)

Summary questions

1 What is meant by the 'molar gas volume'?

2 a How many moles of gas are present in:
   i 36 dm^3 of carbon dioxide, $CO_2(g)$?
   ii 10 000 dm^3 of hydrogen, $H_2(g)$?
   b What volume would be occupied by:
   i 36.0 g of helium, He(g)?
   ii 13.8 g of nitrogen dioxide, $NO_2(g)$?
   c What mass of gas is present in 48 cm^3 of oxygen, O_2? (Take care with the units!)

3 When methane gas burns completely in air, it forms carbon dioxide and water:
   $$CH_4(g) + 2O_2(aq) \rightarrow CO_2(g) + 2H_2O(g)$$
   What volume of oxygen gas is needed to burn 150 dm^3 of methane?

4 Calcium reacts with dilute hydrochloric acid vigorously, giving off hydrogen gas:
   $$Ca(s) + 2HCl(aq) \rightarrow CaCl_2(aq) + H_2(g)$$
   What volume of hydrogen would be produced when 0.80 g of calcium is added to excess dilute acid?

Key points

- A certain volume of gas always contains the same number of gas molecules under the same conditions.

- The volume of 1 mole of any gas at room temperature and pressure is 24 dm^3 (24 000 cm^3).

- You can use the molar gas volume and balanced symbol equations to calculate volumes of gaseous reactants or products.

Chapter summary questions

1 Calculate the mass of 1 mole of each of the following compounds:

  a H_2S **b** SO_2 **c** C_2H_4

  d $NaOH$ **e** Na_2CO_3 **f** $Al_2(SO_4)_3$

  g $NaAl(OH)_4$

(A_r values: H = 1, C = 12, O = 16, Na = 23, Al = 27, S = 32)

2 How many moles of:

  a Ag atoms are there in 27 g of silver?

  b Fe atoms are there in 0.056 g of iron?

  c P_4 molecules are there in 6.2 g of phosphorus?

(A_r values: P = 31, Fe = 56, Ag = 108)

3 a The chemical formula of ethane is C_2H_6. Work out the percentage by mass of carbon in ethane.

  b Work out the mass of hydrogen present in 3.8 g of ethane.

(A_r values: H = 1, C = 12)

4 When aluminium reacts with bromine, 4.05 g of aluminium reacts with 36.0 g of bromine. What is the empirical formula of aluminium bromide?

(A_r values: Al = 27, Br = 80)

5 Balance the following symbol equations:

  a $Na + Cl_2 \rightarrow NaCl$

  b $Al + O_2 \rightarrow Al_2O_3$

  c $Al(OH)_3 \rightarrow Al_2O_3 + H_2O$

  d $Ba(NO_3)_2 \rightarrow BaO + NO_2 + O_2$

  e $C_4H_{10} + O_2 \rightarrow CO_2 + H_2O$

6 In a lime kiln, calcium carbonate is decomposed to calcium oxide:

$$CaCO_3 \rightarrow CaO + CO_2$$

1500 tonnes of calcium carbonate gave 804 tonnes of calcium oxide. Calculate the percentage yield for the process.

(A_r values: C = 12, O = 16, Ca = 40)

7 a Ethene gas (C_2H_4) reacting with steam (H_2O) to form ethanol gas (C_2H_5OH) is a reversible reaction. Write the balanced symbol equation for this reaction, including state symbols.

b If 14.00 g of ethene reacts with excess steam to produce 17.25 g of ethanol, what is the percentage yield of the reaction?

8 Sulfur is mined in Poland and is brought to Britain in ships. The sulfur is used to make sulfuric acid. Sulfur is burnt in air to produce sulfur dioxide. Sulfur dioxide and air are passed over a heated catalyst to produce sulfur trioxide. Water can be added to sulfur trioxide to produce sulfuric acid.

The reactions are:

$$S + O_2 \rightarrow SO_2$$
$$2SO_2 + O_2 \rightleftharpoons 2SO_3$$
$$SO_3 + H_2O \rightarrow H_2SO_4$$

(A_r values: H = 1, O = 16, S = 32)

  a How many moles of sulfuric acid could be produced from one mole of sulfur?

  b Calculate the maximum mass of sulfuric acid that can be produced from 64 kg of sulfur.

  c In an industrial process the mass of sulfuric acid that was produced from 64.00 kg of sulfur was 188.16 kg. Use your answer to part **b** to calculate the percentage yield of this process.

  d Suggest two reasons why the yield of the industrial process was less than the maximum yield.

9 A student placed 12.5 cm³ of potassium hydroxide solution of an unknown concentration into a conical flask using a volumetric pipette. The potassium hydroxide reacted with exactly 20.0 cm³ of 0.50 mol/dm³ hydrochloric acid which was added from a burette.

  a Write the balanced symbol equation, including state symbols, for this reaction.

  b i How many moles of hydrochloric acid are there in 20.0 cm³ of 0.50 mol/dm³ hydrochloric acid?

     ii How many moles of potassium hydroxide will this react with?

  c What is the concentration of the potassium hydroxide solution:

    i in moles per dm³?

    ii in grams per dm³?

     (A_r values: K = 39, O = 16, H = 1)

Dilute hydrochloric acid

Potassium hydroxide solution

Practice questions

1 Use the relative atomic masses in the periodic table in the following calculations.

   a What is the relative formula mass of O_2? (1)

   b What is the relative formula mass of $Ca(OH)_2$? (1)

   c How many moles of sodium hydroxide are there in 20 g of NaOH? (2)

   d What mass of ammonium nitrate is present in 2 mol of NH_4NO_3? (2)

2 The element bromine is formed in the following reaction.

$$HBr + H_2SO_4 \rightarrow H_2O + Br_2 + SO_2$$

   a Write a balanced equation for this reaction. (1)

   b The formula of a bromine compound formerly used in agriculture is CH_3Br.

     What is the percentage by mass of bromine in this compound? (2)

   c Another bromine compound contains 15.2% sodium and 53.0% bromine by mass. The remainder is oxygen.

     Calculate the empirical formula of this compound. (4)

3 An experiment was carried out to investigate the reduction of an oxide of copper.

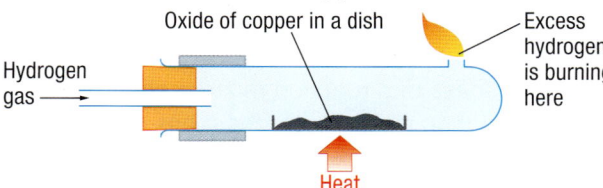

Oxide of copper in a dish

Hydrogen gas

Excess hydrogen is burning here

Heat

   a The diagram shows the hydrogen burning at the end of the tube.

     What would be the risk if the hydrogen were not burnt at the end of the tube? (1)

   b A student observed some drops of a colourless liquid inside the tube shortly after starting to heat it.

     What is the liquid? Outline the reaction in which it formed. (2)

   c Describe how you could be certain that all the oxide had completely changed into copper. (2)

   d A student recorded the following results:

     mass of oxide before heating = 5.72 g
     mass of copper formed   = 5.08 g

     i Calculate the mass of oxygen in the oxide. (1)

     ii Calculate the amount, in moles, of oxygen in the oxide. (1)

     iii Calculate the amount, in moles, of copper formed. (1)

     iv Calculate the empirical formula of the oxide. (2)

4 The symbol equation for a reaction used to make potassium nitrate is:

$$KOH + HNO_3 \rightarrow KNO_3 + H_2O$$

   a What is the simplest ionic equation for this reaction? (1)

   b In an experiment used to make potassium nitrate, it was found that 27.80 cm³ of nitric acid was needed to neutralise 25.00 cm³ of potassium hydroxide solution.

     i Which reactant solution is more concentrated? Give a reason for your choice. (2)

     ii Name the pieces of equipment used to measure the volume of each reactant. (2)

   c In another experiment to make potassium nitrate by the same method, a student recorded these volumes, in cm³, of nitric acid added:

     26.85 27.20 26.10 26.20

     i State which volumes should be used to calculate the average volume of nitric acid added. (1)

     ii Calculate the average volume of nitric acid added. (2)

   d A student recorded the following information from a titration.

     Volume of 0.120 moles/dm³ KOH(aq) = 25.00 cm³

     Volume of HNO_3(aq) = 28.50 cm³

     Calculate the concentration of the nitric acid. (2)

5 A reaction used in industry to manufacture methanol is:

$$CO + 2H_2 \rightleftharpoons CH_3OH$$

A chemist investigated this reaction under different conditions, starting with 56 g of carbon monoxide.

   a What mass of hydrogen is needed to completely react with this mass of carbon monoxide? (2)

   b What is the maximum mass of methanol that can be obtained from this mass of carbon monoxide? (2)

   c In one experiment the chemist obtained 16 g of methanol.

     Calculate the percentage yield of methanol in this experiment. (2)

9.1

How fast?

Learning objectives

After this topic, you should know:

- what is meant by the rate of a chemical reaction
- how to find out the rate of a chemical reaction.

The rate of a chemical reaction tells us how fast reactants turn into products. In our bodies, there are lots of reactions taking place all the time. They happen at the correct rate to supply our cells with what they need, whenever required.

Reaction rate is also very important in the chemical industry. Any industrial process has to make money by producing useful products. This means the right amount of product needed must be made as cheaply as possible. If it takes too long to produce, it will be hard to make a profit when it is sold. The rate of the reaction must be fast enough to make it quickly but safely.

How can you determine the rate of reactions?

Reactions happen at all sorts of different rates. Some are really fast, such as the combustion of chemicals inside a firework exploding. Others are very slow, such as an old piece of iron rusting.

There are two ways to work out the rate of a chemical reaction. You can measure how quickly:

- the reactants are used up as they make products, or
- the products of the reaction are made.

Here are three ways you can make these kinds of measurement in experiments.

Required practical

Measuring the decreasing mass of a reaction mixture

You can measure the rate at which the *mass* of a reaction mixture changes if the reaction gives off a gas. As the reaction takes place, the mass of the reaction mixture decreases. You can measure and record the mass at time intervals which you decide.

Some balances can be attached to a computer to monitor the loss in mass continuously.

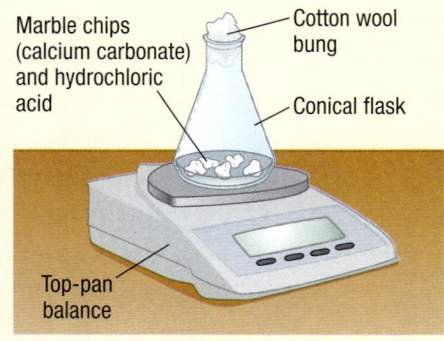

Marble chips (calcium carbonate) and hydrochloric acid — Cotton wool bung — Conical flask — Top-pan balance

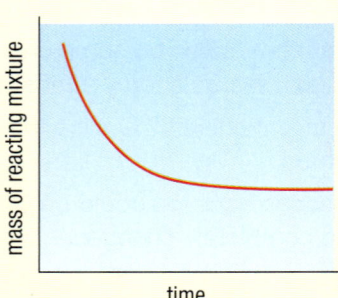

mass of reacting mixture / time

- Why is the cotton wool placed in the neck of the conical flask?
- How would the line on the graph differ if you plot 'Loss in mass' on the vertical axis?

Safety: Wear eye protection. Dilute hydrochloric acid is an irritant.

Study tip

The steeper the line on the graph, the faster the rate of reaction.

Required practical

Measuring the increasing volume of gas given off

If a reaction produces a gas, you can use the gas to find out the rate of reaction. You do this by collecting the gas and measuring the volume given off at time intervals.

■ What are the sources of error when measuring the volume of gas?

Safety: Wear eye protection. Dilute hydrochloric acid is an irritant.

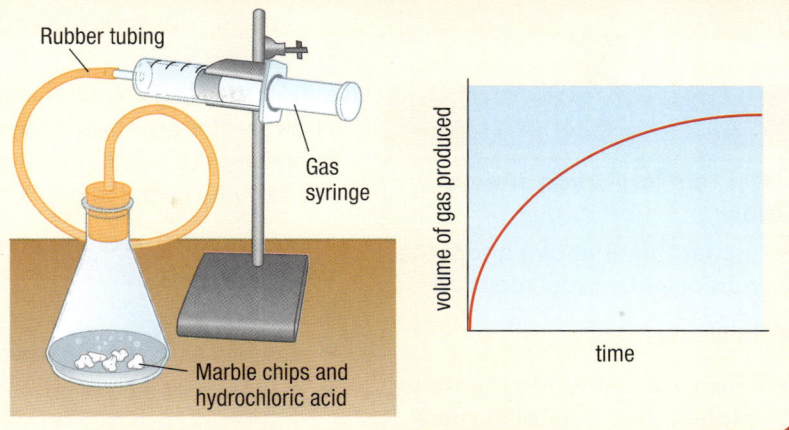

Rubber tubing
Gas syringe
Marble chips and hydrochloric acid

volume of gas produced
time

Required practical

Measuring the decreasing light passing through a solution

Some reactions in solution make a suspension of an insoluble solid (precipitate). This makes the solution go cloudy. You can use this to measure the rate at which the precipitate appears.

The reaction is set up in a flask. Under the flask, put a piece of paper marked with a cross. Then we can record the time taken for the cross to disappear. The shorter the time is, the faster the reaction rate.

Alternatively, you can use a light sensor and data logger. Then measure the amount of light passing through the solution over time, as the graph shows.

■ What are the advantages of using a light sensor rather than the 'disappearing cross' method?

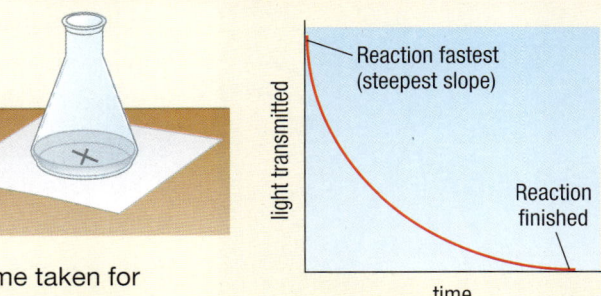

Reaction fastest (steepest slope)
Reaction finished
light transmitted
time

We can summarise these methods of working out the rate of a reaction using this equation:

$$\text{rate of reaction} = \frac{\left[\begin{array}{c}\text{amount of reactant used or} \\ \text{amount of product formed}\end{array}\right]}{\text{time}}$$

Summary questions

1 a Write a word equation and a balanced symbol equation, including state symbols, for the reaction taking place in the practical box on the previous page.
  b In the same experiment, there were some marble chips left when the fizzing stopped. Why was no more gas given off?
  c How can you tell from a graph of the experimental results when the reaction had finished?

2 a Sketch graphs to show the results of:
    i   measuring the mass of products formed in a reaction over time
    ii measuring the mass of reactants remaining in a reaction over time.
  b What does the slope of the line at any particular time on the graphs in part a tell us about the reaction?

Key points

■ You can determine the rate of a chemical reaction by measuring the amount of reactants used up over time.

■ Alternatively, you can measure the rate of reaction by following the amount of products made over time.

■ The slope of the line at any given time on the graphs drawn from such experiments tells us the rate of reaction at that time: the steeper the slope, the faster the reaction.

9.2 Collision theory and surface area

Learning objectives

After this topic, you should know:

- the factors that can affect the rate of a chemical reaction

- collision theory

- how to use collision theory to explain the effect of surface area on reaction rate.

Figure 1 *There is no doubt that the chemicals in these fireworks have reacted. But how can you explain what happens in a chemical reaction?*

Figure 2 *Cooking – an excellent example of controlling reaction rates!*

In everyday life, you often control the rates of chemical reactions without thinking about it – for example, cooking cakes in an oven or revving up a car engine. In chemistry you need to know what affects the rate of reactions. You also need to explain why each factor affects the rate of a reaction.

There are four main factors which affect the rate of chemical reactions:

- temperature
- surface area
- concentration of solutions or pressure of gases
- presence of a catalyst.

Reactions can only take place when the particles (atoms, ions, or molecules) of reactants come together. But the reacting particles do not just have to collide with each other. They also need to collide with sufficient energy to cause a reaction to take place. This is known as **collision theory**.

The minimum amount of energy that particles must have before they can react is called the **activation energy**.

So reactions are more likely to happen between reactant particles if you:

- increase the frequency of reacting particles colliding with each other
- increase the energy they have when they collide.

If you increase the chance of particles reacting, you will also increase the rate of reaction.

Surface area and reaction rate

Imagine lighting a campfire. It is not a good idea to pile large logs together and try to set them alight. You use small pieces of wood to begin with. Doing this increases the surface area of the wood. This results in more wood being exposed to react with oxygen in the air.

When a solid reacts in a solution, the size of the pieces of solid affects the rate of the reaction. The particles inside a large lump of solid are not in contact with the reactant particles in the solution, so they cannot react. The particles inside the solid have to wait for the particles on the surface to react first before they have a chance to react.

In smaller lumps, or in a powder, each tiny piece of solid is surrounded by solution. Many more particles of the solid are exposed to attack at a given time. This means that reactions can take place much more quickly.

Practical

Which burns faster?

Using tongs, try igniting a 2 cm length of magnesium ribbon and time how long it takes to burn.

Take a small spatula tip of magnesium powder and sprinkle it into the Bunsen flame.

■ Write a balanced symbol equation, including state symbols, for the reaction.

■ Explain your observations.

Safety: Make sure you have a heatproof mat under the Bunsen burner and you must wear eye protection.

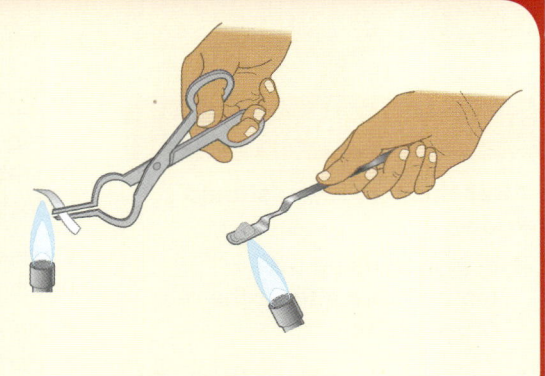

Required practical

Investigating the effect of surface area

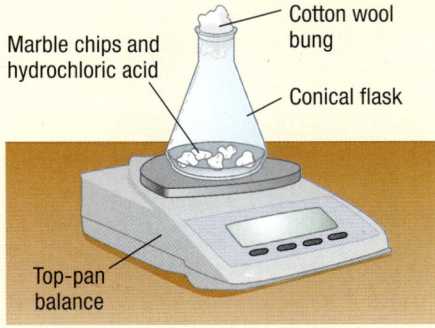

Marble chips and hydrochloric acid
Cotton wool bung
Conical flask
Top-pan balance

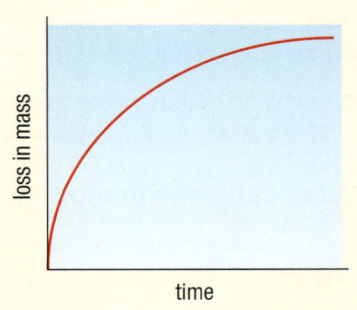

loss in mass

time

In this investigation you will be measuring the mass lost against time for different sizes of marble (calcium carbonate) chips. You need at least two different sizes of marble chips in order to vary the surface area.

■ What variables should you control to make this a fair test?

■ Why does this method of finding out the rate of reaction work?

■ Use the data collected to draw a graph. Explain what the graph shows. (A data logger would help to plot a graph of the results.)

Safety: Wear eye protection.

Summary questions

1 List the factors that can affect the rate of a chemical reaction.

2 **a** Draw a diagram to explain why it is easier for an iron nail that has been cut into small pieces to rust than for a whole iron nail to rust under the same conditions.

  b Why is the term 'under the same conditions' used in part **a**?

3 Explain why the idea of 'activation energy' is an important part of the collision theory used to explain rates of reaction.

4 Explain why the acid in your stomach can help you digest your food more quickly if you chew your food well before you swallow it.

Study tip

Particles collide all the time, but only some collisions lead to reactions.

Increasing the number of collisions in a certain time and the energy of collisions produces faster rates of reaction.

Larger surface area does not result in collisions with more energy but does increase the frequency of collisions.

Key points

■ Particles must collide with a certain minimum amount of energy before they can react.

■ The minimum amount of energy that particles must have in order to react is called the 'activation energy' of a reaction.

■ The rate of a chemical reaction increases if the surface area of any solid reactants is increased. This increases the frequency of collisions between reacting particles.

9.3 The effect of temperature

Learning objectives

After this topic, you should know:

- how increasing the temperature affects the rate of reactions

- how to use collision theory to explain this effect.

Figure 1 *Lowering the temperature will slow down the reactions that make foods go off*

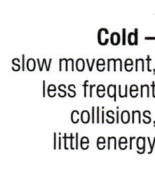

Figure 2 *Moving faster means it is more likely that you'll bump into someone else – and the collision will be harder too!*

When the temperature is increased, it increases the rate of reaction. In our homes, we use fridges and freezers to reduce the temperature and slow down the rate of reactions. When food goes off it is because of chemical reactions, so reducing the temperature slows down these unwanted reactions.

Collision theory tells us why raising the temperature increases the rate of a reaction. There are two reasons:

- Particles collide more frequently.
- Particles collide with more energy.

Particles collide more frequently

When you heat up a substance, energy is transferred to its particles. In solutions and in gases, this means that the particles move around faster. And when particles move around faster they collide more often. Imagine a lot of people walking around in the school playground blindfolded. They may bump into each other occasionally. However, if they start running around, they will bump into each other much more often.

When particles collide more frequently, there are more chances for them to react. This increases the rate of reaction.

Particles collide with more energy

Particles that are moving around more quickly have more energy. This means that any collisions they have are much more energetic. It is like two people colliding when they are running rather than when they are walking.

When you increase the temperature of a reacting mixture, a higher proportion of the collisions will result in a reaction taking place. This is because a higher proportion of particles have energy greater than the activation energy. **This second factor has a greater effect on rate of reaction than the increased frequency of collisions.**

Around room temperature, increasing the temperature of a reaction by 10 °C will cause the rate of the reaction roughly to double.

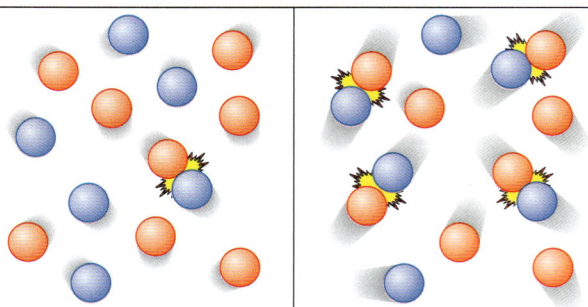

Cold – slow movement, less frequent collisions, little energy

Hot – fast movement, more frequent collisions, more energy

Figure 3 *More frequent collisions, with more energy – both of these factors lead to the increase in the rate of a chemical reaction that is caused by increasing the temperature*

Required practical

The effect of temperature on rate of reaction

Alka-Seltzer tablets are a well-known indigestion remedy. They fizz when added to water. The fizzing is caused by carbon dioxide gas, produced by the reaction of sodium hydrogen carbonate and citric acid, both contained in the tablets. These compounds can come into contact once the tablet is added to water. By varying the temperature of the water, you can measure its effect on the rate of reaction.

- How could you vary the temperature in the investigation?
- How could you measure the mean rate at different temperatures?

Check your ideas with your teacher before you start your investigation.

- Which variables do you have to control to make this a fair test?
- Why is it difficult to get accurate timings in this investigation?
- How can you improve the **precision**, and therefore the accuracy in this case, of any data which has random measurement errors?

Safety: Wear eye protection. Do not raise the temperature above 50 °C

The effect of temperature on the rate of reaction

As one goes up, the other comes down

In the experiment opposite you can measure the time for a cross to disappear as a precipitate forms.

This means that the longer the time, the slower the rate of reaction. There is an inverse relationship between time and rate. So as time increases, rate decreases.

The rate is proportional to $\frac{1}{time}$ (also written as time^{-1}). Therefore, you can plot a graph of temperature against $\frac{1}{time}$ to investigate the effect of temperature on rate of reaction.

The results of an investigation like this can be plotted on a graph (see right).

The graph shows how the time for the reaction to be completed changes with temperature. You could also measure the mean rate of this reaction by measuring either how long it takes to collect a set volume of gas, or the time for the reaction mixture to lose a set mass.

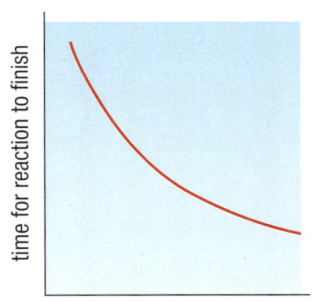

temperature

Summary questions

1 a Why does increasing the temperature increase the rate of a reaction?
  b By how much does a 10 °C rise in temperature increase reaction rate at room temperature?

2 Look at the experiment in the Practical box above.
  a Name the gas given off in the reaction.
  b Describe what happens to the time it takes for the reaction to finish as the temperature increases.
  c Explain your answer to part **b** in detail using collision theory.

3 Water in a pressure cooker boils at a much higher temperature than water in a saucepan because it is under pressure. Explain why food takes longer to cook in a pan than it does in a pressure cooker.

Key points

- Reactions happen more quickly as the temperature increases.

- Increasing the temperature increases the rate of reaction because particles collide more frequently and more energetically. More of the collisions result in a reaction because a higher proportion of particles have energy greater than the activation energy.

9.4

The effect of concentration or pressure

Learning objectives

After this topic, you should know:

- how, and why, increasing the concentration of reactants in solutions affects the rate of reaction

- how, and why, increasing the pressure of reacting gases affects the rate of reaction.

Figure 1 *Limestone statues are damaged by acid rain. This damage happens more quickly as the concentration of the acids in rainwater increases.*

Some of the world's most beautiful buildings are made of limestone or marble. These buildings have stood for centuries. However, they are now crumbling away at a greater rate than before. This is because both limestone and marble are mainly calcium carbonate. This reacts with acids, leaving the stone soft and crumbly. The rate of this reaction has speeded up because the concentration of acids in rainwater has increased.

Increasing the concentration of reactants in a solution increases the rate of reaction because there are more particles of the reactants moving around in the same volume of solution. The more 'crowded' together the reactant particles are, the more likely it is that they will collide. So the more frequent collisions result in a faster reaction.

Increasing the pressure of reacting gases has the same effect. Increased pressure squashes the gas particles more closely together. There are more particles of gas in a given space. This increases the chance that they will collide and react. So, increasing the pressure produces more frequent collisions which speeds up the rate of the reaction.

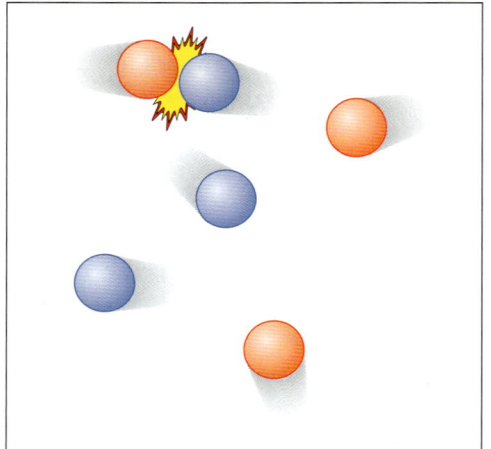

Low concentration, low pressure

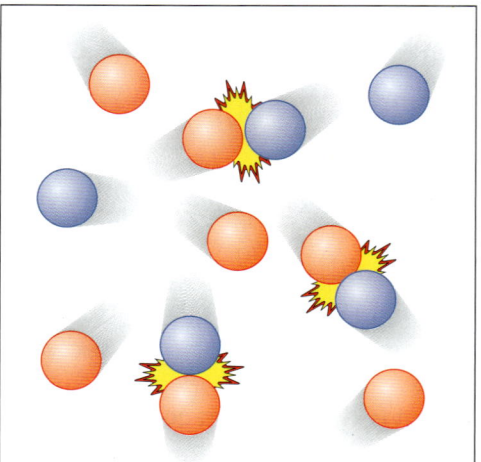

High concentration, high pressure

Figure 2 *Increasing concentration and pressure both mean that particles are closer together. This increases the frequency of collisions between particles, so the reaction rate increases.*

Required practical

Investigating the effect of concentration on rate of reaction

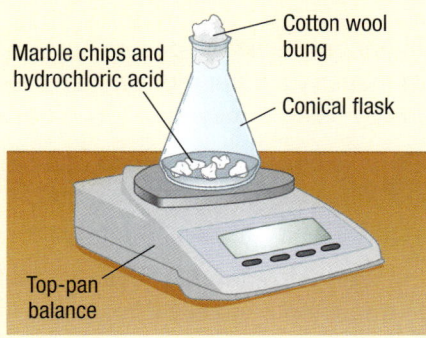

You can investigate the effect of changing concentration by reacting marble chips with different concentrations of hydrochloric acid:

$$CaCO_3(s) + 2HCl(aq) \rightarrow CaCl_2(aq) + CO_2(g) + H_2O(l)$$

Find the rate of reaction by plotting the mass of the reaction mixture over time. The mass will decrease as carbon dioxide gas is given off in the reaction.

- How do you make this a fair test?
- What conclusion can you draw from your results?

Safety: Wear eye protection.

If you plot the results of an investigation like the one above on a graph they look like Figure 3.

The graph shows how the mass of the reaction mixture decreases over time at three different concentrations of acid.

Summary questions

1 Look at the graph in Figure 3.
  a Which line on the graph shows the fastest reaction? How can you tell from the data displayed?
  b Why is the cotton wool bung used in the Practical above?
  c The electronic balance used in the experiment needs a 'high resolution'. What does this mean?

2 You could also follow the reaction described in the Practical above by measuring the volume of gas given off over time. Sketch a graph of volume of gas against time for three different concentrations. Label the three lines as high, medium and low concentration.

3 Explain, in terms of its particles, why the concentration of a gas is proportional to its pressure.

4 Acidic cleaners are designed to remove limescale (calcium carbonate) when they are used neat, that is, undiluted. They do not work as well when they are diluted. Using your knowledge of collision theory, explain why this is.

Study tip

Increasing concentration or pressure does not increase the energy with which the particles collide. However, it does increase the frequency of collisions.

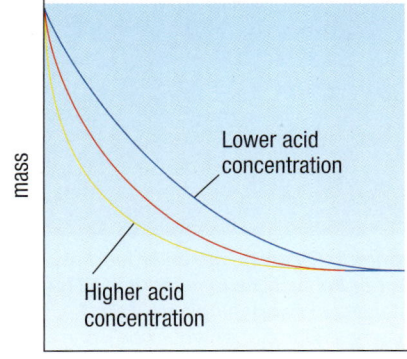

Figure 3 *Investigating the effect of concentration on reaction rate*

Key points

- Increasing the concentration of reactants in solutions increases the frequency of collisions between particles, and so increases the rate of reaction.

- Increasing the pressure of reacting gases also increases the frequency of collisions and so increases the rate of reaction.

9.5

The effect of catalysts

Figure 1 *Catalysts are all around us, in the natural world and in industry. The catalysts in living things are called enzymes. Our planet would be very different without catalysts.*

Sometimes a reaction might only work if it occurs at very high temperatures or pressures. This can cost industry a lot of money. However, it is possible to speed up some reactions by using **catalysts**.

A catalyst is a substance which increases the rate of a reaction. However, it is not changed chemically itself at the end of the reaction. A catalyst is not used up in the reaction so it can be used over and over again.

Different catalysts are needed for different reactions. Many of the catalysts used in industry involve transition metals. Examples include iron, used to make ammonia, and platinum, used to make nitric acid. Catalysts are usually in the form of powders, pellets, or fine gauzes. This gives them the biggest possible surface area and makes them as effective as possible as the reactions they catalyse often involve gases reacting on their surfaces.

Figure 2 *The transition metals platinum and palladium are used in the* **catalytic converters** *in cars*

Advantages of catalysts in industry

Catalysts are often very expensive precious metals. Gold, platinum, and palladium are all costly but are the most effective catalysts for particular reactions. However, it is often cheaper to use a catalyst than to pay for the extra energy needed without one. To get the same rate of reaction without a catalyst would require higher temperatures and/or pressures.

So catalysts save money and help the environment. This is because using high temperatures and pressures often involves burning fossil fuels. So operating at lower temperatures and pressures conserves these **non-renewable** resources. It also stops more carbon dioxide entering the atmosphere, helping to combat climate change.

Not only does a catalyst speed up a reaction, but also it does not get used up in the reaction, so a tiny amount of catalyst can be used to speed up a reaction over and over again.

Required practical

Investigating catalysis

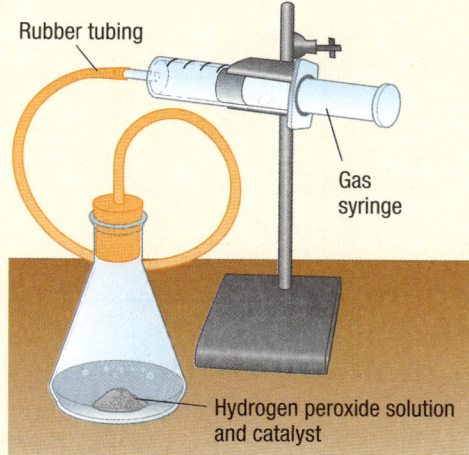

Rubber tubing

Gas syringe

Hydrogen peroxide solution and catalyst

You can investigate the effect of different catalysts on the rate of a reaction. You will look at hydrogen peroxide solution decomposing:

$$2H_2O_2(aq) \rightarrow 2H_2O(l) + O_2(g)$$

The reaction produces oxygen gas. Collect this in a gas syringe using the apparatus shown above.

You can investigate the effect of many different substances on the rate of this reaction. Examples include manganese(IV) oxide, MnO_2, and potassium iodide, KI.

- State the independent variable in this investigation.

A table of the time taken to produce a certain volume of oxygen can then tell us which catalyst makes the reaction go fastest.

- What type of graph would you use to show the results of your investigation? Why?

Safety: Wear eye protection. If the syringe is glass take care to not expel the piston. Manganese(IV) oxide is harmful.

?? Did you know … ?

The catalysts used in chemical plants eventually become 'poisoned' so that they don't work any more. This happens because impurities in the reaction mixture combine with the catalyst and stop it working properly.

Study tip

Catalysts change only the rate of reactions. They do not change the products.

Summary questions

1 How is a catalyst affected by a chemical reaction it speeds up?

2 Solid catalysts used in chemical processes are often shaped as tiny beads or cylinders with holes through them. Why are they made in these shapes?

3 Why is the number of moles of catalyst needed to speed up a chemical reaction very small compared with the number of moles of reactants?

4 Do some research to find out four industrial processes that make products using catalysts. Write a word equation for each reaction and name the catalyst used.

5 Explain why catalysts are so useful in the chemical industry.

Key points

- A catalyst speeds up the rate of a chemical reaction but is not used up itself during the reaction. It remains chemically unchanged.

- Different catalysts are needed for different reactions.

- Catalysts are used whenever possible in industry to increase rates of reaction and reduce energy costs.

Chapter summary questions

1 A student investigated the reaction between magnesium and dilute hydrochloric acid. He started by reacting a piece of magnesium ribbon in 1.0 mol/dm³ hydrochloric acid.

a Describe **one** method the student could use to measure the rate of this reaction.

b Suggest **three** ways in which the student could increase the rate of this reaction.

c i The student decided to investigate the effect of *temperature* on rate of reaction. Write an outline plan of how he should set up his investigation.

ii What would you expect the student to find out from the data he collected?

iii Explain the expected pattern in his results using collision theory.

2 Two students investigated the reaction of some marble chips with dilute nitric acid.

Time (minutes)	Investigation A Mass of gas produced (g)	Investigation B Mass of gas produced (g)
0.0	0.00	0.00
0.5	0.56	0.28
1.0	0.73	0.36
1.5	0.80	0.39
2.0	0.82	0.41
2.5	0.82	0.41

a Which gas was produced in the reaction? What is the test for this gas?

b i Marble chips contain calcium carbonate. Write a word equation for the reaction investigated.

ii Now write a balanced symbol equation, including state symbols, for the reaction.

c The students were investigating the effect of concentration on rate of reaction. How do you think the students got the data for their table above?

d Plot a graph of these results, with time on the *x*-axis.

e After 30 seconds, how does the rate of reaction in Investigation B compare with the rate of reaction in Investigation A?

f How does the final mass of gas produced in Investigation B compare with that produced in Investigation A?

g From the results, what can you say about the initial concentration of the acids in Investigations A and B?

h Explain the data obtained using collision theory.

3 A pair of students are studying the effect of surface area on rate of reaction. They use zinc foil reacting with dilute sulfuric acid.

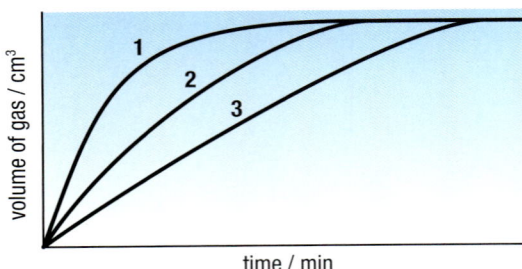

a What did the students investigating the reaction have to do to make it a *fair test*?

b Which gas was given off in the reaction? How could you test for this gas?

c Write a balanced symbol equation, including state symbols, for the reaction under investigation.

d Which line (1, 2, or 3) shows results for the zinc with the largest surface area (given the same mass of zinc in each test)?

e Which sized pieces of zinc would react most slowly?

f The students doing the experiments also tried reacting the same mass of *powdered* zinc with the acid. Describe what their results would look like on the graph above.

g Use collision theory to explain the results in this investigation.

4 a Define the word 'catalyst'.

b Hydrogen peroxide (H_2O_2) solution decomposes to form water and oxygen gas. Write a balanced symbol equation, including state symbols, for this reaction.

c The reaction is catalysed by some metal oxides. You are provided with oxides of copper, lead, manganese, and iron. Describe how you can test which is the best catalyst for the decomposition of hydrogen peroxide, collecting quantitative (measured) data to support your conclusion.

d How can you get a pure dry sample of the metal oxides from the mixture left at the end of the reaction?

Practice questions

1 Some students investigated the reaction between calcium carbonate and nitric acid. They used this apparatus to collect the gas given off when dilute nitric acid was added to marble chips of different sizes.

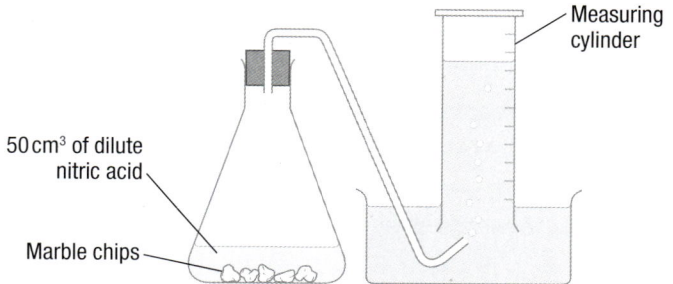

They recorded the volume of gas every 10 seconds until no more gas was collected.

The graph of their results is shown below.

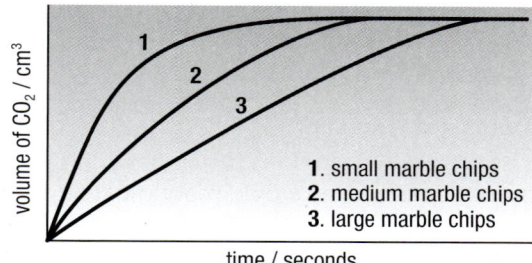

1. small marble chips
2. medium marble chips
3. large marble chips

a The students all used the same mass of marble chips.

State **two** other factors that should have been kept constant to ensure a fair test. (2)

b Reaction 1 was the fastest of the three reactions. State how the graph shows this. (2)

c Why do small marble chips react faster than large marble chips? (1)

d Another group of students investigated the same reaction using equal masses of medium marble chips but using different concentrations of dilute nitric acid in each experiment. Student 1 used the original nitric acid. The other students added water to decrease the concentrations of the acid in their experiments. They recorded the time taken to collect 100 cm³ of gas.

The table shows their results.

Student	Acid concentration	Time in seconds
1	100%	11
2	80%	16
3	60%	22
4	40%	32
5	20%	61

i Suggest which student's recorded results are least likely to be accurate. Give a reason for your choice. (2)

ii One student calculated the rate of reaction using the expression

$$rate = \frac{1}{time}$$

What is the rate of the experiment carried out by student 1? (2)

iii Another student plotted a graph of rate against concentration. She drew a straight line of best fit through the origin.
State the relationship between rate and concentration shown by this line. (2)

iv Explain why the rate of student 2's reaction was double that of student 4's reaction. (3)

2 The reaction between sodium thiosulfate and hydrochloric acid can be used to investigate reaction rates. The equation for the reaction is:

$Na_2S_2O_3(aq) + 2HCl(aq) \rightarrow 2NaCl(aq) + H_2O(l) + SO_2(aq) + S(s)$

a Identify the product that is used in measuring the rate, giving a reason for your choice. (2)

b Some students measured the rate of this reaction in a series of experiments. All the experiments were done in the same way, except that the temperatures were different. The students noticed that the rate was a lot faster at higher temperatures.

One student suggested that this was because there were more frequent collisions between the particles.

i Why do the particles collide more frequently? (1)

ii What is the other reason why the rate is faster at higher temperatures? (1)

3 Many reactions in industry are carried out at high temperatures and using catalysts.

a What is a catalyst? (2)

b Some expensive metals, such as platinum, are used as catalysts.

Explain why it is worthwhile to use expensive metals as catalysts. (1)

c Platinum is used as a catalyst in car exhausts.

Suggest why it is used in thin layers rather than as lumps. (1)

10.1

Reversible reactions

Learning objectives

After this topic, you should know:

- what a reversible reaction is

- how to represent reversible reactions.

In most chemical reactions, the reactants react completely to form the products. You show this by using an arrow pointing *from* the reactants *to* the products:

$$A + B \quad \rightarrow \quad C + D$$
reactants   products

But in some reactions the products can react together to make the original reactants again. This is called a **reversible reaction**.

A reversible reaction can go in both directions so we use two 'half-arrows' in the equation. One arrow points in the forwards direction and one backwards:

$$A + B \rightleftharpoons C + D$$

The substances on the left-hand side of the equation are still called the 'reactants' and those on the right-hand side the 'products'. So it is important that you write down the equation of the reversible reaction you are referring to when you use the words 'reactants' and 'products'. If the equation is written as:

$$C + D \rightleftharpoons A + B$$

the reactants are C and D, and the products are A and B!

Examples of reversible reactions

Have you ever tried to neutralise an alkaline solution with an acid? It is very difficult to get a solution which is exactly neutral. You can use an indicator to tell when just the right amount of acid has been added. An indicator forms compounds that are different colours in acidic solutions and in alkaline solutions.

Litmus is a complex molecule. It is represented as HLit (where H is hydrogen). HLit is red. If you add alkali, HLit turns into the Lit^- ion by losing an H^+ ion. Lit^- is blue. If you then add more acid, blue Lit^- changes back to red HLit and so on.

$$HLit(aq) \rightleftharpoons H^+(aq) + Lit^-(aq)$$
Red litmus   Blue litmus

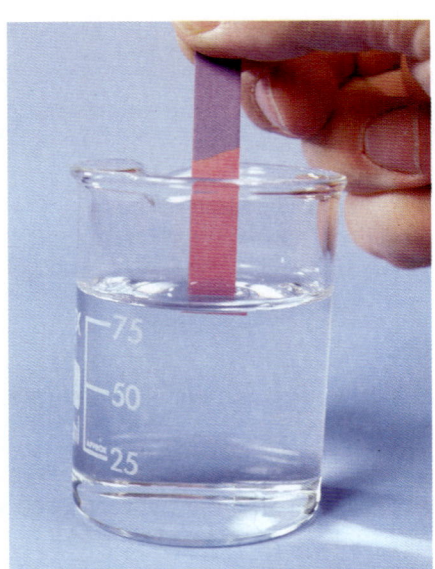

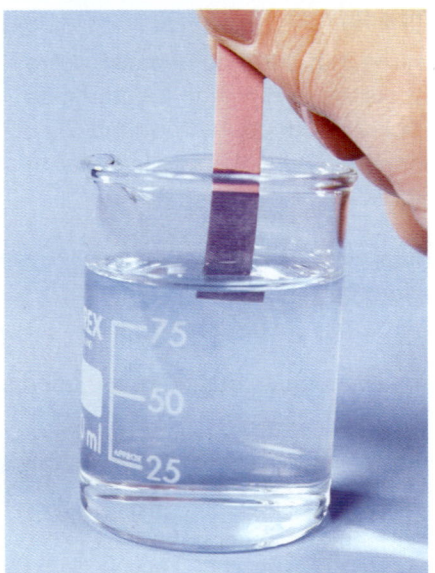

Figure 1 *Indicators undergo reversible reactions, changing colour to show us whether solutions are acidic or alkaline*

Practical

Changing colours

Use litmus solution, dilute hydrochloric acid, and dilute sodium hydroxide solution to show the reversible reaction described here.

- Explain the changes you see when adding acid and alkali to litmus.

Safety: Wear eye protection.

Other reversible reactions involve salts and their water of crystallisation. For example:

$CuSO_4 \cdot 5H_2O$	$\rightleftharpoons$	$CuSO_4$	+	$5H_2O$
hydrated	$\rightleftharpoons$	anhydrous	+	water
copper(II) sulfate (blue)		copper(II) sulfate (white)		

Practical

Heating ammonium chloride

Gently heat a small amount of ammonium chloride (harmful) in a test tube with a loose plug of mineral wool. Use test tube holders or clamp the test tube at an angle. Make sure you warm the bottom of the tube.

- What do you see happen inside the test tube?
- Explain the changes.

Safety: Wear eye protection. Take care if you are asthmatic.

When ammonium chloride is heated, a reversible reaction takes place.

Ammonium chloride breaks down on heating. It forms ammonia gas and hydrogen chloride gas. This is an example of thermal decomposition:

$$\text{ammonium chloride} \xrightarrow{\text{heat}} \text{ammonia} + \text{hydrogen chloride}$$
$$NH_4Cl(s) \longrightarrow NH_3(g) + HCl(g)$$

The two hot gases rise up the test tube. When they cool down near the mouth of the tube they react with each other (see Figure 2). The gases re-form ammonium chloride again. The white ammonium chloride solid forms on the inside of the glass:

$$\text{ammonia} + \text{hydrogen chloride} \longrightarrow \text{ammonium chloride}$$
$$NH_3(g) + HCl(g) \longrightarrow NH_4Cl(s)$$

The reversible reaction can be shown as:

$$\text{ammonium chloride} \rightleftharpoons \text{ammonia} + \text{hydrogen chloride}$$
$$NH_4Cl(s) \rightleftharpoons NH_3(g) + HCl(g)$$

links

For more about reversible reactions, see 11.3 'Energy and reversible reactions'.

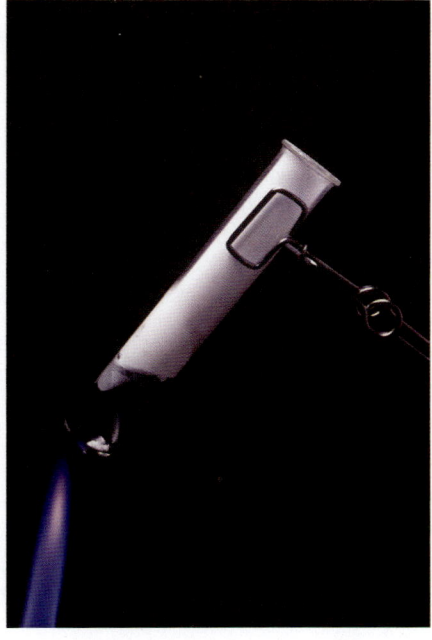

Figure 2 *An example of a reversible reaction – heating ammonium chloride*

Summary questions

1. What do chemists mean by the term 'a *reversible* chemical reaction'?

2. Phenolphthalein is an indicator. It is colourless in acid and pure water but is pink-purple in alkali. In a demonstration a teacher started with a beaker containing a mixture of water and phenolphthalein. In two other beakers she had different volumes of acid and alkali. The acid and alkali had the same concentration.
   She then poured the mixture into the beaker containing $2\,cm^3$ of sodium hydroxide solution. Finally she poured the mixture into a third beaker with $5\,cm^3$ of hydrochloric acid in it.
   Describe what you would observe happening in the demonstration.

3. You can represent the phenolphthalein indicator in question **2** as HPhe. Assuming it behaves like litmus, write a balanced symbol equation to show its reversible reaction in acid and alkali. Show the colour of HPhe and Phe⁻ under their formulae in your equation.

4. Thermochromic materials change colours at different temperatures. The change is reversible. Give a use or potential use for thermochromic materials in the home.

5. Find out about two reversible reactions that can be used to test for the presence of water. Give the positive result for each test, including a balanced symbol equation.

Key points

- In a reversible reaction the products of the reaction can react to make the original reactants.

- You can show a reversible reaction using the $\rightleftharpoons$ sign.

10.2

Chemical equilibrium

Learning objectives

After this topic, you should know:

- what it means when a reaction is 'at equilibrium'
- how the position of equilibrium can be affected in a reversible reaction.

Some reactions are reversible. The products formed can react together to make the original reactants again:

$$A + B \rightleftharpoons C + D$$

So what happens when you start with just reactants in a reversible reaction?

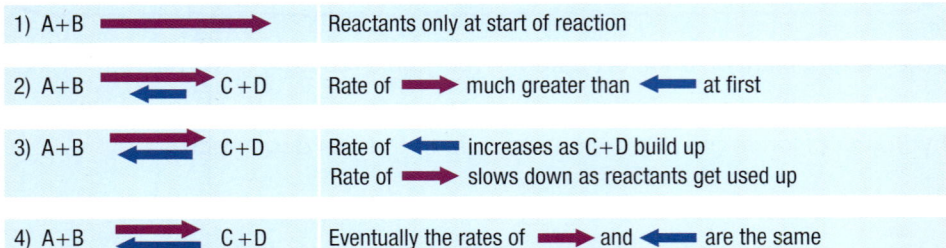

1) A+B ⟶ Reactants only at start of reaction

2) A+B ⟶⟵ C+D Rate of ⟶ much greater than ⟵ at first

3) A+B ⟶⟵ C+D Rate of ⟵ increases as C+D build up
Rate of ⟶ slows down as reactants get used up

4) A+B ⟶⟵ C+D Eventually the rates of ⟶ and ⟵ are the same

In a **closed system**, no reactants, products or energy can get in or out. So in a reversible reaction, as the concentration of products builds up, the rate at which they react to re-form reactants increases. As this starts to happen, the rate of the forward reaction is decreasing. This is because the concentration of reactants is decreasing from its original maximum value. Eventually both forward and reverse reactions are going at the same rate, but in opposite directions.

When this happens the reactants are making products at the same rate as the products are making reactants. Overall, there is *no change* in the amount of products and reactants. We say that the reaction has reached **equilibrium**.

At equilibrium, the rate of the forward reaction equals the rate of the reverse reaction. As the forward and reverse reactions are continuously taking place, we sometimes say that there is a state of 'dynamic' equilibrium.

⬤⬤ **links**

For more information on reversible reactions, look back at 10.1 'Reversible reactions'.

Figure 2 *The situation at equilibrium is just like running up an escalator which is going down – if you run up as fast as the escalator goes down, you will get nowhere!*

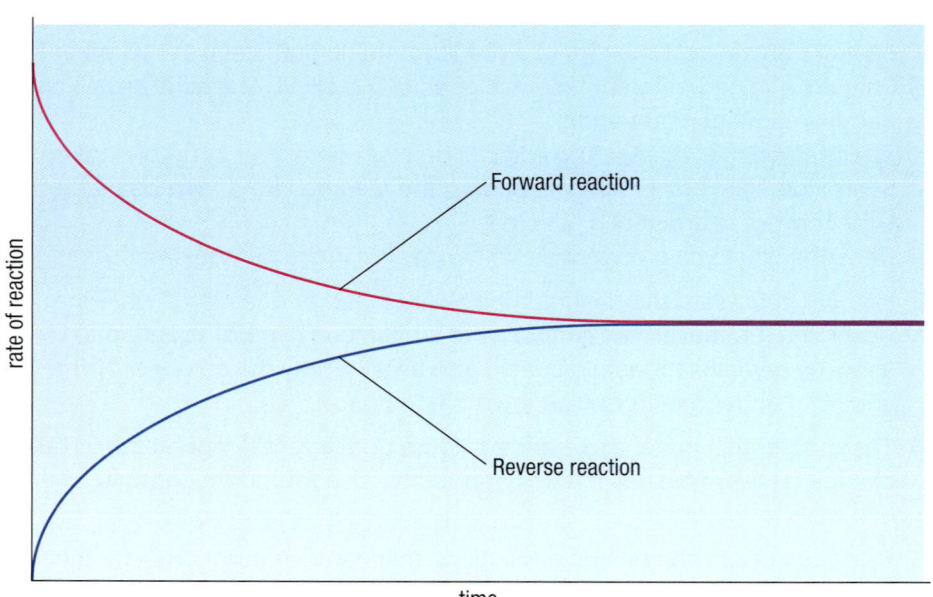

Figure 1 *In a reversible reaction at equilibrium, the rate of the forward reaction is the same as the rate of the reverse reaction*

One example of a reversible reaction is the reaction between iodine monochloride (ICl) and chlorine gas (Figure 3). Iodine monochloride is a brown liquid, whilst chlorine is a yellowish green gas. These substances can react together to make yellow crystals of iodine trichloride (ICl$_3$).

When there is plenty of chlorine gas the forward reaction makes iodine trichloride crystals, which are quite stable. But, if you lower the concentration of chlorine gas the rate of the forward reaction decreases. The reverse reaction becomes the faster of the two reactions. This starts turning more iodine trichloride back to iodine monochloride and chlorine, until equilibrium is established again.

It is possible to change the relative amounts of the reactants and products in a reacting mixture by changing the conditions. This is very important in the chemical industry. In a process with a reversible reaction, engineers need conditions that give as much product as possible.

However, there are always other economic and safety factors to consider when chemists manipulate reversible reactions in industry.

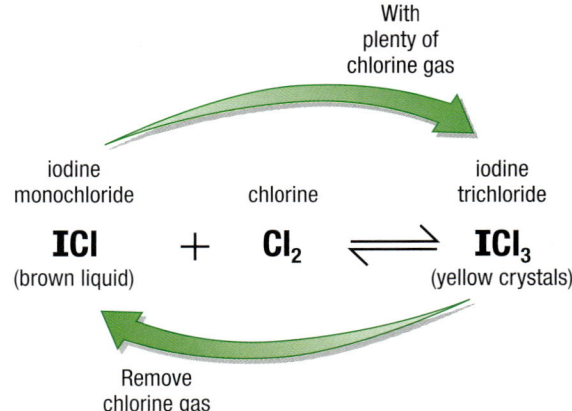

∞ links

For more information about economic and safety issues associated with reversible reactions in industry, see 10.4 'Making ammonia – the Haber process' and 10.5 'The economics of the Haber process'.

Figure 3 *This reacting mixture can be changed by adding or removing chlorine from the mixture*

Summary questions

1 How does the rate of the reverse reaction compare with the rate of the forward reaction in a reaction at equilibrium?

2 What can you say about the concentration of reactants and products in a reversible reaction at equilibrium?

3 Explain why chemists describe chemical equilibrium as 'dynamic' as opposed to 'static'.

4 An equilibrium mixture is set up in a closed system with iodine monochloride, chlorine gas and iodine trichloride. In order to make more iodine trichloride, would you pump more chlorine gas into the mixture or remove chlorine gas? Explain your answer.

Key points

■ In a reversible reaction the products of the reaction can react to re-form the original reactants.

■ In a closed system, the rate of the forward and reverse reactions are equal at equilibrium.

■ Changing the reaction conditions can change the amounts of products and reactants in a reaction mixture at equilibrium.

10.3 Altering conditions

After this topic, you should know:

- how changing the pressure affects reversible reactions involving gases

- how changing the temperature affects reversible reactions.

Pressure and equilibrium

You have seen how changing concentration can affect a reversible reaction at equilibrium. In general, you will find that the position of equilibrium shifts as if trying to cancel out any change in conditions. Think about increasing the concentration of a reactant. This will cause the position of equilibrium to shift to the right, in favour of the products, in order to reduce the concentration of that reactant. It opposes the change introduced.

If a reversible reaction involves changing numbers of gas molecules, altering the pressure can also affect the equilibrium mixture. In many reversible reactions, there are more molecules of gas on one side of the equation than on the other. By changing the pressure at which these reactions are carried out, it's possible to change the amount of products that are made. Look at the table below:

If the forward reaction produces *more* molecules of gas …	If the forward reaction produces *fewer* molecules of gas …
… an increase in pressure decreases the amount of products formed.	… an increase in pressure increases the amount of products formed.
… a decrease in pressure increases the amount of products formed.	… a decrease in pressure decreases the amount of products formed.

The following reversible reaction gives an example of how pressure affects the equilibrium:

$$2NO_2(g) \rightleftharpoons N_2O_4(g)$$
brown gas   pale yellow gas

In this reaction you can see from the balanced symbol equation that we have:

2 molecules of gas on the left-hand side of the equilibrium equation and
1 molecule of gas on the right-hand side.

Imagine that the pressure in the reaction vessel is increased. The position of equilibrium will shift to reduce the pressure. It will move in favour of the reaction that produces fewer gas molecules. In this case that is to the right, in favour of the forward reaction. So more N_2O_4 gas will be made. The colour of the gaseous mixture will get lighter.

> **Study tip**
>
> Changing the pressure affects the equilibrium only if there are different numbers of molecules of gases on each side of the balanced equation.

links

For information on the effect of concentration on a reversible reaction, look back at 10.2 'Chemical equilibrium'.

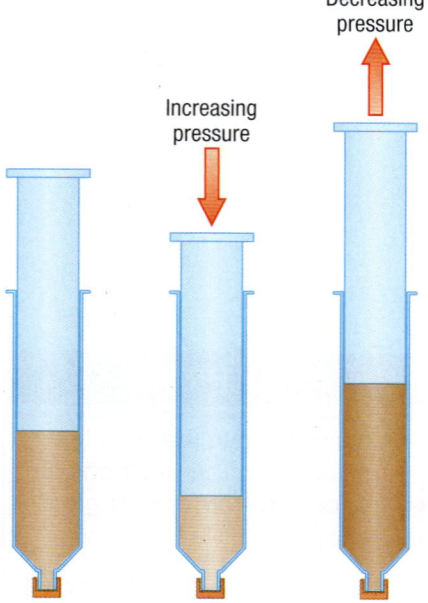

Decreasing pressure

Increasing pressure

Figure 1 *The effect of changing the pressure on $2NO_2(g) \rightleftharpoons N_2O_4(g)$*

links

For information on how changes in pressure are useful in industry, look at 10.4 'Making ammonia – the Haber process' and 10.5 'The economics of the Haber process'.

Temperature and equilibrium

When you have a closed system, nothing is added or taken away from the reaction mixture. In a closed system, the relative amounts of the reactants and products in a reversible reaction at equilibrium depend on the temperature.

By changing the temperature, you can plan to get more of the products and less of the reactants. Look at the table below:

If the forward reaction is exothermic …	If the forward reaction is endothermic …
… an increase in temperature decreases the amount of products formed.	… an increase in temperature increases the amount of products formed.
… a decrease in temperature increases the amount of products formed.	… a decrease in temperature decreases the amount of products formed.

This can be seen when looking at the reaction involving $NO_2(g)$ and $N_2O_4(g)$ again.

$$2NO_2(g) \rightleftharpoons N_2O_4(g)$$

The forward reaction is exothermic, releasing energy, so the reverse reaction is endothermic, absorbing the same amount of energy.

As the temperature is increased, the equilibrium shifts as if to try to reduce the temperature. The reaction that is endothermic (taking in energy) will cool it down. So in this case the reverse reaction is favoured and more NO_2 is formed.

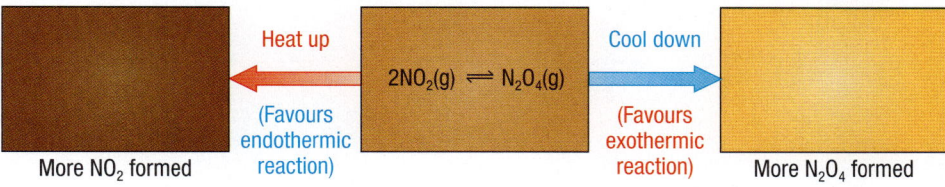

Heat up
(Favours endothermic reaction)

$2NO_2(g) \rightleftharpoons N_2O_4(g)$

Cool down
(Favours exothermic reaction)

More NO_2 formed

More N_2O_4 formed

Figure 3 *The effect of changing the temperature on $2NO_2(g) \rightleftharpoons N_2O_4(g)$*

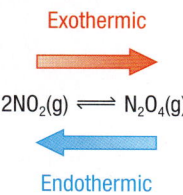

Exothermic

$2NO_2(g) \rightleftharpoons N_2O_4(g)$

Endothermic

Figure 2 *Energy changes in the reversible reaction $2NO_2(g) \rightleftharpoons N_2O_4(g)$*

Demonstration

The effect of temperature and pressure on equilibrium

Using a gas syringe, your teacher can show an equilibrium mixture of NO_2 and N_2O_4 gases.

You can see the effect of changing temperature by using warm water and ice.

Pushing and pulling the plunger of the syringe can change the pressure.

■ Explain your observations.

Safety: Take care if you are asthmatic. The teacher should be wearing chemical splash-proof eye protection as these gases are 'very toxic'.

Summary questions

1 How does increasing the pressure affect the amount of products formed in a reaction that produces a larger volume (more molecules) of gas?

2 Look at the reversible reaction below:

$$H_2O(g) + C(s) \rightleftharpoons CO(g) + H_2(g)$$

The forward reaction is endothermic.
How will the amount of hydrogen gas formed change by increasing the temperature?

3 a Look at Figure 1 on the previous page. How will increasing the pressure affect the colour of a mixture of NO_2 and N_2O_4 gases? Explain your answer.

   b Look at Figure 3 above. What will happen to the colour of a mixture of NO_2 and N_2O_4 gases if the temperature is increased? Explain your answer.

4 Explain what effect increasing the pressure would have on the equilibrium mixture below:

$$H_2(g) + I_2(g) \rightleftharpoons 2HI(g)$$

Key points

■ Pressure can affect reversible reactions involving gases at equilibrium. Increasing the pressure favours the reaction that forms fewer molecules of gas. Decreasing the pressure favours the reaction with the greater number of molecules of gas formed.

■ It is possible to change the relative amount of products formed at equilibrium by changing the temperature at which we carry out a reversible reaction.

■ Increasing the temperature favours the endothermic reaction. Decreasing the temperature favours the exothermic reaction.

10.4

Making ammonia – the Haber process

Humans need plants for food, and as a way of maintaining oxygen in the air. Plants need nitrogen to grow. Although nitrogen gas makes up nearly 80 per cent of the air, most plants cannot use it directly.

Instead, plants absorb soluble nitrates from the soil through their roots. When farmers harvest crops the nitrogen in plants is lost. This is because the plants do not die and decompose to replace the nitrogen in the soil. So farmers need to put nitrogen back into the soil for the next year's crops to maximise food yield.

Nowadays this is usually done by adding nitrate fertilisers to the soil. These fertilisers are made using a process devised by a German chemist called Fritz Haber.

Figure 1 *Plants are surrounded by nitrogen in the air. They cannot use this nitrogen, and rely on soluble nitrates in the soil instead. These are supplied by spreading fertiliser on the soil.*

The Haber process

The Haber process provides a way of turning nitrogen in the air into ammonia.

$$\text{nitrogen gas} + \text{hydrogen gas} \rightleftharpoons \text{ammonia gas}$$

Scientists can use ammonia in many different ways. The most important of these is to make fertilisers.

The raw materials for the production of ammonia are:
- nitrogen from the air
- hydrogen, which we get mainly from natural gas (which contains methane, CH_4).

The nitrogen and hydrogen are purified. Then they are passed over an iron catalyst at a high temperature (about 450 °C) and a high pressure (about 200 atmospheres). The product of this reversible reaction is ammonia.

- The reaction used in the Haber process is reversible. This means that the ammonia made breaks down again into nitrogen and hydrogen.
- The ammonia is removed by cooling the gases so that the ammonia liquefies. It can then be separated from the unreacted nitrogen gas and hydrogen gas.

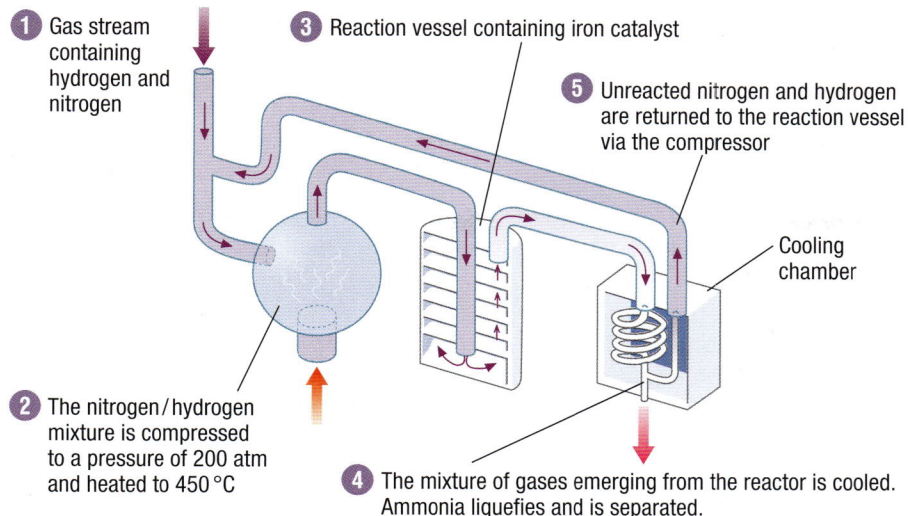

① Gas stream containing hydrogen and nitrogen

③ Reaction vessel containing iron catalyst

⑤ Unreacted nitrogen and hydrogen are returned to the reaction vessel via the compressor

Cooling chamber

② The nitrogen/hydrogen mixture is compressed to a pressure of 200 atm and heated to 450 °C

④ The mixture of gases emerging from the reactor is cooled. Ammonia liquefies and is separated.

Figure 2 *The Haber process*

- The unreacted nitrogen and hydrogen gases are recycled back into the reaction mixture. They are then re-compressed and heated before returning to the reaction vessel. There they have a chance to react again on the surface of the iron catalyst.

$$N_2(g) + 3H_2(g) \overset{\text{iron catalyst}}{\rightleftharpoons} 2NH_3(g)$$

The Haber process is conducted in carefully chosen conditions to give a reasonable yield of ammonia as quickly as possible.

 Did you know …?

Fritz Haber's work on the Haber process took place just before the First World War. When the war started, he was involved in the first chemical weapons used in warfare. His wife, also a chemist, opposed his work on chemical weapons. She committed suicide before the war ended.

∞ links

For more information about the choice of conditions, see 10.5 'The economics of the Haber process'.

Summary questions

1 What are the main raw materials used to get the nitrogen gas and hydrogen gas needed for the production of ammonia in industry?

2 a Write a balanced symbol equation, including state symbols, for the production of ammonia in the Haber process.
  b State the conditions chosen for the Haber process, including the catalyst used.

3 a How is ammonia separated from the reaction mixture in the Haber process?
  b What happens to unreacted nitrogen and hydrogen that leaves the reaction vessel?

4 Draw a flowchart to show how the Haber process is used to make ammonia.

Key points

- Ammonia is an important chemical for making other products, including fertilisers.

- Ammonia is made from nitrogen and hydrogen in the Haber process.

- The Haber process is carried out under conditions of about 450 °C and 200 atmospheres pressure, using an iron catalyst.

- Any unreacted nitrogen and hydrogen are recycled back into the reaction vessel in the Haber process.

10.5

The economics of the Haber process

After this topic, you should know:

- why a pressure of about 200 atmospheres is chosen for the Haber process
- why a temperature of about 450 °C is chosen for the Haber process.

The effect of pressure

You have just seen how ammonia is made in the Haber process. Nitrogen and hydrogen react to make ammonia in a reversible reaction:

$$N_2(g) + 3H_2(g) \rightleftharpoons 2NH_3(g)$$

As the balanced equation above shows, there are four molecules of gas on the left-hand side of the equation (N_2 and $3H_2$), but on the right-hand side there are only two molecules of gas ($2NH_3$). This means that the volume of the reactants is greater than the volume of the products. So, an increase in pressure will tend to shift the position of equilibrium to the right, producing more ammonia.

To get the maximum possible yield of ammonia, the pressure must be as high as possible. But very high pressures need lots of energy to compress the gases. Very high pressures also need expensive reaction vessels and pipes. They have to be strong enough to withstand very high pressures, otherwise there is always the danger of an explosion. To avoid the higher costs of building a stronger chemical plant, the Haber process uses a pressure of 200 atmospheres. This is a good *compromise*. This pressure gives a lower yield than even higher pressures would, but it does reduce the expense and helps produce a reasonable rate of reaction between the gases.

Figure 1 *It is very expensive to build chemical plants that operate at very high pressures*

The effect of temperature

The effect of temperature on the Haber process is more complicated than the effect of pressure. The forward reaction is exothermic:

$$N_2(g) + 3H_2(g) \rightleftharpoons 2NH_3(g) \qquad \Delta H = -93\,kJ/mol$$

So lowering the temperature would increase the amount of ammonia in the reaction mixture at equilibrium. This happens because the forward reaction to form ammonia releases energy to the surroundings and raises the temperature of the surroundings.

However, at a low temperature, the rate of the reaction would be very slow because the gas molecules would collide less often and less energetically. To make ammonia commercially, scientists must get the reaction to go quickly. They do not want to waste time waiting for the ammonia to be produced.

This is achieved in the Haber process by another compromise. A reasonably high temperature of 450 °C is used to get the reaction going at a reasonable rate, even though this reduces the yield of ammonia. Look at the graph in Figure 2.

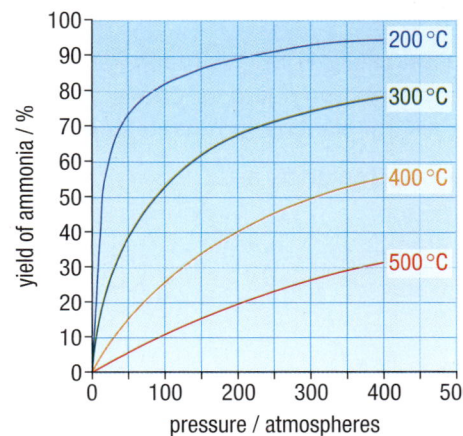

Figure 2 *The conditions for the Haber process are a compromise between getting a reasonable yield of ammonia and getting the reaction to take place at a fast enough rate*

A lower temperature can also reduce the effectiveness of the iron catalyst.

The effect of a catalyst

An iron catalyst can also be used to speed up the reaction. A catalyst speeds up the rate of both the forward and reverse reactions by the same amount. Therefore it does not affect the actual yield of ammonia but it does produce ammonia more quickly, which is an important economic consideration in industry.

Study tip

Remember that the yield is the amount of product obtained compared with (i.e., divided by) the amount obtained if all of the reactants were used up to make product.

??? Did you know ... ?

The Haber process is sometimes called the Haber–Bosch process, since Fritz Haber found out how to make ammonia from nitrogen and hydrogen but Carl Bosch carried out the work to find the best conditions for the reaction. Bosch and his team carried out 6500 experiments to find the best catalyst for the reaction.

Summary questions

1 Explain the effect of increasing the pressure in the Haber process on:
   a the yield of ammonia from the reversible reaction
   b the rate of production of ammonia.

2 Why is an iron catalyst used in the Haber process?

3 Look at Figure 2.
   a What is the approximate yield of ammonia at a temperature of 500 °C and 400 atmospheres pressure?
   b What is the approximate yield of ammonia at a temperature of 500 °C and 100 atmospheres pressure?
   c What is the approximate yield of ammonia at a temperature of 200 °C and 400 atmospheres pressure?
   d What is the approximate yield of ammonia at a temperature of 200 °C and 100 atmospheres pressure?
   e Given your answers to parts **a–d**, explain why the Haber process is carried out at around 200 atmospheres and 450 °C.

Key points

- The Haber process uses a pressure of around 200 atmospheres to increase the amount of ammonia produced.

- Although higher pressures would produce higher yields of ammonia, they would make the chemical plant too expensive to build and run.

- A temperature of about 450 °C is used for the reaction. Although lower temperatures would increase the yield of ammonia, it would be produced too slowly.

10.6

The Contact process

Learning objectives

After this topic, you should know:

■ the three stages in the Contact process

■ why particular reaction conditions are chosen for the second stage in the process.

Many of the important reactions in industry are reversible. This is a challenge for industrial chemists. They have to choose conditions that will produce as much product as possible in the shortest possible time. This will maximise profits for the chemical company. However, safety and environmental aspects of the manufacturing process also have to be taken into account before deciding on the actual operating conditions.

Sulfuric acid, H_2SO_4, is made in the **Contact process**. It is one of the most important manufactured chemicals as it is used to make paints, pigments, fertilisers, plastics, synthetic fibres, and dyes.

The raw materials are:

■ sulfur

■ air

■ water.

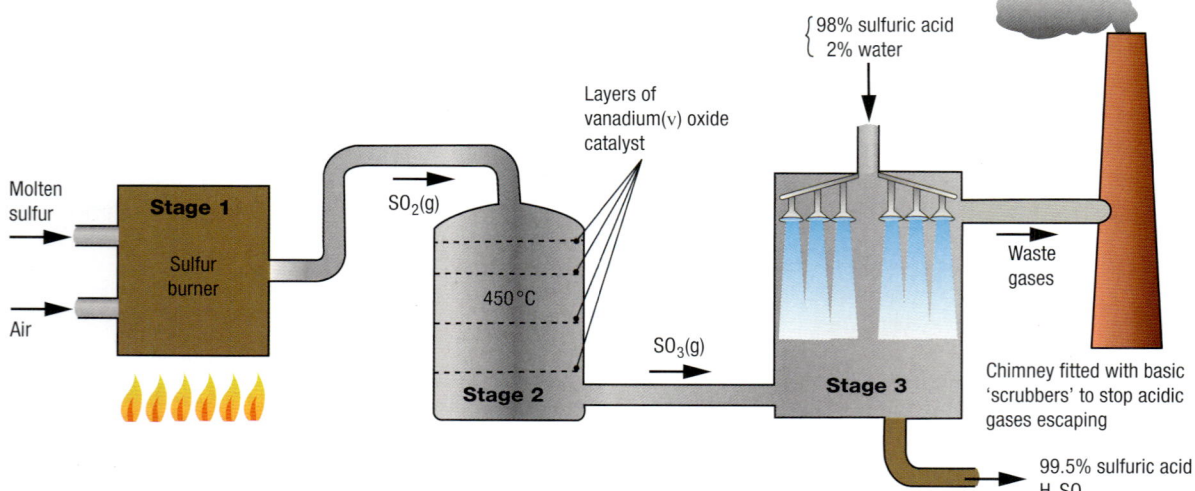

Figure 1 *The Contact process for making sulfuric acid is a three-stage process*

Stage 1

In the first stage, sulfur is turned into sulfur dioxide gas. The sulfur is a yellow solid at room temperature. It can be imported into the UK on ships from Poland or the USA, where underground deposits of sulfur are found. It can also be extracted from impurities in crude oil or natural gas. However, natural gas from the North Sea contains very little sulfur, so it cannot be used.

Molten sulfur is burnt in air:

$$S(l) \quad + \quad O_2(g) \quad \rightarrow \quad SO_2(g)$$
sulfur   oxygen   sulfur dioxide

Stage 2

This is the most important step because the reaction is reversible.

The sulfur dioxide gas made in Stage 1 is turned into sulfur trioxide gas:

$$2SO_2(g) \quad + \quad O_2(g) \quad \rightleftharpoons \quad 2SO_3(g) \qquad \Delta H = -197 \, kJ/mol$$
sulfur dioxide   oxygen   sulfur trioxide

Chemists want as much of the SO_3 gas made in the forward reaction as they can, as quickly as possible. Notice that the forward reaction is exothermic. So a low temperature would favour the forward reaction (as this would shift the equilibrium to raise the temperature up again). But if the temperature is too low, the rate of reaction will be too slow. A compromise of 450 °C is chosen. You still get a good yield (97%) of SO_3 at this temperature, and at a reasonable rate.

So what about the pressure chosen? Count the molecules of gas on either side of the equation. How could you get more SO_3? As there are three molecules of gas on the left-hand side of the equation and only two molecules of gas on the right-hand side, increasing the pressure will favour more SO_3. However, with a 97% yield already, it is not worth spending much money compressing the gases. The gases are compressed slightly above atmospheric pressure to push them through the pipes and reaction vessels. So the pressure is only just above 1 atmosphere or 100 kPa.

A catalyst, vanadium(v) oxide, is used to speed up the rate of reaction. The V_2O_5 is supported on porous, daisy-shaped cylinders of silica to maximise its surface area in each layer of the reaction vessel.

Figure 2 V_2O_5 catalyst on porous support

Stage 3

In the final stage sulfur trioxide is converted into sulfuric acid. The sulfur trioxide gets absorbed into a mixture of 98% sulfuric acid and 2% water. In effect, the sulfur trioxide reacts with the water:

$$SO_3(g) \quad + \quad H_2O(l) \quad \rightarrow \quad H_2SO_4(l)$$
$$\text{sulfur trioxide} \qquad \text{water} \qquad \text{conc. sulfuric acid}$$

Summary questions

1 Give the word equation and balanced symbol equation, including state symbols, for the reactions that take place in each of the three stages in the Contact process.

2 a State the conditions chosen for Stage 2 of the Contact process.
  b Explain how the efficiency of the catalyst is maximised.
  c Explain the conditions chosen for Stage 2 of the Contact process.

3 Assuming 100% conversion at each stage in the Contact process, calculate the mass of sulfuric acid in kilograms that can be made from each kilogram of sulfur burnt. Give your answer to 3 significant figures.

??? Did you know …?

The exothermic reactions in each stage of the Contact process transfer energy to the surroundings. This energy is not wasted, but is used to heat the gases in Stage 2.

Key points

■ The Contact process is used to manufacture sulfuric acid, H_2SO_4.

■ In the first stage, sulfur is burnt and oxidised to sulfur dioxide gas.

■ In the second stage, sulfur dioxide is oxidised by oxygen gas to make sulfur trioxide in a reversible reaction that takes place at 450 °C and atmospheric pressure, in the presence of a vanadium(v) oxide catalyst.

■ In the third stage, in effect sulfur trioxide reacts with water to make sulfuric acid.

Chapter summary questions

1 Here is an example of a reaction mixture in a state of dynamic equilibrium:

$$ICl(l) \quad + \quad Cl_2(g) \quad \rightleftharpoons \quad ICl_3(s)$$
brown liquid　　yellow/green gas　　yellow solid

The equilibrium can be set up in a glass U-tube:

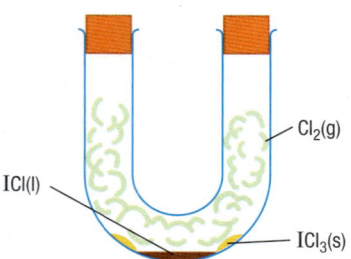

a The equilibrium mixture contains the three states of matter. What does this statement mean?

b The pressure inside the U-tube is increased. What happens to the equilibrium mixture? Explain your answer.

c What would happen to the equilibrium mixture if the chlorine gas was allowed to escape? Explain your answer.

2 A chemical reaction can make product C from reactants A and B. Under the reaction conditions, A, B, and C are all gases.
A, B, and C react in the proportions 1:1:2. The reaction is carried out at 250°C and 100 atmospheres. The reaction is reversible, and it is endothermic in the forward direction.

a Write a balanced symbol equation, including state symbols, for this reversible reaction.

b How would increasing the pressure affect:
  i the amount of C formed?
  ii the rate at which C is formed?

c How would increasing the temperature affect:
  i the amount of C formed?
  ii the rate at which C is formed?

3 In the manufacture of sulfuric acid, sulfur dioxide is oxidised to form sulfur trioxide:

$$2SO_2(g) + O_2(g) \rightleftharpoons 2SO_3(g)$$

This reversible reaction is exothermic from left to right. State and explain the effect of each of the following conditions on the rate and yield of this reaction.

a Increase in temperature

b Increase in pressure

c Use of a catalyst

4 When nitric acid is made in industry, ammonia is oxidised to nitrogen monoxide.

$$4NH_3(g) + 5O_2(g) \rightleftharpoons 4NO(g) + 6H_2O(g)$$

The reaction is exothermic in the forward direction. Ammonia and air, at about 7 atmospheres pressure, is passed over a platinum catalyst at 900°C.
Explain how changing the conditions used could increase the yield of nitrogen monoxide.

5 The table below gives data on the yield of ammonia at equilibrium in the reversible reaction with nitrogen and hydrogen:

| Ammonia present at equilibrium (%) | | | | | |
| Pressure (atmospheres) | Temperature (°C) | | | | |
	100	200	300	400	500
10	88	51	15	4	1
25	92	64	27	9	3
50	94	74	39	15	6
100	97	82	52	25	11
200	98	89	67	39	18
400	99	95	80	55	32
1000	99.9	98	93	80	57

a i What is the effect of increasing the pressure on the yield of ammonia?
  ii What is the effect of increasing the temperature on the yield of ammonia?
  iii Explain how you used the table to judge the effect of changing the temperature and the pressure on the yield of ammonia.

b Explain why the conditions in the Haber process (200 atmospheres pressure and a temperature of 450°C) are described as a 'compromise'.

c How does the use of an iron catalyst affect the yield of ammonia?

6 Here is the equation for the reaction used in the Haber process:

$$N_2(g) + 3H_2(g) \overset{\underset{\text{Fe(s)}}{\text{catalyst}}}{\rightleftharpoons} 2NH_3(g) \quad \Delta H = -93\,kJ/mol$$

Draw an energy level diagram for this reaction, showing the reaction pathway with and without the use of the iron catalyst. (See Topic 11.6).

Practice questions

1 If a reversible reaction occurs in a closed container it can reach equilibrium.

  a State **two** features of a reversible reaction that has reached equilibrium. (2)

  b A gaseous oxide of nitrogen has the formula N_2O_4. It decomposes into another gaseous oxide of nitrogen when it is heated without changing the pressure. This decomposition is endothermic. The progress of this reaction can be checked using the differences in colour.

$$N_2O_4(g) \rightleftharpoons 2NO_2(g)$$
$$\text{pale yellow} \qquad \text{brown}$$

    i How does the colour of the reaction mixture change when the temperature increases? (1)

    ii Explain how the information in the question helps to predict the colour change. (2)

  c A mixture of the two oxides of nitrogen is placed in a gas syringe and the pressure is increased.

    Predict what happens to the colour of the mixture. Explain your answer by reference to the equation in part **b**. (3)

2 Ammonia is manufactured by the Haber process. This process is shown in the flow diagram.

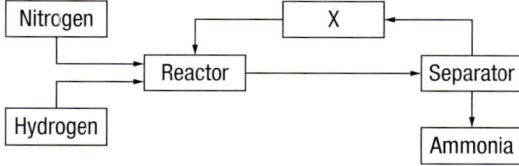

  a **i** What is the source of the nitrogen? (1)

    ii What is the source of the hydrogen? (1)

  b State **three** conditions used in the reactor. (3)

  c What substances are represented by X in the diagram? (1)

  d What condition is used in the separator? (1)

  e Write a symbol equation for the reaction that occurs in the reactor. (2)

3 A company builds a new factory to produce ammonia.

The new factory uses the same process as the old factory but with these changes in conditions:

- The temperature is decreased by 100 °C in the new factory.
- The pressure is increased by 100 atmospheres in the new factory.

  a Predict how each change affects the yield of ammonia in this reaction in the new factory. Explain your answers. (6)

  b How does the catalyst affect the rate of reaction? (1)

  c Why does the catalyst not affect the yield of the reaction? (2)

4 Some ammonium chloride was heated in a test tube as shown in the diagram:

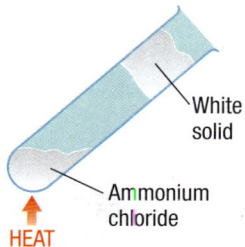

White solid

Ammonium chloride

HEAT

  a When the ammonium chloride is heated gently it decomposes according to this equation:

$$NH_4Cl \rightarrow NH_3 + HCl$$

    The diagram shows that a white solid forms near the open end of the test tube.

    i Name the two gases in the tube between the ammonium chloride and the white solid. (2)

    ii Suggest what the white solid is. (1)

    iii Deduce the equation to show its formation. (1)

  b A student heated the tube strongly and did not observe the formation of the white solid.

    Suggest an explanation for this. (1)

5 The manufacture of sulfuric acid is carried out in three stages:

Stage 1 – burning sulfur
Stage 2 – oxidation of sulfur dioxide
Stage 3 – conversion of sulfur trioxide to sulfuric acid.

  a Name the industrial process used to make sulfuric acid. (1)

  b In this process, the first stage involves the combustion of sulfur. If 256 tonnes of sulfur are burnt in air, what is the maximum mass of sulfur dioxide gas obtained for Stage 2? (2)

  c **i** Write the balanced symbol equation for the reaction that takes place in Stage 2. (2)

    ii Name the catalyst used in Stage 2. (1)

  d Assuming 100% conversion of the 256 tonnes of sulfur in Stage 1 into sulfuric acid, calculate the maximum mass of sulfuric acid that could be made. (2)

Exothermic and endothermic reactions

Learning objectives

After this topic, you should know:

- that energy is transferred to or from the surroundings in chemical reactions
- some examples of exothermic and endothermic reactions.

Whenever chemical reactions take place, energy is always transferred as chemical bonds are broken and new ones are made.

Some reactions transfer energy **from** the reacting chemicals **to** their surroundings. These are called **exothermic** reactions. The energy transferred from the reacting chemicals often heats up the surroundings. This means that you can measure a rise in temperature as the reaction happens.

Some reactions transfer energy **from** the surroundings **to** the reacting chemicals. These are called **endothermic** reactions. As they take in energy from their surroundings, these reactions cause a drop in temperature as they happen.

Exothermic reactions

Burning fuels are an obvious example of exothermic reactions. For example, when methane (in natural gas) burns it gets oxidised and releases energy.

Neutralisation reactions between acids and alkalis are also exothermic. You can easily measure the rise in temperature using simple apparatus (see the experiment on the next page).

Scientists use the term 'enthalpy change' to describe the transfer of energy in chemical reactions at constant pressure. As the reactions carried out in the lab are at constant pressure, the energy we are considering in these reactions is enthalpy, symbol H. Any temperature change measured when a reaction takes place is a result of enthalpy changes between the reactants and products of the reaction. The enthalpy change is given the symbol ΔH. The products of exothermic reactions have less enthalpy content than the reactants so numerical values, measured in kilojoules per mole, are given a negative sign. For example:

$$CH_4(g) + 2O_2(g) \rightarrow CO_2(g) + 2H_2O(l) \qquad \Delta H = -890 \text{ kJ/mol}$$

Figure 1 *When a fuel burns in oxygen, energy is transferred to the surroundings. You usually don't need a thermometer to know that there is a temperature change!*

∞ links

For more information about enthalpy changes and how to represent them, see 11.6 'Energy level diagrams'.

Figure 2 *All warm-blooded animals rely on exothermic reactions in **respiration** to keep their body temperatures steady*

Endothermic reactions

Endothermic reactions are much less common than exothermic ones.

Thermal decomposition reactions are endothermic. An example is the decomposition of calcium carbonate. When heated, it forms calcium oxide and carbon dioxide. This reaction only takes place if you keep heating the calcium carbonate strongly. The calcium carbonate needs to absorb energy from the surroundings to be broken down.

The products have more enthalpy content than the reactants in endothermic reactions, so numerical values are given a positive sign. For example:

$$CaCO_3(s) \rightarrow CaO(s) + CO_2(g) \qquad \Delta H = +178\,kJ/mol$$

Figure 3 *When you eat sherbet you can feel an endothermic reaction. Sherbet dissolving in the water in your mouth takes in energy. It provides a slight cooling effect.*

Practical

Investigating energy changes

You can use very simple apparatus to investigate the energy changes in reactions. Often you do not need to use anything more complicated than a polystyrene cup and a thermometer.

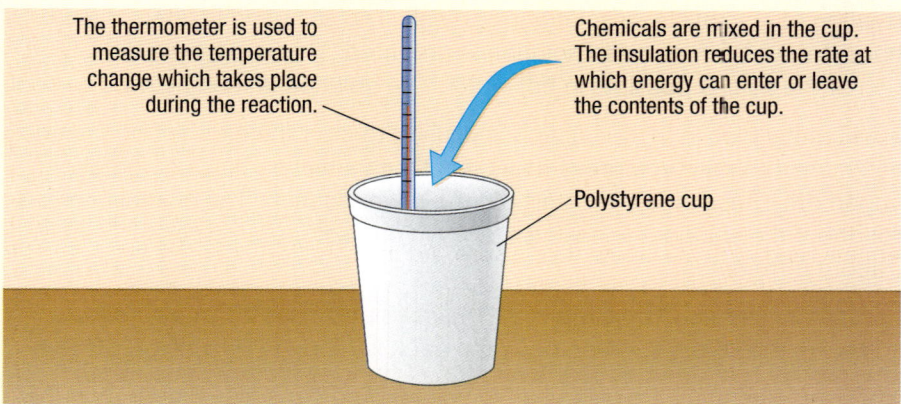

The thermometer is used to measure the temperature change which takes place during the reaction.

Chemicals are mixed in the cup. The insulation reduces the rate at which energy can enter or leave the contents of the cup.

Polystyrene cup

■ State two ways in which you could make the data you collect more accurate.

Safety: Wear eye protection.

Summary questions

1. **a** What do you call a reaction that releases energy to its surroundings?
   b What do you call a reaction that absorbs energy from its surroundings?
   c Give **two** examples of exothermic reactions.
   d Give an example of an endothermic reaction.

2. Potassium nitrate dissolving in water is an endothermic process. Explain what you would feel if you held a beaker of water in your hand as you stirred in potassium nitrate.

3. Two solutions are added together and the temperature changes from 19 °C to 27 °C. Explain what you can say about the enthalpy change in this reaction, including the sign given to ΔH.

4. The enthalpy change for the thermal decomposition of 16.8 g of magnesium carbonate is +23.4 kJ. (A_r values: C = 12, O = 16, Mg = 24)
   a Write a balanced symbol equation, including state symbols, for the reaction.
   b Calculate the value of ΔH for the thermal decomposition in kilojoules per mole.

Study tip

Remember that exothermic reactions involve energy EXiting (leaving) the reacting chemicals, so the surroundings get hotter.

In endothermic reactions energy is transferred INTO (sounds like 'endo'!) the reacting chemicals, so the surroundings get colder.

Key points

■ Energy may be transferred to or from the reacting substances in a chemical reaction.

■ A reaction in which energy is transferred from the reacting substances to their surroundings is called an exothermic reaction.

■ In exothermic reactions, the change in enthalpy, ΔH, has a negative value.

■ A reaction in which energy is transferred to the reacting substances from their surroundings is called an endothermic reaction.

■ In endothermic reactions, the change in enthalpy, ΔH, has a positive value.

11.2

Using energy transfers from reactions

Learning objectives

After this topic, you should know:

- how to make use of the energy from exothermic reactions

- how to use the cooling effect of endothermic reactions.

Figure 1 *Here is a hand warmer based on the recrystallisation of sodium ethanoate*

Warming up

Chemical hand and body warmers can be very useful. These products use exothermic reactions to warm you up. People can take hand warmers to places where they know they will get very cold. For example, spectators at outdoor sporting events in winter can warm their hands up. People usually use the body warmers to help ease aches and pains.

Some hand warmers can only be used once. An example of this type uses the energy released in the oxidation of iron. Iron turns into hydrated iron(III) oxide in an exothermic reaction. The reaction is similar to rusting. Sodium chloride (common salt) is used as a catalyst. This type of hand warmer is disposable. It can be used only once but it lasts for hours.

Other hand warmers can be reused many times. These are based on the formation of crystals from solutions of a salt. The salt used is often sodium ethanoate. A supersaturated solution is prepared. This is done by dissolving as much of the salt as possible in hot water. The solution is then allowed to cool.

A small metal disc in the plastic pack is used to start the exothermic change (see Figure 1). When you press this a few times small particles of metal are scraped off. These 'seed' (or start off) the crystallisation. The crystals spread throughout the solution, releasing energy in an exothermic change. They work for about 30 minutes.

To reuse the warmer, you simply put the solid pack into boiling water to re-dissolve the crystals. When cool, the pack is ready to be activated again.

Exothermic reactions are also used in self-heating cans (see Figure 2 on the next page). The reaction used to release the energy is usually:

$$\text{calcium oxide} + \text{water} \rightarrow \text{calcium hydroxide}$$

You press a button in the base of the can. This breaks a seal and lets the water and calcium oxide mix. Coffee is available in these self-heating cans.

Development took years and cost millions of pounds. Even then, over a third of the can was taken up with the reactants to release energy. Also, in some early versions, the temperature of the coffee did not rise high enough in cold conditions.

Activity

Hot food

Mountaineers and explorers can take 'self-heating' foods with them on their journeys. One example uses the energy released when calcium oxide reacts with water to heat the food.

Design a self-heating, disposable food container for stew.

- Draw a labelled diagram of your container and explain how it works.
- What are the safety issues involved in using your product?

Cooling down

Endothermic processes can be used to cool things down. For example, chemical cold packs usually contain ammonium nitrate and water (see Figure 3). When ammonium nitrate dissolves it absorbs energy from its surroundings, making them colder. These cold packs are used as emergency treatment for sports injuries. The coldness reduces swelling and numbs pain.

The ammonium nitrate and water (sometimes present in a gel) are kept separate in the pack. When squeezed or struck the bag inside the water pack breaks releasing ammonium nitrate. The instant cold packs work for about 20 minutes.

They can only be used once but are ideal where there is no ice available to treat a knock or strain.

The same endothermic change can also be used to chill cans of drinks.

TO HEAT CONTAINER

Turn container UPSIDE DOWN before opening and follow instructions.

STEP 4
HOT SPOT turns from pink to white when beverage is hot. (6–8 minutes)

STEP 5
Once hot, shake 5 to 10 seconds then twist lid to align opening with pull-tab. Open and enjoy.

TWIST WHEN HOT

STEP 3
coloured water drains into the activation chamber.
Wait 5 SECONDS and turn can right side up.
5 Seconds

STEP 2
Place container on flat surface. Using thumb, FIRMLY push button DOWNWARD until internal foil seal tears and

STEP 1
PULL off tamper-proof metal bottom

Figure 2 *Development of this self-heating can in the USA took about 10 years*

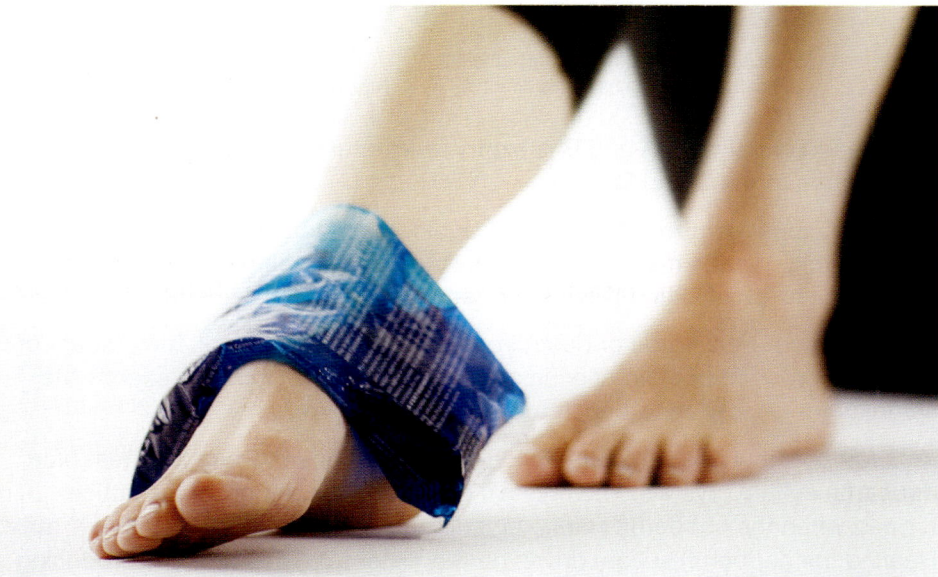

Figure 3 *Instant cold packs can be applied as soon as an injury occurs to minimise damage*

Summary questions

1 a Describe the chemical reaction which takes place in a disposable hand warmer.
  b Describe how a re-usable hand warmer works.
  c Give an advantage and a disadvantage of each type of hand warmer.
  d Name **one** use of an exothermic reaction in the food industry.

2 a Give **two** uses of endothermic changes.
  b Which endothermic change is often used in cold packs?
  c The solid used in cold packs is often ammonium nitrate.
    i Give the formula of ammonium nitrate.
    ii State another use of ammonium nitrate.

3 a Which solid is usually used in the base of self-heating coffee cans?
  b Write a balanced symbol equation, including state symbols, for the reaction of water with the solid in part **a**.
  c Why is it essential that the coffee stays out of contact with the solid in part **a**?

Key points

- Exothermic changes can be used in hand warmers and self-heating cans. Crystallisation of a supersaturated solution is used in reusable warmers. However, disposable, one-off warmers can give off heat for longer.

- Endothermic changes can be used in instant cold packs for sports injuries.

11.3

Energy and reversible reactions

Energy changes are involved in reversible reactions too. Let's consider an example.

Figure 1 shows a reversible reaction where A and B react to form C and D. The products of this reaction (C and D) can also react to form A and B again.

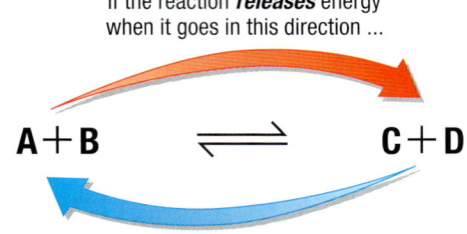

If the reaction **releases** energy when it goes in this direction ...

$$A + B \rightleftharpoons C + D$$

... it will **absorb** exactly the same amount of energy when it goes in this direction

Figure 1 *A reversible reaction*

If the reaction between A and B is exothermic, energy will be released when the reaction forms C and D.

If C and D then react to make A and B again, the reaction must be endothermic. What's more, it must absorb exactly the same amount of energy as it released when C and D were formed from A and B.

Energy cannot be created or destroyed in a chemical reaction. The amount of energy released in one direction in a reversible reaction must be exactly the same as the energy absorbed in the opposite direction.

You can see how this works if you look at what happens when you heat blue copper(II) sulfate crystals. The crystals contain water as part of the lattice formed when the copper(II) sulfate crystallised. The copper sulfate is **hydrated**. Heating the copper(II) sulfate drives off the water from the crystals, producing white **anhydrous** ('without water') copper(II) sulfate. This is an endothermic reaction.

$$\begin{array}{ccc} CuSO_4 \cdot 5H_2O & \rightleftharpoons & CuSO_4 + 5H_2O \\ \text{hydrated} & \rightleftharpoons & \text{anhydrous} + \text{water} \\ \text{copper(II) sulfate (blue)} & & \text{copper(II) sulfate (white)} \end{array}$$

When water is added to anhydrous copper(II) sulfate it forms hydrated copper(II) sulfate. The colour change in the reaction, from white to blue, is a useful test for the presence of water. The reaction in this direction is exothermic. In fact, so much energy may be produced that you may see steam rising as the water boils.

Figure 2 *Hydrated copper(II) sulfate and white anhydrous copper(II) sulfate*

Practical

Energy changes in a reversible reaction

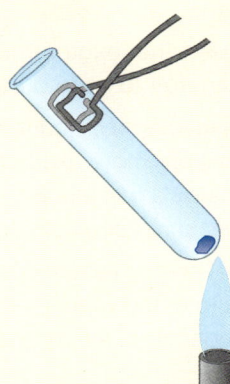

Figure 3 *Heating hydrated copper(II) sulfate*

Gently heat a few copper(II) sulfate crystals in a test tube. Observe the changes. When the crystals are completely white allow the tube to cool to room temperature (this takes several minutes). Add two or three drops of water from a dropper and observe the changes. Carefully feel the bottom of the test tube.

- Explain the changes you have observed.

You can repeat this with the same solid, as it is a reversible reaction, or try with other hydrated crystals, such as cobalt(II) chloride. Some are not so colourful but the changes are similar.

Safety: Wear eye protection. Avoid skin contact with cobalt(II) chloride (toxic). Copper salts are harmful.

You can soak filter paper in cobalt(II) chloride solution and allow it to dry in an oven. The blue paper that is produced is called cobalt(II) chloride paper. The paper turns pale pink when water is added to it, so it can be used as an indicator for water.

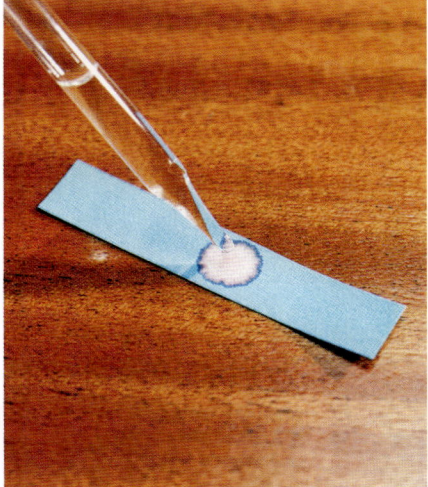

Figure 4 *Blue cobalt(II) chloride paper turns pink when water is added*

Summary questions

1 a How does the enthalpy change for a reversible reaction in one direction compare with the enthalpy change for the reaction in the opposite direction?
   b What can anhydrous copper(II) sulfate be used to test for?
   c Why does blue cobalt(II) chloride turn pink if left out in the open air?
   d When water is added to blue cobalt(II) chloride is energy released or absorbed?

2 A reversible reaction gives out 50 kilojoules (kJ) of energy in the forward reaction. In this reaction W and X react to give Y and Z.
   a Write an equation to show the reversible reaction.
   b What can you say about the energy transfer in the reverse reaction?

3 Blue cobalt(II) chloride crystals turn pink when they become damp. The formula for the two forms can be written as $CoCl_2 \cdot 2H_2O$ and $CoCl_2 \cdot 6H_2O$.
   a How many moles of water will combine with 1 mole of $CoCl_2 \cdot 2H_2O$?
   b Write a balanced chemical equation for the reaction, which is reversible.
   c How can pink cobalt(II) chloride crystals be changed back to blue cobalt(II) chloride crystals?

Key points

- In reversible reactions, one reaction is exothermic and the other is endothermic.

- In any reversible reaction, the amount of energy released when the reaction goes in one direction is exactly equal to the energy absorbed when the reaction goes in the opposite direction.

11.4

Comparing the energy released by fuels

You have already seen that exothermic reactions release energy. When we burn a fuel, we use this energy to keep ourselves warm or we use it for transport.

Not all fuels release the same amount of energy when they burn. Some combustion reactions are more exothermic than others. It is often very important to know how much energy a fuel releases when it burns. In a school chemistry lab it is very difficult to measure accurately the energy released by fuels when they burn.

The actual amount of energy released by a burning fuel in an experiment is related to the rise in temperature of the water in a calorimeter (a glass or metal container). The larger the rise in temperature, the more energy has been released (see Figure 1).

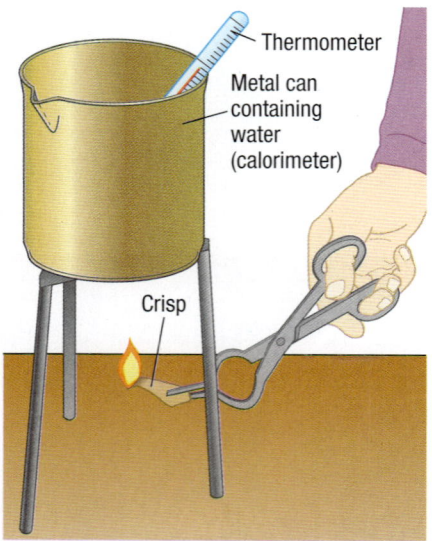

Figure 1 *The energy released by fuels and foods when they burn can be compared using a simple calorimeter. The energy content of food is often given in units called calories. However, the joule is the usual scientific unit of energy.*

∞ links

For more information on exothermic reactions, look back at 11.1 'Exothermic and endothermic reactions'.

Calculating the energy released

There is an equation you can use to work out the energy released, measured in joules (J):

$$\frac{\text{energy}}{\text{released}} = \frac{\text{mass of water}}{\text{heated}} \times \frac{\text{specific heat capacity}}{\text{of water}} \times \frac{\text{rise in}}{\text{temperature}}$$

This is sometimes written as:

$$Q = mc\Delta T$$

Where:

Q is the energy released by the fuel (in joules, J)

m is the mass of water heated in the calorimeter (1 cm^3 of water has a mass of 1 g)

c is the specific heat capacity of water (this is the amount of energy needed to raise 1 g of water by 1 °C)

ΔT is the rise in temperature (i.e., the final temperature of the water minus the initial temperature).

Worked example

In an experiment, a burning fuel raised the temperature of 50 cm^3 of water from 16 °C to 41 °C. How much energy was released by the fuel? (4.2 J of energy raises the temperature of 1 g of water by 1 °C.)

Solution

Write down the data given:

Mass of water heated in the calorimeter = 50 g
Specific heat capacity of water = 4.2 J/g °C
Rise in temperature of the water = 41 °C − 16 °C = 25 °C

Now use the data in the equation $Q = mc\Delta T$ (which will always be provided for you in the exam):

$$Q = mc\Delta T$$
$$Q = 50 \times 4.2 \times 25 \text{ J}$$
$$Q = \textbf{5250 J (or 5.25 kJ)}$$

It is useful to be able to compare the energy content of different fuels. Their combustion data is given as the number of joules (or kilojoules) of energy released per gram or per mole of fuel burnt.

In the experiment described in the worked example, the fuel was weighed before and after burning. It was found that the mass had decreased by 0.2 g. So the energy given out in the experiment will be multiplied by the number of 0.2 g there are in 1 g.

$$\text{energy released per gram} = 5.25 \times \frac{1}{0.2}\,\text{kJ/g}$$
$$= \mathbf{26.25\,kJ/g}$$

Suppose you know that the relative formula mass of the fuel is 46. You can now also work out the energy released per mole. One mole of the fuel has a mass of 46 g. So the energy given out in the experiment will be multiplied by the number of 0.2 g there are in 46 g:

$$\text{energy released per mole} = 5.25 \times \frac{46}{0.2}\,\text{kJ/mol}$$
$$= \mathbf{1207.5\,kJ/mol}$$

And as it is an exothermic reaction you can write:

$$\Delta H = \mathbf{-1207.5\,kJ/mol}$$

Study tip

Measurements using simple calorimeters are not accurate because of energy losses, but they can be used to compare the amounts of energy released.

Practical

Comparing the energy released when fuels burn

You can compare the energy released by different fuels when they burn. The fuels are burnt to heat water in a copper can or a glass beaker. Measure the temperature changes produced by different fuels, and then compare the energy they release when they burn. Use the worked example to help with your calculations.

■ What variables must you control in order to compare the energy released by different fuels?

Safety: Wear eye protection. Take care with burning fuels.

Summary questions

1 A simple calorimeter was used to compare the energy released by three different fuels. The results were as follows:

	Mass of fuel burnt (g)	Volume of water (cm³)	Initial temperature (°C)	Final temperature (°C)
Fuel A	0.24	160	18.0	28.0
Fuel B	0.18	100	18.0	26.0
Fuel C	0.27	150	18.0	27.0

a Using the equation $Q = mc\,\Delta T$, calculate the energy released by each fuel in the three tests. (4.2 J of energy raises the temperature of 1 g of water by 1 °C.)

b Calculate the energy released per gram for each fuel.

c Arrange the fuels in order of the amount of energy they release per gram.

d The relative formula mass of Fuel A is 48, Fuel B is 42, and Fuel C is 58. What is the energy released by Fuels A, B, and C in kilojoules per mole (kJ/mol)?

e Now arrange the fuels in order of the amount of energy they release per mole.

f Comment on the accuracy of the figures for the amount of energy released per mole of fuel burnt that were calculated in part d.

Key points

■ When fuels and food react with oxygen, energy is released in an exothermic reaction. The unit of energy is the joule (but calories are still often used for the energy content of foods).

■ A simple calorimeter can be used to compare the energy released by different fuels or different foods in a school chemistry lab.

■ Use the equation $Q = mc\,\Delta T$ to calculate the amount of energy transferred from a burning fuel or food to the water in a calorimeter.

11.5 Energy transfers in solutions

Energy changes in chemical reactions

Learning objectives

After this topic, you should know:

- how to measure the energy change for a reaction that takes place in solution.

You can use a simple calorimeter to measure temperature changes in reactions involving solutions. You can use a polystyrene cup as a simple calorimeter with a thermometer to monitor reactions of solids with solutions, or between solutions.

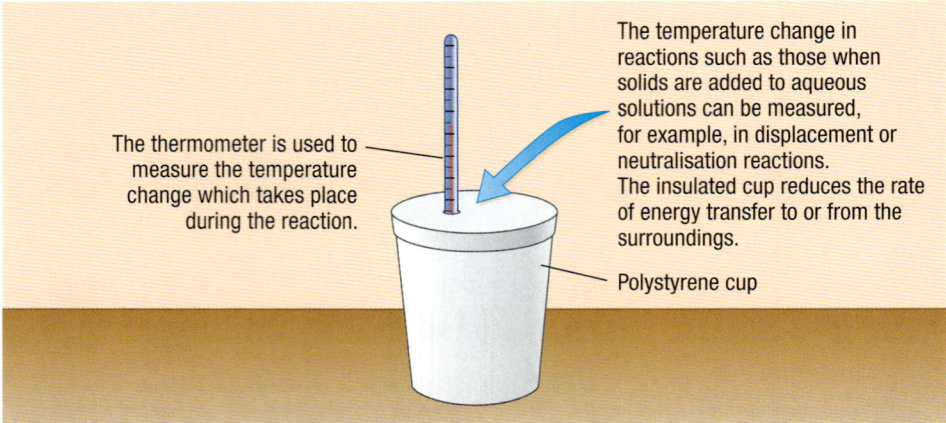

The thermometer is used to measure the temperature change which takes place during the reaction.

The temperature change in reactions such as those when solids are added to aqueous solutions can be measured, for example, in displacement or neutralisation reactions. The insulated cup reduces the rate of energy transfer to or from the surroundings.

Polystyrene cup

Figure 1 *A simple calorimeter used to measure energy changes in solution. Polystyrene is a good thermal insulator, which helps to reduce energy transfer through the sides of the container during reactions. A lid on the calorimeter reduces energy transfer to the surroundings even further.*

You have seen how to calculate actual energy changes using:

$$Q = mc\,\Delta T$$

In words:

energy released = mass of water heated × specific heat capacity of water × change in temperature

In these calculations, assume that the solutions behave like water. So 1 cm³ of solution has a mass of 1 g. Assume also that solutions have a specific heat capacity of 4.2 J/g °C. Therefore 4.2 J of energy raises the temperature of 1 g of solution by 1 °C. Look at the worked example on the next page.

Study tip

You do not have to learn the equation $Q = mc\,\Delta T$
In the exam this equation will be provided for you, but you should know how to use it.

Practical

Measuring energy changes in reactions

You can use a simple polystyrene calorimeter to work out the energy changes in the following:

- iron filings + copper sulfate solution (a displacement reaction)
- magnesium ribbon + dilute hydrochloric acid
- sodium hydroxide solution + dilute hydrochloric acid (a neutralisation reaction)
- dissolving potassium nitrate, anhydrous copper sulfate, and other salts in water

Safety: Wear eye protection. Copper sulfate is harmful. Sodium hydroxide, if equal to or less than 0.4 mol/dm³, is irritant. Potassium nitrate is oxidising. Dilute hydrochloric acid is an irritant.

Worked example

A simple calorimeter is used to measure the energy change in the reaction:

$$A + B \rightarrow C$$

$60\,cm^3$ of a solution containing 0.1 moles of A is mixed with $40\,cm^3$ of a solution containing 0.1 moles of B. The temperature of the two solutions before mixing was 19.6 °C. After mixing them, the maximum temperature reached was 26.1 °C. Calculate the enthalpy change ($\triangle H$) of the reaction.

Solution

Step 1 – calculate the temperature change:

$$\text{temperature change} = 26.1\,°C - 19.6\,°C$$
$$= 6.5\,°C$$

Step 2 – calculate the energy change:

$$Q = mc\,\Delta T$$

The mass of solution heated up in the reaction is $60\,g + 40\,g = 100\,g$

$$\text{energy change} = 100\,g \times 4.2\,J/g\,°C \times 6.5\,°C$$
$$= 2730\,J$$
$$= 2.73\,kJ$$

This is the energy change when 0.1 moles of reactants A and B are mixed. So when 1 mole of reactants are mixed there will be 10 times as much energy released (1 mole is 10×0.1 moles)

$$= 2.73\,kJ \times 10$$
$$= 27.3\,kJ$$

So this experiment gives the energy change for the reaction:

$$A + B \rightarrow C$$

as 27.3 kJ/mol.

The temperature rises so the reaction is exothermic. Therefore:

$$\Delta H = -27.3\,\textbf{kJ/mol}$$

Summary questions

1 Why is polystyrene a good material for a simple calorimeter used for measuring temperature changes of reactions in solution?

2 A student added $25\,cm^3$ of dilute hydrochloric acid to $25\,cm^3$ of sodium hydroxide solution in a polystyrene calorimeter. She recorded a temperature rise of 5.5 °C. There were 0.050 moles of both the acid and the alkali used in the experiment.
   a Write a balanced symbol equation, including state symbols, for the reaction that takes place.
   b What is this type of reaction called?
   c Calculate the concentration of the solutions used in this experiment in moles per decimetre cubed (mol/dm^3).
   d Using $Q = mc\,\Delta T$, work out the energy change in the experiment. (Specific heat capacity of the solution is 4.2 J/g °C.)
   e What would be the change in enthalpy, ΔH, for this reaction in kJ per mole?

Study tip

The energy released by an exothermic reaction heats up the water (and its container). The temperature rise is proportional to the amount of energy released.

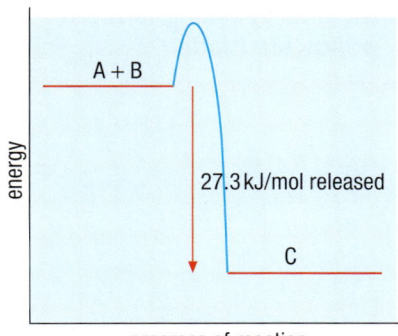

Figure 2 *Here is the energy level diagram for the reaction $A + B \rightarrow C$*

∞ links

For more information on energy level diagrams, see 11.6 'Energy level diagrams'.

Key points

- You can calculate the energy change for reactions in solution by measuring the temperature change and using the equation:

$$Q = mc\,\Delta T$$

(The equation will be given in the exam.)

- Neutralisation and displacement reactions are both examples of reactions that you can use this technique for.

11.6 Energy level diagrams

Learning objectives

After this topic, you should know:

- what energy level diagrams show us
- how catalysts affect the activation energy of a reaction
- the energy changes when bonds are broken
- the energy changes when bonds are made.

You can find out more about what is happening in a particular reaction by looking at its energy level diagram. These diagrams show us the relative amounts of energy contained in the reactants and the products. This energy is measured in kilojoules per mole (kJ/mol).

Exothermic reactions

Figure 1 shows the energy level diagram for an exothermic reaction. The products are at a lower energy level than the reactants. Therefore when the reactants form the products, energy is released.

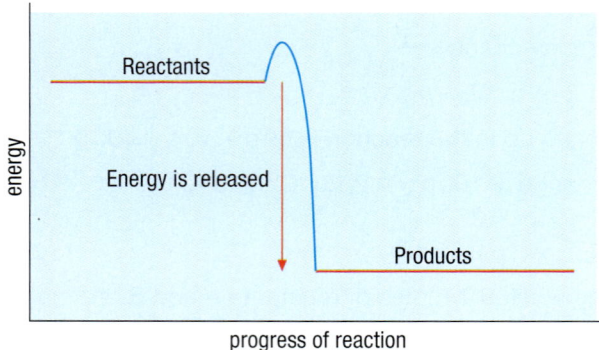

Figure 1 *The energy level diagram for an exothermic reaction*

The difference between the energy levels of the reactants and the products is the energy change during the reaction, measured in kJ/mol.

The difference in energy between the products and the reactants is released to the surroundings. Therefore in exothermic reactions the temperature of the surroundings increases. The surroundings get hotter and the value of ΔH for the reaction has a negative sign.

Endothermic reactions

Figure 2 shows the energy level diagram for an endothermic reaction.

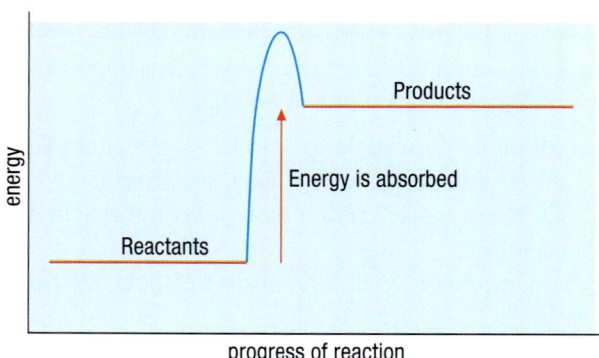

Figure 2 *The energy level diagram for an endothermic reaction*

Here the products are at a higher energy level than the reactants. As the reactants react to form products, energy is absorbed from the surroundings. The temperature of the surroundings decreases because energy is taken in during the reaction. The surroundings get colder. The products are at a higher energy level than the reactants, and the value of ΔH for the reaction is given a positive sign.

links

For more information about the enthalpy change, ΔH, in chemical reactions, look back at 11.1 'Exothermic and endothermic reactions'.

Activation energy and catalysis

Think back to your work on rates of reaction in Chapter 9. There you learnt about the collision theory of reactions. This stated that there is a minimum amount of energy needed before colliding particles can react. This energy needed to start a reaction is called the activation energy. This can be shown on the energy level diagrams. Look at Figure 3.

Catalysts can increase the rate of a reaction. The way they do this is to provide an alternative pathway to the products, which has a lower activation energy. This is shown on the energy level diagram in Figure 4. This means that a higher proportion of reactant particles now have enough energy to react when they collide.

Bond breaking and bond making

Think about what happens as a chemical reaction takes place. You can think of the chemical bonds between the atoms or ions in the reactants being broken. Then new chemical bonds can be formed to make the products.

- Energy has to be supplied to break chemical bonds. This means that breaking bonds is an **endothermic** process. Energy is taken in from the surroundings.
- But, when new bonds are formed, energy is released. Making bonds is an **exothermic** process.

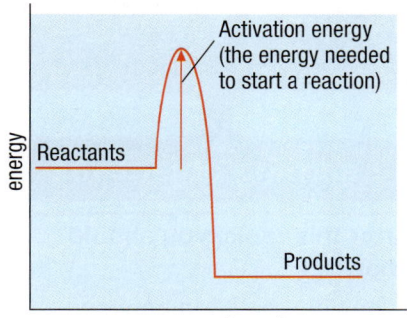

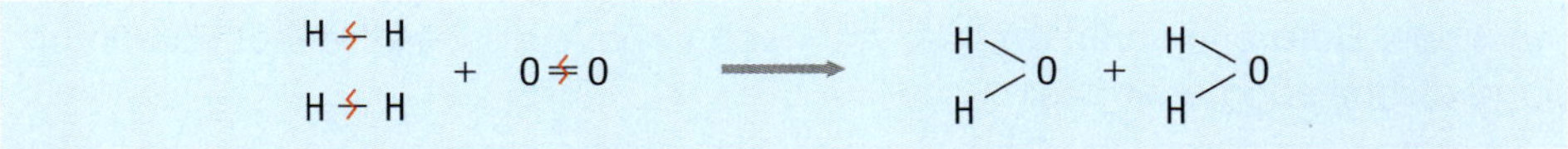

Figure 3 *The minimum amount of energy needed to start a reaction is called its activation energy*

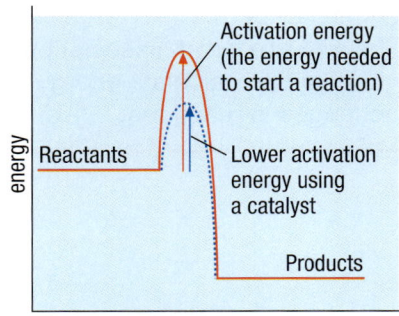

Figure 4 *A catalyst provides a different reaction pathway with a lower activation energy, so that a higher proportion of reactant particles have enough energy to react*

$$H-H + H-H + O=O \longrightarrow \overset{H}{\underset{H}{>}}O + \overset{H}{\underset{H}{>}}O$$

Figure 5 *Hydrogen and oxygen react to make water. The bonds between hydrogen atoms and between oxygen atoms have to be broken so that bonds between oxygen atoms and hydrogen atoms in water can be formed.*

Study tip

Remember that **B**reaking bonds a**B**sorbs energy, fo**R**ming bonds **R**eleases energy.

Summary questions

1 Draw energy level diagrams for the following reactions:
 a $H_2(g) + Cl_2(g) \rightarrow 2HCl(g)$
   ΔH for the reaction is -184 kJ/mol.
 b $H_2(g) + I_2(g) \rightarrow 2HI(g)$
   ΔH for the reaction $+26.5$ kJ/mol.

2 a Draw an energy level diagram for an endothermic reaction, $X \rightarrow Y + Z$, including its activation energy.
 b Now show, on your energy level diagram drawn in part **a**, the effect of adding a catalyst.
 c Explain how a catalyst increases the rate of a reaction.

3 a Explain why bond breaking is an endothermic process.
 b Using a diagram like Figure 5 above, show the bonds being broken and new bonds made in the reaction between methane (CH_4) and oxygen which produces carbon dioxide and water.
 c List the number and type of each bond broken and formed in your answer to part **b**.

Key points

- You can show the relative difference in the energy of reactants and products on energy level diagrams.

- Catalysts lower the activation energy so a greater proportion of reactant particles have enough energy to react.

- Bond breaking is endothermic, whereas bond making is exothermic.

11.7 Bond dissociation energy calculations

11.7

Learning objectives

After this topic, you should know:

- how the balance between bonds breaking in the reactants and bonds being made in the products affect the overall energy change of a reaction

- how to use bond dissociation energies to calculate energy changes in reactions.

Making and breaking bonds

There is always a balance between the energy needed to break bonds and the energy released when new bonds are made in a reaction. This is what decides whether the reaction is endothermic or exothermic.

- In some reactions the energy released when new bonds are formed (as the products are made) is more than the energy needed to break the bonds in the reactants. These reactions transfer energy to the surroundings. They are **exothermic**.

- In other reactions the energy needed to break the bonds in the reactants is more than the energy released when new bonds are formed in the products. These reactions transfer energy from the surroundings to the reacting chemicals. They are **endothermic**.

Bond dissociation energy

The energy needed to break the bond between two atoms is called the **bond dissociation energy** for that bond.

Bond dissociation energies are measured in kJ/mol. You can use bond dissociation energies to work out the enthalpy change (ΔH) for many chemical reactions. Before you can do this, you need to have a list of the most common bond dissociation energies:

Bond	Bond dissociation energy (kJ/mol)	Bond	Bond dissociation energy (kJ/mol)
C–C	347	H–Cl	432
C–O	358	H–O	464
C–H	413	H–N	391
C–N	286	H–H	436
C–Cl	346	O=O	498
Cl–Cl	243	N≡N	945

To calculate the energy change for a chemical reaction you need to work out:

1 how much energy is needed to break the chemical bonds in the reactants

2 how much energy is released when the new bonds are formed in the products.

It is very important to remember that the data in the table is the energy required for *breaking* bonds. However, the energy released as these bonds are formed is the same (see Figure 1).

For example, the bond dissociation energy for a C—C bond is +347 kJ/mol (an endothermic change). This means that the energy released *forming* a C—C bond is –347 kJ/mol (an exothermic change).

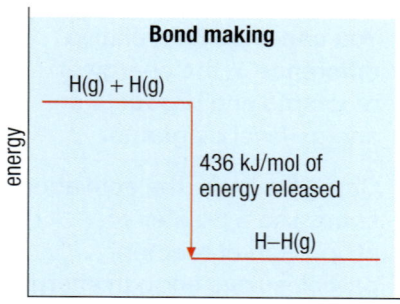

Bond breaking

H(g) + H(g)

436 kJ/mol of energy absorbed

H–H(g)

progress of reaction

Bond making

H(g) + H(g)

436 kJ/mol of energy released

H–H(g)

progress of reaction

Figure 1 *Breaking and making a particular bond always involves the same amount of energy, but the change in enthalpy will have opposite signs (+ for bond breaking and – for bond making)*

Worked example

Ammonia is made from nitrogen and hydrogen in the Haber process. The balanced symbol equation for this reaction is:

$$N_2(g) + 3H_2(g) \rightleftharpoons 2NH_3(g)$$

Calculate the overall energy change for the forward reaction using bond dissociation energies.

Solution

This equation tells us that you need to break the bonds in 1 mole of nitrogen molecules and three moles of hydrogen molecules in this reaction. Look at Figure 2.

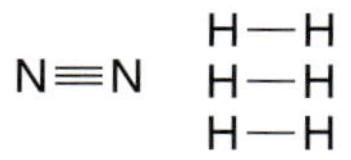

Nitrogen molecules are held together by a triple bond (written like this: N≡N). This bond is very strong. Using data from the table, its bond energy is 945 kJ/mol.

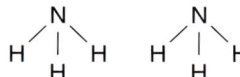

Figure 2 *These bonds are broken in the forward reaction*

Hydrogen molecules are held together by a single bond (written like this: H—H). From the table, the bond energy for this bond is 436 kJ/mol. The energy needed to break 1 mole of N≡N and 3 moles of H—H bonds is 945 + (3 × 436) kJ = **+2253 kJ** (given a plus sign as this is an endothermic process).

When these atoms form ammonia (NH_3), six new N—H bonds are made as two moles of NH_3 are formed (see Figure 3). The bond energy of the N—H bond is 391 kJ/mol.

Figure 3 *These bonds are made in the forward reaction*

Energy released when six moles of N—H bonds are made = 6 × 391 kJ = **−2346 kJ** (given a minus sign as this is an exothermic process).

Figure 4 shows the overall energy change for the forward reaction, as written.

So *the overall enthalpy change (△H)*
= (+2253 kJ) + (−2346 kJ)
= **−93 kJ** (This is the energy *released* in the forward reaction as written above.)

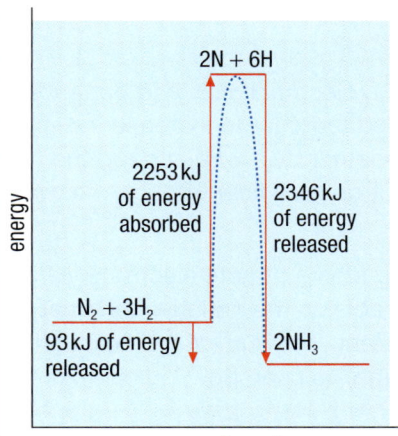

Figure 4 *The formation of ammonia. The energy released, 93 kJ, is for the formation of 2 moles of ammonia as shown in the balanced equation. So if you wanted to know the energy change for the reaction per mole of ammonia formed, it would release exactly half this, that is, 46.5 kJ/mol.*

Key points

- In chemical reactions, energy must be supplied to break the bonds between atoms in the reactants.

- When new bonds are formed between atoms in a chemical reaction, energy is released.

- In an exothermic reaction, the energy released when new bonds are formed is greater than the energy absorbed when bonds are broken.

- In an endothermic reaction, the energy released when new bonds are formed is less than the energy absorbed when bonds are broken.

- You can calculate the overall energy change in a chemical reaction using bond energies.

Summary questions

1. If the energy required to break bonds is greater than the energy released when bonds are made, will the reaction be exothermic or endothermic?

2. What is meant by the 'bond dissociation energy' of a chemical bond?

3. Write balanced symbol equations and calculate the energy changes for the following chemical reactions:
   a. hydrogen + chlorine → hydrogen chloride
   b. hydrogen + oxygen → water
   (Use the bond energies supplied in the table on the previous page.)

11.8 Chemical cells and batteries

Learning objectives

After this topic, you should know:

- how to interpret data on chemical cells in terms of the relative reactivity of different metals
- how to evaluate the use of chemical cells when given information
- how to plan and carry out an investigation of the voltage produced by simple cells using different metals.

Figure 1 *There are many different types of electrical cells and batteries used in mobile electrical appliances*

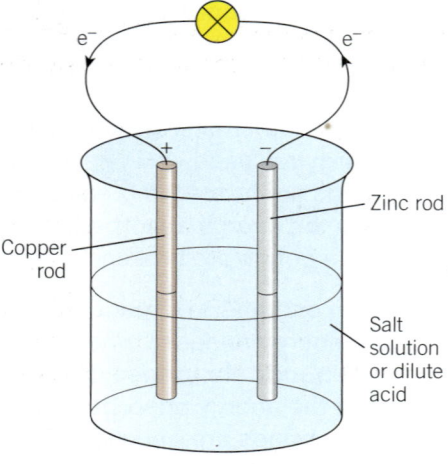

Figure 2 *An electrical cell made from zinc and copper. The electrons flow from the more reactive metal (zinc) to the less reactive metal (copper). So zinc acts as the negative terminal of the cell, providing electrons to the external circuit.*

Did you realise that the electrical cells and batteries used in many mobile appliances rely on the differing reactivity of metals? You have seen in Chapter 4 how metals can be extracted from their ores and how the method used is related to the metal's position in the reactivity series.

You also used this order of reactivity to predict displacement reactions, as a more reactive metal will displace a less reactive metal from its compounds.

For example:

$$Zn(s) + CuSO_4(aq) \rightarrow ZnSO_4(aq) + Cu(s)$$

The sulfate ions do not change in the displacement reaction above. They are spectator ions.

So you can leave them out of the equation and write an ionic equation:

$$Zn(s) + Cu^{2+}(aq) \rightarrow Zn^{2+}(aq) + Cu(s)$$

You can think of this redox reaction as two half equations.

One will represent reduction:

$$Cu^{2+}(aq) + 2e- \rightarrow Cu(s)$$

The Cu^{2+} ions are reduced to Cu.

The other will be an oxidation reaction:

$$Zn(s) \rightarrow Zn^{2+}(aq) + 2e-$$

The Zn atoms are oxidised to Zn^{2+} ions.

You can use this difference in reactivity to make electrical cells and batteries. (A battery is made up of two or more cells joined together to increase the voltage available.) If you join the two metals together by a wire and dip them into an electrolyte, such as a salt solution, electrons will flow through the wire from the zinc to the copper.

In a simple cell like that shown in Figure 2, zinc atoms donate electrons via the connecting wire to the copper(II) ions, so zinc acts as the negative terminal of the cell. The flow of electrons is an electric current. The current will flow in the circuit until one of the reactants, Zn(s) or $Cu^{2+}(aq)$, is used up. You can use the current to light a lamp.

Remember that there is a tendency for any metal atom to give away electrons and form a positive ion. The greater the tendency to form their positive ion, the more reactive the metal is. So the copper(II) ions accept electrons from the zinc atoms and change into copper atoms.

In general:

The greater the difference in reactivity between the two metals used, the higher the voltage produced.

You can test this out in the following experiment. Place a voltmeter in the external circuit. The voltage reading will give you a measure of the difference in reactivity between the two metals used in the cell. You can think of this as the difference in the metals' tendency to give away electrons. The larger the voltage, the bigger the difference in their power as reducing agents.

Practical

Investigating chemical cells

Use this apparatus to investigate the voltage produced by different metals paired with magnesium.

You can also compare metals against zinc, iron, copper, nickel, and tin in your electrical cells.

- Record and display your data using a suitable table.
- Then draw your conclusion.
- You can extend your investigation to find out if any other factors affect the voltage produced, besides the two metals used.
- Or you can check predictions of the voltage produced by other pairs of metals using the data in your table.

Safety: Wear eye protection.

Voltmeter

Magnesium ribbon

Other metal tested

Salt solution

The first mass-produced batteries made were called primary cells, and improved versions are still manufactured today. They cannot be recharged. The dry cell, with electrodes made of zinc and carbon, are non-rechargeable, as are the popular, modern alkaline cells that can produce a larger voltage. Once one of the reactants has run out, the cell stops working and should be disposed of in a recycling centre.

Other cells can be recharged and used over and over again. In the recharging process, the battery is connected to a power supply that reverses the reactions that occur at each electrode when the cell is discharging. This regenerates the original reactants.

Summary questions

1 Why is it not possible to make an electrical cell using two electrodes made of zinc metal?

2 An electrical cell is made using the metals iron and zinc as the two electrodes, and a salt solution as the electrolyte.
   a Draw a diagram of the electrical cell set up with a lamp in the external circuit.
   b Which metal is reduced in the cell?
   c Explain which metal will act as the negative terminal of the cell. Include two half equations in your answer.

3 Using Figure 3, explain two disadvantages of the zinc–carbon cell compared with some other more modern cells.

links

To revise the reactivity series see 4.3 'The reactivity series'.

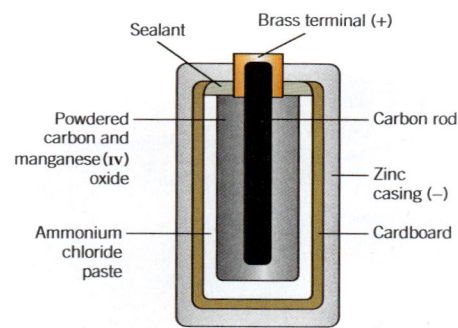

Sealant Brass terminal (+)

Powdered carbon and manganese(IV) oxide

Carbon rod

Zinc casing (–)

Ammonium chloride paste

Cardboard

Figure 3 *A zinc–carbon dry cell produces a voltage of about 1.5 V, but cannot be recharged and is prone to leakage if left in an appliance for a long period of time*

Key points

- **Metals tend to lose electrons and form positive ions.**

- **When two metals are dipped in a salt solution and joined by a wire, the more reactive metal will donate electrons to the less reactive metal. This forms a simple electrical cell.**

- **The greater the difference in reactivity between the two metals, the higher the voltage produced by the cell.**

11.9 Fuel cells

Learning objectives

After this topic, you should know:

- that hydrogen–oxygen fuel cells can be used to run vehicles

- that when in use, the only waste product of hydrogen–oxygen fuel cells is water, reducing pollution compared with the use of fossil fuels

- the advantages and disadvantages of using hydrogen as a fuel.

Figure 1 *A Mercedes-Benz London bus runs on fuel cells*

??? Did you know ... ?

Hydrogen refuelling stations tend to have no roof. Then if there is a leak of hydrogen, it will escape upwards into the atmosphere. This reduces the risk of an explosion.

Much of the world's pollution is caused by the increasing numbers of vehicles on our roads. Some people think that the best solution would be to move away from fossil fuels, especially crude oil, and build a society which relies on hydrogen-based fuels.

A more efficient use of the energy from oxidising a fuel is in a **fuel cell**. For example, hydrogen–oxygen fuel cells are fed with hydrogen and oxygen which produce water. Most of the energy released in the reaction is transferred to electrical energy. This can be used to run a vehicle. However, we need a constant supply of hydrogen to run the fuel cell.

Scientists are aware that replacing engines powered by fossil fuels with 'cleaner' energy sources could have great benefits. Therefore they have developed many types of fuel cell and hydrogen-powered engines. The challenge is to match the performance, convenience, and price of petrol or diesel cars.

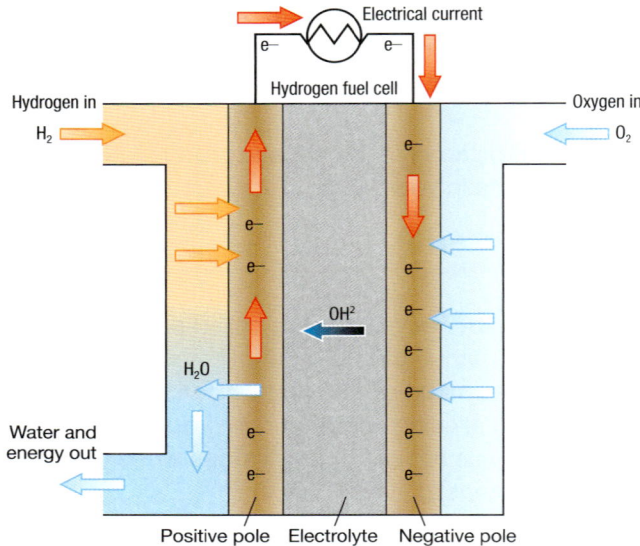

Figure 2 *A hydrogen fuel cell which has an alkaline electrolyte. Notice that the only waste product is water when the cell is in use.*

Hydrogen – a fuel for the future

Scientists are very interested in developing hydrogen as a fuel for combustion engines that can be used in vehicles. It burns well with a very clean flame as there is no carbon in the fuel:

$$\text{hydrogen} \quad + \quad \text{oxygen} \quad \rightarrow \quad \text{water}$$
$$2H_2(g) \quad + \quad O_2(g) \quad \rightarrow \quad 2H_2O(g)$$

As you can see in the equation, water is the only product in the combustion of hydrogen. There are no pollutants made when hydrogen burns and no extra carbon dioxide is added to the air. Therefore, it would help tackle global warming and climate change. Not only that, water is potentially a huge natural source of hydrogen. The hydrogen can be obtained from water by electrolysis. However, the electricity needed to break down the water into hydrogen and oxygen must be supplied by a renewable energy source to conserve fossil fuels and control carbon dioxide emissions.

Problems with hydrogen

However, there are problems that need to be solved before hydrogen becomes a common fuel.

- When mixed with air and ignited it is explosive. So there are safety concerns in case of leaks, or accidents, in vehicles powered by hydrogen.
- Vehicles normally run on liquid fuels but hydrogen is a gas. Therefore, it takes up a much larger volume than liquid fuels so storage becomes an issue. We can use high-pressure cylinders but these also have safety problems in crashes.

Oxidising hydrogen in fuel cells is much more efficient than burning it so this could be the way forward. These cells are used in spacecraft and can be used to run a vehicle. However, a constant supply of hydrogen is needed to run the fuel cell and there are advantages and disadvantages.

Advantages of hydrogen fuel cells

- They do not need to be electrically recharged.
- There are no pollutants produced.
- They can be designed in a range of sizes for different uses.

Disadvantages of hydrogen fuel cells

- Hydrogen is highly flammable.
- Hydrogen is sometimes produced for the cell by non-renewable sources.
- Hydrogen is difficult to store.

Summary questions

1 Copy and complete the sentences using the words below:

water   fuel   oxygen   carbon   combustion   carbon dioxide

Hydrogen can be used as an alternative fuel to burn in _____ engines in cars. It can also power electrical cars that run on _____ cells.

In both these uses, hydrogen reacts with _____ to form _____ . This helps to reduce levels of _____ in the atmosphere, reducing the fuel's _____ footprint compared with petrol or diesel.

2 Your family decides to buy an electric car that needs regular recharging by plugging it into an electrical socket in your garage. Why would this not necessarily mean that you had found a way to get around without adding to the carbon dioxide in the air? How could it run without contributing to global warming?

3 a Explain why hydrogen is potentially a pollution-free fuel. Include a balanced symbol equation in your answer.
  b Explain two problems that have to be overcome before hydrogen is used as an everyday fuel.

Key points

- Much of the world relies on fossil fuels. However, they are non-renewable and they cause pollution. Alternative fuels, such as hydrogen, need to be found soon.

- Hydrogen can be burnt in combustion engines or used in fuel cells to power vehicles. However, it does have problems to overcome in its use as a fuel, such as safety and storage issues.

Chapter summary questions

1 Two solutions are mixed and react in an endothermic reaction. When the reaction has finished, the reaction mixture is allowed to stand until it has returned to its starting temperature.

   a Sketch a graph of temperature (y-axis) against time (x-axis) to show how the temperature of the reaction mixture changes.

   b Label the graph clearly and explain what is happening wherever you have shown that the temperature is changing.

   c In the experiment 15 cm³ of one solution was mixed with 25 cm³ of the other solution, and once mixed the temperature fell from 18 °C to 15 °C. Use $Q = mc\,\Delta T$, where $c = 4.2$ J/g °C (i.e., 4.2 J of energy raises the temperature of 1 g of water by 1 °C), to calculate the energy absorbed in the experiment.

2 a Draw an energy level diagram to show the exothermic reaction between nitric acid and sodium hydroxide, including its activation energy.

   b Draw an energy level diagram to show the endothermic change when ammonium nitrate dissolves in water, including its activation energy.

   c Show an energy level diagram, including its activation energy, for the exothermic breakdown of hydrogen peroxide into water and oxygen. Then add to your diagram to show what happens when the same reaction is catalysed by manganese(IV) oxide.

3 When you eat sugar you break it down eventually to produce water and carbon dioxide.

   a Copy and complete the balanced symbol equation:

$$C_{12}H_{22}O_{11} + O_2 \rightarrow$$

   b Why must your body *supply* energy in order to break down a sugar molecule?

   c When you break down sugar in your body, energy is released. Explain where this energy comes from in terms of the bonds in molecules.

   d A person can get about 1700 kJ of energy by breaking down 100 g of sugar.
If a heaped teaspoon contains 5 g of sugar, how much energy does this release when broken down by their body?

4 Hydrogen peroxide has the structure H—O—O—H. It decomposes slowly to form water and oxygen.

$$2H_2O_2(aq) \rightarrow 2H_2O(l) + O_2(g)$$

The table shows the bond dissociation energies for different types of bond.

Bond	Bond dissociation energy (kJ/mol)
H—O	464
H—H	436
O—O	144
O=O	498

   a Use the bond dissociation energies to calculate the enthalpy change, ΔH, for the decomposition of hydrogen peroxide as shown above.

   b Explain the sign (+ or –) in your answer to part **a** in terms of bond breaking and making.

5 Bomb calorimeters are used in research laboratories to calculate the energy in food.

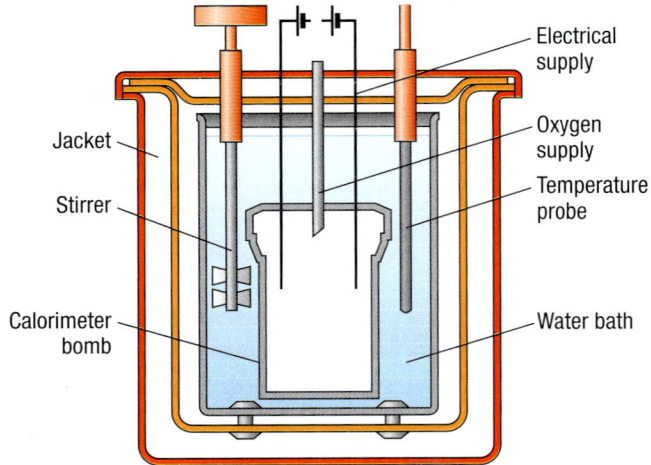

As you can see from this diagram, it is sealed and has its own supply of oxygen. There is a supply of electricity to produce a spark that will ignite the food. The increase in temperature of the water around the bomb is measured.

   a Using a bomb calorimeter, a food scientist burnt a sample of a sugar. There were 250 cm³ of water in the calorimeter and its temperature went from 18 °C to 44 °C.
Use $Q = mc\,\Delta T$, where $c = 4.2$ J/g °C (i.e., 4.2 J of energy raises the temperature of 1 g of water by 1 °C), to calculate the energy released by the burning sugar.

   b The scientists burnt 0.01 moles of the sugar in the bomb calorimeter. How much energy does this sugar release in kJ per mole?

Practice questions

1 Most chemical reactions are exothermic. Some are endothermic.

Some equations are shown with a ΔH value.

a **i** What is the difference between these types of reaction? (2)

    ii What is the meaning of ΔH? (1)

    iii How can you decide from the ΔH value whether a reaction is exothermic or endothermic? (1)

b The progress of a reaction can be represented on an energy level diagram.

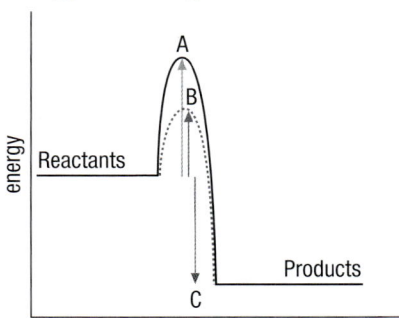

The diagram represents a reaction that can be done with or without a catalyst.

  i How can you decide from this diagram that the reaction is exothermic? (1)

  ii What name is used for the energies represented by arrows A and B? (1)

  iii Why is arrow B shorter than arrow A? (1)

  iv What is the enthalpy change for the reaction? (1)

c The following equation represents a reversible reaction.

$$A + B \rightleftharpoons C + D \qquad \Delta H = +31 \text{ kJ/mole}$$

What is the value of ΔH for the following reaction?

$$C + D \rightleftharpoons A + B \qquad (1)$$

2 Some students investigated the energy released when a liquid fuel burns.

They used the following apparatus.

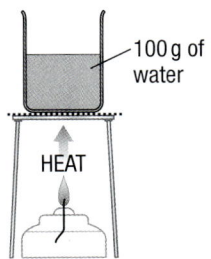

100 g of water

HEAT

They obtained these results.

Student	Mass of fuel burnt in grams	Start temperature of water in °C	End temperature of water in °C
1	0.32	16.5	29.0
2	0.38	17.0	32.0
3	0.35	18.0	32.5

a **i** What temperature rise did student 1 obtain? (1)

    ii Which student obtained the biggest increase in temperature? (1)

b The amount of energy released can be calculated using this expression:

$$\text{energy} = \begin{array}{c}\text{mass of}\\\text{water}\end{array} \times \begin{array}{c}\text{specific heat}\\\text{capacity of}\\\text{water}\end{array} \times \begin{array}{c}\text{temperature}\\\text{rise}\end{array}$$

  i Use this expression to calculate the energy, in joules, released in student 3's experiment. Assume that it takes 4.18 J to raise the temperature of 1 g of water by 1 °C. (3)

  ii Another student used the same apparatus and obtained a value of 6200 J by burning 0.40 g of fuel.

    Calculate the energy released in kJ/gram. (3)

c Why is the calculated value of energy released in these experiments much less than the true value? (1)

3 The energy change in a reaction can be calculated using values of bond energy. The combustion of methane can be represented as follows:

$$\text{H}-\overset{\displaystyle\text{H}}{\underset{\displaystyle\text{H}}{\text{C}}}-\text{H} + 2\,\text{O}{=}\text{O} \longrightarrow \text{O}{=}\text{C}{=}\text{O} + 2\,\text{H}-\text{O}-\text{H}$$

Use bond energy values from the tables in your calculations.

Bond	Bond energy in kJ/mole	Bond	Bond energy in kJ/mole
C–H	412	C=O	743
O=O	496	H–O	463

a Calculate the energy needed to break the bonds in the reactants. (2)

b Calculate the energy given out when the bonds in the products are formed. (2)

c What is the enthalpy change for the combustion of methane? (1)

12.1 Hydrocarbons

Learning objectives

After this topic, you should know:

- what crude oil is made up of
- what alkanes are
- how to represent alkanes.

Figure 1 *The price of nearly everything you buy is affected by oil because the cost of moving goods to the shops affects the price you pay for them*

So far, some of the 21st century's most important chemicals come from crude oil. These chemicals play a major part in our lives. We use them as fuels to run our cars, to warm our homes, and to make electricity.

Fuels are important because they keep us warm and on the move. So when oil prices rise, it affects us all. Countries that produce crude oil can affect the whole world economy by the price they charge for their oil.

Crude oil

Crude oil is a dark, smelly liquid. It is a **mixture** of many different chemical compounds. A mixture contains two or more elements or compounds that are not chemically combined together. Nearly all of the compounds in crude oil are compounds containing only hydrogen and carbon atoms. These compounds are called **hydrocarbons**.

Crude oil straight from the ground is not much use. There are too many substances in it, all with different boiling points. Before we can use crude oil, it must be separated into different substances which contain molecules with a similar number of carbon atoms and have similar boiling points. These are known as **fractions**. Because the properties of substances do not change when they are mixed (as opposed to reacted), engineers can separate mixtures of substances in crude oil by **fractional distillation**. Fractional distillation separates liquids with different boiling points.

Demonstration

Fractional distillation of crude oil

Mixtures of liquids can be separated using fractional distillation. This can be done in the lab on a small scale. Heat the crude oil mixture so that it boils. The different fractions vaporise between different ranges of temperature. Collect the vapours by cooling and condensing them.

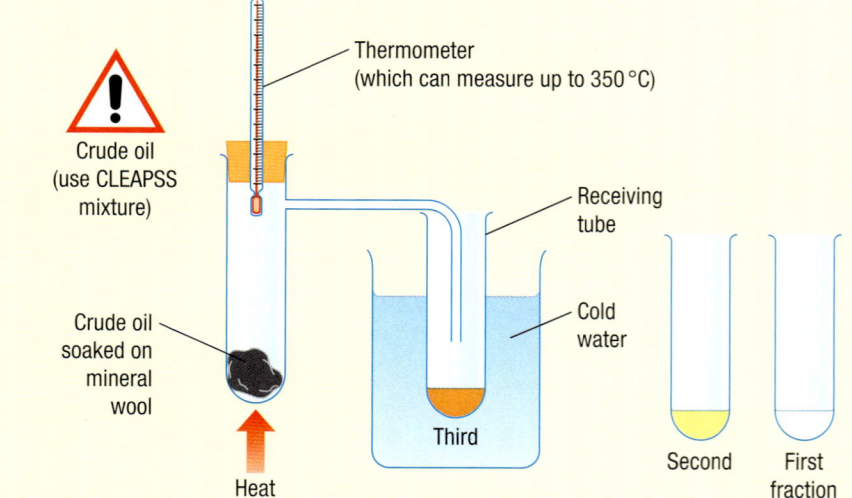

Figure 2 *The fractional distillation of crude oil in the lab*

- What colour are the first few drops of liquid collected?

Alkanes

Alkanes are **saturated hydrocarbons**. All the carbon–carbon bonds are single covalent bonds. This means that they contain as many hydrogen atoms as possible in each molecule. No more hydrogen atoms can be added.

Most of the hydrocarbons in crude oil are **alkanes**. You can see some examples of alkane molecules in Figure 3. Notice how all their names end in '–ane'. The first part of their names tells you how many carbon atoms are in their molecules.

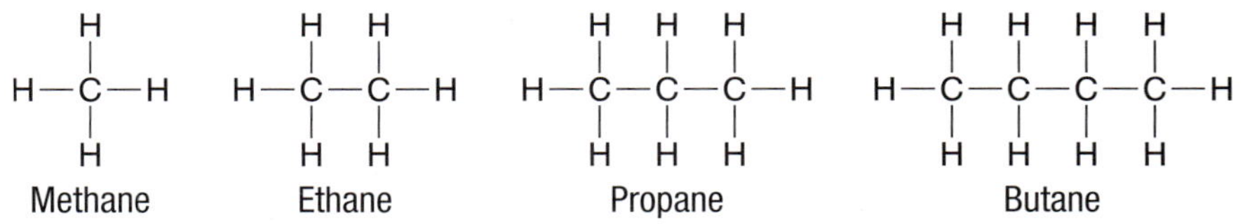

Figure 3 *You can represent alkanes like this, showing all of the atoms in the molecule. They are called 'displayed formulae'. The line drawn between two atoms in a molecule e.g. C – H or C – C, as shown above, represents a single covalent bond.*

Look at the molecular formulae of the first five alkane molecules below:

$$CH_4 \text{ (methane)}$$

$$C_2H_6 \text{ (ethane)}$$

$$C_3H_8 \text{ (propane)}$$

$$C_4H_{10} \text{ (butane)}$$

$$C_5H_{12} \text{ (pentane)}$$

Can you see a pattern in the molecular formulae of the alkanes? You can write the general formula for alkane molecules like this:

$$C_nH_{2n+2}$$

This means that 'for every n carbon atoms there are $(2n + 2)$ hydrogen atoms' in an alkane molecule. For example, if an alkane molecule contains 12 carbon atoms its formula will be $C_{12}H_{26}$.

∞ **links**

For information on covalent bonding, look back at 2.3 'Covalent bonding'.

Summary questions

1 a What is crude oil?
 b Why is oil so important to society?
 c Why can crude oil be separated using distillation?

2 Engineers drill crude oil from beneath the ground or seabed. Why is this crude oil not very useful as a product itself?

3 a Write the general formula of the alkanes.
 b Write the molecular formulae of the alkanes which have six to ten carbon atoms. Then find out their names.

4 a Draw the displayed formula of octane, whose molecules have eight carbon atoms.
 b How many hydrogen atoms are there in an alkane molecule which has 22 carbon atoms?
 c How many carbon atoms are there in an alkane molecule which has 32 hydrogen atoms?
 d Why are alkanes described as 'saturated hydrocarbons'?

Key points

- Crude oil is a mixture of many different compounds.

- Most of the compounds in crude oil are hydrocarbons – they contain only hydrogen and carbon atoms.

- Alkanes are saturated hydrocarbons. They contain as many hydrogen atoms as possible in their molecules.

- The general formula of an alkane is:
$$C_nH_{2n+2}$$

12.2 Fractional distillation of crude oil

Learning objectives

After this topic, you should know:

- how the volatility, viscosity, and flammability of hydrocarbons are affected by the size of their molecules

- how to separate crude oil into fractions

- the properties that make a fraction useful as a fuel.

The properties of hydrocarbons

The properties of hydrocarbons depend on the size of their molecules. Some have molecules that are quite small, with relatively few carbon atoms in short chains. These short-chain molecules make up the hydrocarbons that tend to be most useful. These hydrocarbons make good fuels as they ignite easily and burn well, with less smoky flames than hydrocarbons made up of larger molecules. They are very **flammable** (see Figure 1 below). Other hydrocarbons have lots of carbon atoms in their molecules, and may have branches or side-chains.

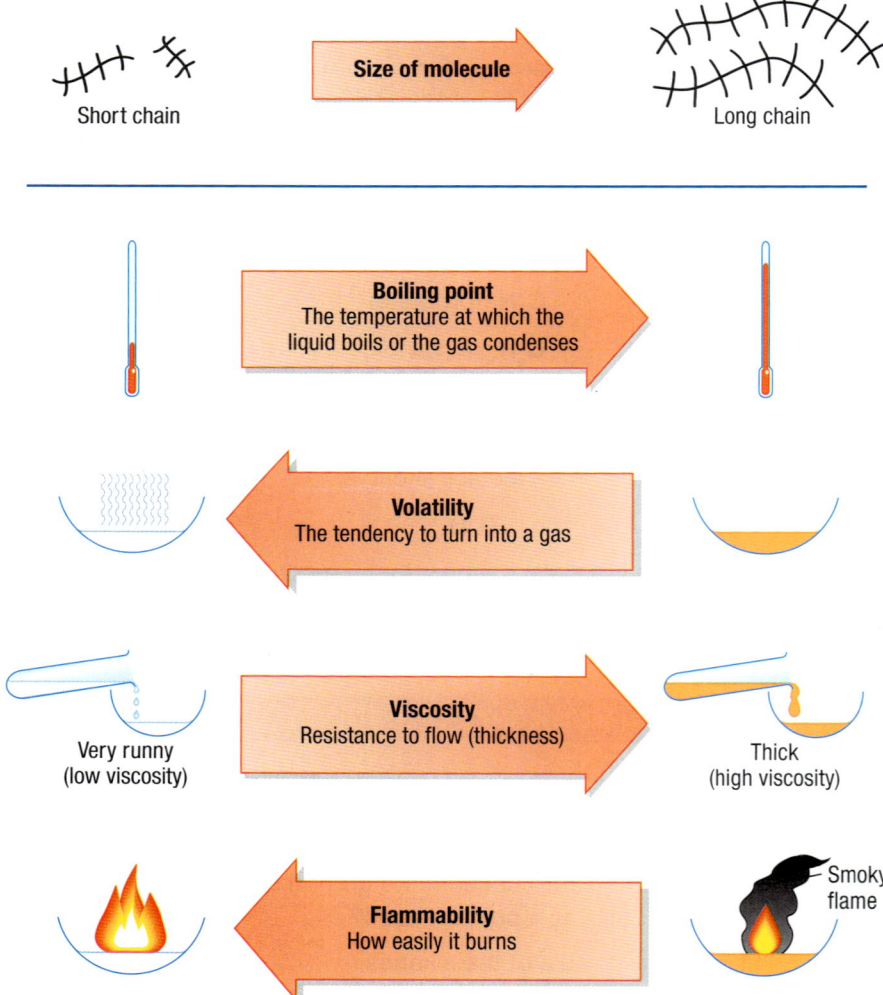

Figure 1 *The properties of hydrocarbons depend on the chain-length of their molecules*

Demonstration

Comparing fractions

Your teacher might compare the **viscosity** and flammability of some fractions (mixtures of hydrocarbons with similar boiling points) that are obtained from crude oil.

The boiling point of a hydrocarbon depends on the size of its molecules. Engineers can use the differences in boiling points in fractional distillation to separate the hydrocarbons in crude oil into fractions.

Fractional distillation of crude oil

Engineers separate out crude oil into hydrocarbons with similar boiling points, called fractions. This process is called **fractional distillation**. Each hydrocarbon fraction contains molecules with similar numbers of carbon atoms. Each of these fractions boils at a different temperature range because of the different sizes of the molecules in it.

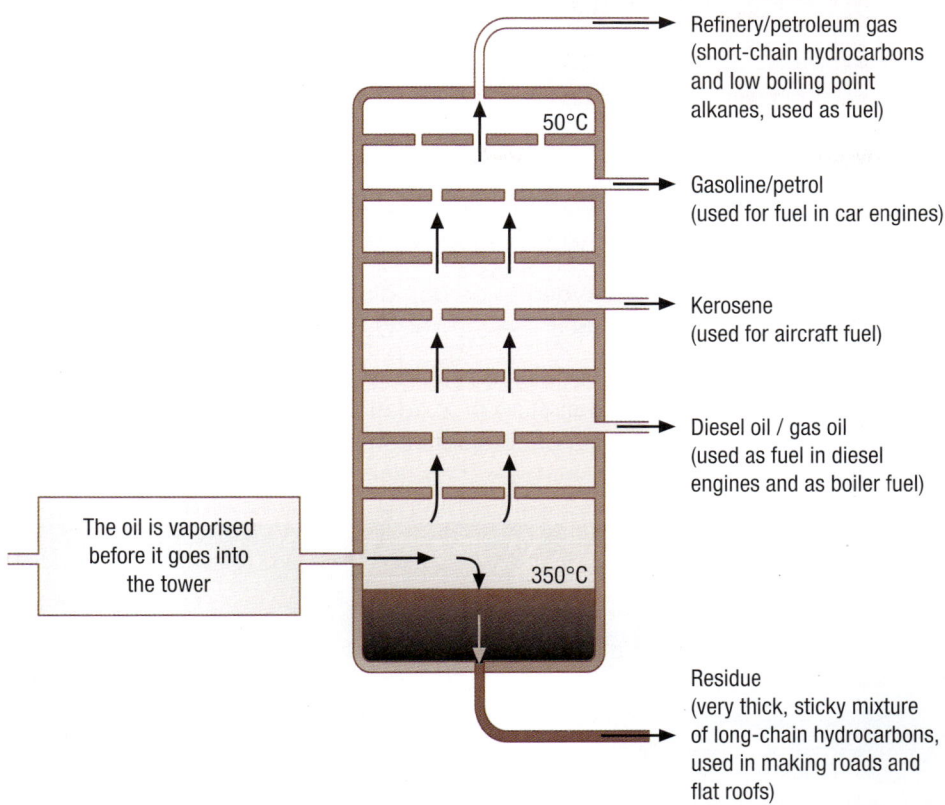

Refinery/petroleum gas (short-chain hydrocarbons and low boiling point alkanes, used as fuel)

50°C

Gasoline/petrol (used for fuel in car engines)

Kerosene (used for aircraft fuel)

Diesel oil / gas oil (used as fuel in diesel engines and as boiler fuel)

The oil is vaporised before it goes into the tower

350°C

Residue (very thick, sticky mixture of long-chain hydrocarbons, used in making roads and flat roofs)

Figure 2 *Scientists and engineers use fractional distillation to separate the mixture of hydrocarbons in crude oil into fractions. Each fraction contains hydrocarbons with similar boiling points.*

Crude oil is heated and fed in near the bottom of a tall tower (a fractionating column) as hot vapour. The column is kept very hot at the bottom and much cooler at the top. The temperature decreases going up the column. The gases condense when they reach the temperature of their boiling points. The different fractions are collected as liquids at different levels. The fractions are collected from the column in a continuous process.

Hydrocarbons with the smallest molecules have the lowest boiling points. They are collected at the cool top of the column. At the bottom of the column, the fractions have high boiling points. They cool to form very thick liquids or solids at room temperature.

Once collected, the fractions need more processing before they can be used.

?? Did you know … ?

2014 figures show that Venezuela contains 17.5% of the world's crude oil reserves, with Saudi Arabia containing 15.7% of total proved resources.

There are many different types of crude oil. For example, crude oil from Venezuela contains many long-chain hydrocarbons. It is very dark and thick and is called 'heavy' crude. Other countries, such as Nigeria and Saudi Arabia, produce crude oil which is much paler in colour and runnier. This is 'light' crude.

Figure 3 *An oil refinery at night*

Summary questions

1 a How does the size of a hydrocarbon molecule affect:
   i   the boiling point?
   ii the volatility?
   iii the viscosity (thickness) of a hydrocarbon?
  b A hydrocarbon catches fire very easily. Is it likely to have molecules with long hydrocarbon chains or short ones? Will it give off lots of black smoke when it burns?

2 Make a table to summarise how the properties of hydrocarbons depend on the size of their molecules.

3 Explain the steps involved in the fractional distillation of crude oil.

Key points

■ Crude oil is separated into fractions using fractional distillation.

■ The properties of each fraction depend on the size of the hydrocarbon molecules in it.

■ Lighter fractions make better fuels as they ignite more easily and burn well, with cleaner (less smoky) flames.

12.3 Burning fuels

Learning objectives

After this topic, you should know:

- the products of combustion when fuels are burnt in a good supply of air, or in a limited supply of air
- the pollutants produced when fuels are burnt.

The lighter fractions from crude oil are very useful as fuels. When hydrocarbons burn in plenty of air they release lots of energy.

For example, when propane burns:

$$\text{propane} + \text{oxygen} \rightarrow \text{carbon dioxide} + \text{water}$$
$$C_3H_8 + 5O_2 \rightarrow 3CO_2 + 4H_2O$$

The carbon and hydrogen in the fuel are **oxidised** completely when they burn like this. Remember that one definition of 'oxidation' means adding oxygen in a chemical reaction. The products of complete combustion are carbon dioxide and water.

Figure 1 *On a cold day you can often see the water produced when fossil fuels burn*

Practical

Products of combustion

You can test the products given off when a hydrocarbon burns as shown in Figure 2.

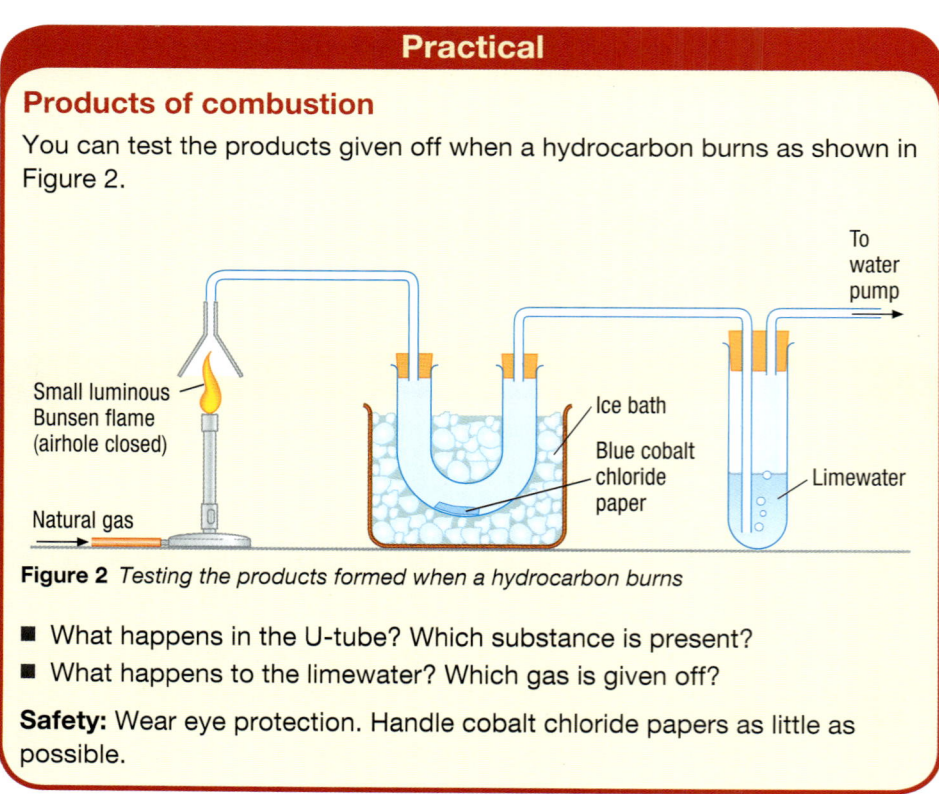

Small luminous Bunsen flame (airhole closed)

Natural gas

Ice bath

Blue cobalt chloride paper

Limewater

To water pump

Figure 2 *Testing the products formed when a hydrocarbon burns*

- What happens in the U-tube? Which substance is present?
- What happens to the limewater? Which gas is given off?

Safety: Wear eye protection. Handle cobalt chloride papers as little as possible.

Pollution from fuels

All fossil fuels – oil, coal, and natural gas – produce carbon dioxide and water (vapour) when they burn in plenty of air. But as well as hydrocarbons, these fuels also contain other substances. Impurities containing sulfur found in fuels cause us major problems.

All fossil fuels contain at least some sulfur. This reacts with oxygen when the fuel is burnt. It forms a gas called **sulfur dioxide** (SO_2). This acidic gas is toxic. It is bad for the environment, as it is a cause of acid rain which damages trees, as well as killing animal life in lakes. The sulfur impurities can be removed from a fuel *before* the fuel is burnt. This happens in petrol and diesel for cars. In power stations, sulfur dioxide is also removed from the waste or 'flue' gases by reacting it with basic calcium oxide or calcium hydroxide.

When fuel is burnt in a car engine, even more pollution can be produced.

- When any fuel containing carbon is burnt, it makes carbon dioxide. Carbon dioxide is the main greenhouse gas in the air. It absorbs energy released as radiation from the surface of the Earth. Most scientists think that this is causing **climate change**, which affects temperatures around the world and might result in more severe weather events and the flooding of low-lying land as sea levels rise.
- When there is not enough oxygen inside an engine, it results in **incomplete** (or partial) **combustion**. Instead of all the carbon in the fuel turning into carbon dioxide, **carbon monoxide** gas (CO) is also formed. Carbon monoxide is a toxic gas. It is colourless and odourless. Your red blood cells pick up this gas and carry it around in your blood instead of oxygen.
- The high temperature inside an engine also allows the nitrogen and oxygen in the air to react together. This reaction makes **nitrogen oxides**. These are toxic and can trigger some people's asthma. They also cause acid rain.
- Diesel engines burn hydrocarbons with bigger molecules than petrol engines. When these large molecules react with oxygen in an engine they do not always burn completely. Tiny solid particles containing carbon (soot) and unburnt hydrocarbons are produced. These **particulates** get carried into the air. They travel into the upper atmosphere, reflecting sunlight back into space, causing **global dimming**. Scientists also think that they may damage the cells in our lungs and even cause cancer.

Figure 3 *A combination of many cars in a small area and the right weather conditions can cause smog to be formed. This is a mixture of SMoke and fOG.*

Key points

- When hydrocarbon fuels are burnt in plenty of air the carbon and hydrogen in the fuel are completely oxidised. They produce carbon dioxide and water.

- Sulfur impurities in fuels burn to form sulfur dioxide, which can cause acid rain. Sulfur can be removed from fuels before burning them or sulfur dioxide can be removed from flue gases.

- Changing the conditions in which hydrocarbon fuels are burnt can change the products made.

- In insufficient oxygen, poisonous carbon monoxide gas is formed. We can also get particulates of carbon (soot) and unburnt hydrocarbons, especially if the fuel is diesel. These can cause global dimming.

- At the high temperatures in engines, nitrogen from the air reacts with oxygen to form oxides of nitrogen. These cause breathing problems and can also cause acid rain.

Summary questions

1. a What are the products of the complete combustion of a hydrocarbon?
   b When fossil fuels burn, which element present in impurities can produce sulfur dioxide?

2. a Which pollution problem does sulfur dioxide gas contribute to?
   b Which other non-metal oxides released from cars also cause this pollution problem?

3. Explain how a sulfur dioxide, b nitrogen oxides, and c particulates, are produced when fuels burn in vehicles.

4. a Natural gas is mainly methane (CH_4). Write a balanced symbol equation, including state symbols, for the complete combustion of methane.
   b When natural gas burns in a faulty gas heater it can produce carbon monoxide (and water). Write a balanced symbol equation, including state symbols, to show this reaction.
   c Why is carbon monoxide dangerous?

12.4

Alternative fuels

Learning objectives

After this topic, you should know:

- what biofuels are
- the advantages and disadvantages of using biofuels.

Biofuels

Biofuels are fuels that are made from plant or animal products (renewable resources). Examples are **biodiesel**, which is made from oils extracted from plants, and biogas, which is generated from animal waste. Biofuels will become more and more important as our supplies of crude oil diminish.

Figure 1 *This coach runs on biodiesel*

Advantages of biodiesel and other biofuels

There are advantages in using biodiesel and other biofuels as a fuel.

- Biodiesel is much less harmful to animals and plants if it is spilled than the diesel obtained from crude oil.
- When biodiesel is burnt in an engine it burns much more cleanly, reducing the particulates emitted. It also makes very little sulfur dioxide.
- As **non-renewable** fossil fuels, such as crude oil, run out, their price will increase and biofuels will become cheaper to use.
- Another really big advantage over fossil fuels is the fact that the crops used to make biofuels absorb carbon dioxide gas as they grow. So biofuels from plants are in theory 'CO_2 neutral'. That means the amount of carbon dioxide given off when the fuel burns is balanced by the amount absorbed as the plants it was made from grew. Therefore, biofuels make little contribution to the carbon dioxide in our atmosphere. We say that they have a 'reduced carbon footprint' as fuels compared with petrol and diesel.

However, we cannot claim that all biofuels make a *zero* contribution to carbon dioxide emissions. We should really take into account the CO_2 released when:

- fertilising and harvesting the crops
- extracting and processing the biofuel
- transporting the plant material and biofuel.

Disadvantages of biofuels

There are also disadvantages in using biofuels.

- The use of large areas of farmland to produce biofuel instead of food could pose problems. If people start to rely on oil-producing crops for their fuel, farmers will turn land once used for food crops to land for growing biofuel crops.
- This could result in famine in poorer countries if the price of staple food crops rises as demand overtakes supply. Forests, which absorb lots of carbon dioxide, might also be cleared to grow the biofuel crops if they get more popular.
- People are also worried about endangered species because of the destruction of their habitats, such as tropical forests, cleared to grow biofuel crops.

Plants absorb CO_2 as they grow

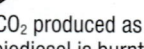

Converted to biodiesel

CO_2

CO_2 produced as biodiesel is burnt

Figure 2 *Cars that run on biodiesel produce very little CO_2 overall, as CO_2 is absorbed by the plants used to make the biofuel*

Using ethanol as a biofuel

Another biofuel is ethanol. It can be made by fermenting sugar solution, extracted from sugar beet or sugar cane, using yeast (whose enzymes provide a natural catalyst). The **fermentation** is a slow batch process that takes place in large vats, at temperatures between 20 °C and 35 °C. It produces yields of up to 15% ethanol. The ethanol is then distilled from the aqueous solution for use as a biofuel. This fermentation process is:

$$\text{glucose} \rightarrow \text{ethanol} + \text{carbon dioxide}$$
$$C_6H_{12}O_6(aq) \rightarrow 2C_2H_5OH(aq) + 2CO_2(g)$$

Ethanol made by fermentation is known as 'bioethanol'.

As with other biofuels, the ethanol gives off carbon dioxide (a greenhouse gas) when it burns:

$$C_2H_5OH(l) + 3O_2(g) \rightarrow 2CO_2(g) + 3H_2O(g)$$

However, this should be balanced against the sugar cane absorbing CO_2 gas during photosynthesis.

Ethanol can also be made in industry from ethene, made when large hydrocarbons found in crude oil are cracked at high temperatures. (See page 173.)

Look at the reaction below used to convert ethene into ethanol:

phosphoric
acid catalyst

$$C_2H_4(g) + H_2O(g) \rightleftharpoons C_2H_5OH(g)$$

ethene   steam   high pressure   ethanol

Making ethanol from ethene relies on crude oil supplies. As you know, crude oil is a fossil fuel that is running out (non-renewable). The process can be run continuously. Although the reaction is reversible, unreacted ethene is recycled back into the reaction vessel.

In countries like Brazil, that have no natural oil supplies of their own, ethanol is manufactured on a large scale by fermentation. They have a climate that is ideal for growing sugar cane (renewable). So although the fermentation process is a slow reaction, it is the cheapest way for them to make ethanol to be used as a fuel.

The ethanol has to be made in batches and left to ferment, but can only produce yields of up to 15% ethanol in solution. However, the energy requirements cost less as the reaction mixture ferments at just warm temperatures (between 20 °C and 35 °C), and atmospheric pressure. Ethanol is distilled off, then used to run some Brazilian cars. This helps preserve fossil fuel supplies, as petrol and ethene come from crude oil. It also reduces carbon dioxide released into the atmosphere. On the other hand, lots of land that could be used for food crops are set aside for biofuel crops.

Figure 3 *Ethanol can be made from sugar cane. The growing and harvesting of the sugar cane crop is labour-intensive work.*

⊙⊙ links

For more information on ethanol, see 14.2 'Properties and uses of alcohols'.

Summary questions

1 Why is burning ethanol a better choice of fuel than petrol if you want to reduce carbon dioxide emissions?

2 Explain where the energy in biodiesel comes from.

3 a Explain why ethanol can be described as a renewable fuel or a non-renewable fuel.

 b Explain why shifting from fossil fuels to biofuels could possibly cause food shortages.

Key points

- Biofuels are a renewable source of energy that could be used to replace some fossil fuels.

- Biodiesel can be made from vegetable oils.

- There are advantages, and some disadvantages, in using biofuels.

- Ethanol is also a biofuel as it can be made from the sugar in plants.

Chapter summary questions

1 a i Copy and complete this general formula for the alkanes:

$$C_nH$$

ii What is the formula of the alkane with 18 carbon atoms?

iii What is the formula of the alkane with 18 hydrogen atoms?

b Look at the boiling points in this table:

Alkane	Number of carbon atoms	Boiling point in °C
methane	1	−161
ethane	2	−88
propane	3	−42
butane	4	−0.5
pentane	5	
hexane	6	69

Draw a graph of the alkanes' boiling points (vertical axis) against the number of carbon atoms (horizontal axis).

c What is the general pattern you see from your graph?

d Use your graph to predict the boiling point of pentane.

2 a The alkanes are all 'saturated hydrocarbons'.
i What is a hydrocarbon?
ii What does 'saturated' mean when describing an alkane?

b i Give the name and formula of this alkane:

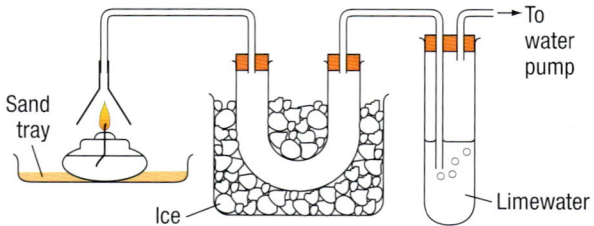

ii What do the letters represent in this displayed formula?
iii What do the lines between the letters represent?

3 One alkane, A, has a boiling point of 344 °C and another, B, has a boiling point of 126 °C.

a Which one will be collected nearer the top of a fractionating column in an oil refinery? Explain your choice.

b Which one will be the better fuel? Explain your choice.

c State the differences you would expect between A and B in terms of their:
i viscosity
ii volatility.

4 a What problem is associated with the increased levels of carbon dioxide gas in the atmosphere?

b i What gas is given off from fossil fuel power stations that can cause acid rain?
ii Give **two** ways of stopping this acidic gas getting into the atmosphere.
iii Name the other cause of acid rain which comes from car engines and how it arises.

5 Two students are testing the products formed when ethanol burns. They set up the experiment below:

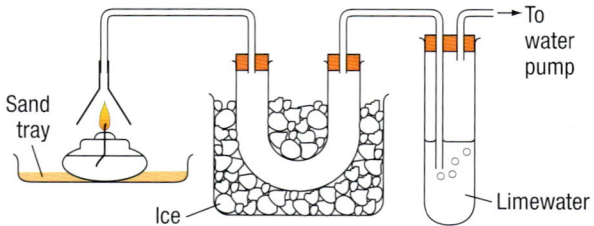

a Why do they put ice around the U-tube?

b How can they test for the substance formed in the U-tube?

c Explain what happens to the limewater.

d There is a small amount of carbon dioxide in the air. How can they show that the carbon dioxide they test for is not just the result of the carbon dioxide in the air?

e Write a word equation and balanced symbol equation for the complete combustion of ethanol.

f The students wanted to test the products of combustion of a hydrocarbon. Wax is made of hydrocarbons. How would the students change their experiment to carry out this test?

g Write a word equation and balanced symbol equation for the complete combustion of propane gas.

h Propane is used in portable gas heaters but there must be good ventilation in the room. Which toxic gas will be produced when the propane burns if there is a poor supply of air in the room?

6 a Which one of these fuels could be termed a 'biofuel'? Explain your choice.
Hydrogen   Butane   Ethanol   Diesel   Petrol

b Explain what is meant by the statement 'Biodiesel is potentially CO_2 neutral'.

c Why is the use of hydrogen as a fuel one way to tackle the problem of climate change?

d State **two** problems with the use of hydrogen as a fuel.

e Write a word equation and a balanced symbol equation to show the combustion of hydrogen.

f What piece of equipment can turn energy from the reaction in part **e** into electrical energy?

Practice questions

1 Crude oil is a mixture of many compounds, most of which are saturated hydrocarbons. Most of these hydrocarbons are alkanes.

a i What is a hydrocarbon? (2)

ii What is the meaning of the term 'saturated'? (1)

b The formulae of the first two alkanes are CH_4 and C_2H_6

i What is the general formula of the alkanes? (1)

ii What is the molecular formula of the alkane with three carbon atoms in each molecule? (1)

iii Draw the displayed formula of the alkane with four carbon atoms in each molecule. (1)

c The compounds in crude oil are separated into fractions by fractional distillation. This process involves converting most of the crude oil into vapour, and then cooling it. The fraction that reaches the top of the fractionating column contains the smallest molecules.

i Name the changes of state involved in this process of fractional distillation. (2)

ii State **two** properties of the liquid fraction that leaves the column near the lowest part. (2)

2 Many of the smaller molecules obtained from crude oil are used as fuels.

a Name the **two** products of complete combustion of a hydrocarbon fuel. (2)

b Write the symbol equation for the complete combustion of methane. (2)

c If there is not enough oxygen, then a hydrocarbon may undergo incomplete combustion.

Name one solid product and one gaseous product that could form by incomplete combustion. (2)

d The combustion of hydrocarbon fuels can produce small amounts of oxides of nitrogen.

i What condition is needed for oxides of nitrogen to form? (1)

ii What is the main environmental problem caused by oxides of nitrogen? (1)

e Hydrocarbon fuels may contain small amounts of sulfur, which can release sulfur dioxide into the atmosphere during combustion.

i What is done to prevent the emission of sulfur dioxide from a car exhaust? (1)

ii What is done to prevent the emission of sulfur dioxide from a power station? (1)

3 Alternative fuels are increasingly used to replace fuels obtained from crude oil. Biofuels are one type of alternative fuel.

a One common biofuel that has been used for many centuries is wood.

Why is wood described as a biofuel? (1)

b Ethanol is another biofuel.

State the main source of ethanol and name the method used to convert it into ethanol. (2)

c i State **one** reason why biofuels are being increasingly used. (1)

ii State the main disadvantages of using large areas of land for biofuel production. (1)

d Hydrogen is another alternative fuel.

i State why the combustion of hydrogen is less likely to cause climate change than burning a hydrocarbon fuel. (1)

ii State **two** problems in using hydrogen as a fuel in cars. (2)

13.1

Cracking hydrocarbons

Learning objectives

After this topic, you should know:

■ how smaller, more useful hydrocarbon molecules are made from larger, less useful ones in crude oil

■ what alkenes are and how they differ from alkanes.

Figure 1 *In an oil refinery, huge crackers like this are used to break down large hydrocarbon molecules into smaller ones*

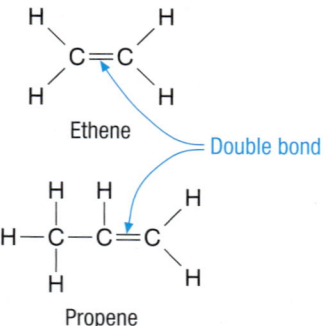

Figure 3 *The displayed formula of a molecule of ethene (C_2H_4) and a molecule of propene (C_3H_6). These are both alkenes – each molecule has a carbon–carbon (C=C) double bond in it.*

Some of the heavier fractions that are produced in the fractional distillation of crude oil are not very useful. The hydrocarbons in them are made up of large molecules. They are thick liquids or solids with high boiling points. They are difficult to vaporise and do not burn easily – so they are poor fuels. Yet the main demand from crude oil is for fuels. Fortunately these large hydrocarbon molecules can be broken down in a process called **cracking**.

The process takes place at an oil refinery in a steel vessel called a cracker. In the cracker, a heavy fraction distilled from crude oil is heated to vaporise the hydrocarbons. The vapour is then either passed over a hot catalyst or mixed with steam and heated to a very high temperature. The hydrocarbons are cracked as thermal decomposition reactions take place. The large molecules split apart to form smaller, more useful ones.

An example of cracking

Decane, $C_{10}H_{22}$, is a medium-sized alkane molecule. When heated to 500 °C with a catalyst, it breaks down. One of the molecules produced is pentane, which is used in petrol.

Figure 2 *Pentane (C_5H_{12}) can be used as a fuel. This is the displayed formula of pentane.*

We also get propene and ethene, which can be used to produce other chemicals.

$$C_{10}H_{22} \xrightarrow{\text{500 °C + catalyst}} C_5H_{12} + C_3H_6 + C_2H_4$$

decane   pentane   propene   ethene

This reaction is an example of thermal decomposition.

Notice how this cracking reaction produces different types of molecules. One of the molecules is pentane. The first part of its name tells us that it has five carbon atoms (*pent-*). The last part of its name (*-ane*) shows that it is an alkane. Like all other alkanes, pentane is a saturated hydrocarbon. Its molecules have as much hydrogen as possible in them.

The other molecules in this reaction have names that end slightly differently. They end in *-ene*. This type of molecule is called an **alkene**. The different ending tells us that these molecules are **unsaturated**. Unsaturated compounds contain at least one **double bond** between their carbon atoms. Look at Figure 3. You can see that these alkenes have one C=C double covalent bond and have the general formula C_nH_{2n}.

A simple experiment like the one on the next page shows that alkenes burn (but not as well as equivalent small alkanes which are used as fuels). Alkenes also react with bromine water, which is orange in colour. The products of this reaction are colourless so this gives you a good test to see if a compound contains a C=C double bond:

Positive test for an unsaturated hydrocarbon: orange bromine water turns colourless.

Contains required practical

Cracking

Medicinal paraffin is a mixture of hydrocarbon molecules. You can crack it by heating it and passing the vapour over hot pieces of broken pot. The broken pot acts as a catalyst.

- Why must you remove the end of the delivery tube from the water before you stop heating?

If you carry out this practical, collect at least two test tubes of gas.

Required practical: Test the gas (or another unsaturated hydrocarbon) by shaking it with a few drops of bromine water (harmful).

- Record your observations and conclusions.

Safety: Avoid cold water being sucked back into the hot tube by removing the delivery tube from the water trough before you stop heating. Wear eye protection. Some gases collected may be harmful or even toxic.

Diagram labels: Delivery tube; Gaseous product; Ceramic wool soaked in medicinal paraffin; Heat; Broken pot (catalyst); Safety valve; Water

Making ethanol from ethene (hydration)

Ethanol for industrial use as a fuel or solvent can be made from ethene gas instead of by fermentation. Ethene is the main by-product made in cracking. Ethene gas can react with steam to make ethanol.

$$\text{ethene} + \text{steam} \xrightarrow{\text{catalyst}} \text{ethanol}$$
$$C_2H_4 + H_2O \rightleftharpoons C_2H_5OH$$

This reaction is called **hydration**. The reaction requires energy to heat the gases and to generate a high pressure. The reaction is reversible so ethanol can break down back into ethene and steam. So unreacted ethene and steam are recycled over the concentrated phosphoric acid catalyst.

∞ links

For more information on fermentation to make ethanol, look back to 12.4 'Alternative fuels'.

Summary questions

1. **a** Why is cracking so important?
   b How can large hydrocarbon molecules be cracked in an oil refinery?

2. Cracking a hydrocarbon makes two new hydrocarbons, A and B. When bromine water is added to A, nothing happens. Bromine water added to B turns from an orange solution to colourless.
   a i Which hydrocarbon, A or B, is unsaturated?
   ii What is an 'unsaturated' hydrocarbon?
   iii What do you call the type of unsaturated hydrocarbon formed in cracking?
   b i Which hydrocarbon, A or B, is used as a fuel?
   ii What do you call the type of saturated hydrocarbon formed in cracking?
   c What type of reaction is cracking an example of?

3. An alkene molecule with one double bond contains 12 carbon atoms. How many hydrogen atoms does it have? Write down its formula.

4. Dodecane (an alkane with 12 carbon atoms) can be cracked into octane (with 8 carbon atoms) and ethene. Write a balanced symbol equation for this reaction.

Key points

- Large hydrocarbon molecules can be split up into smaller molecules by passing the vapours over a hot catalyst or by mixing them with steam and heating them to a very high temperature.

- Cracking produces saturated hydrocarbons which are used as fuels and unsaturated hydrocarbons (called alkenes).

- Alkenes (and other unsaturated compounds containing carbon–carbon double bonds) react with yellow-orange bromine water, turning it colourless.

13.2 Making polymers from alkenes

Learning objectives

After this topic, you should know:
- how monomers make polymers
- how alkenes undergo polymerisation reactions
- how to represent polymerisation reactions by an equation.

Figure 1 *All of these products were manufactured using chemicals originating from crude oil*

The fractional distillation of crude oil and cracking produces a large range of hydrocarbons. These are very important to our way of life. Products based on crude oil are all around us. It is difficult to imagine life without them.

Hydrocarbons are our main fuels. We use them in our transport and at home to cook and for heating. We also use them to generate electricity in oil-fired and gas-fired power stations.

Then there are the chemicals made from crude oil. They are used to make things ranging from cosmetics to explosives. But one of the most important ways that chemicals from crude oil are used is to make plastics.

Plastics

Plastics are made up of huge molecules made from lots of small molecules joined together. We call the small molecules **monomers**. We call the huge molecules they make **polymers**. (*Mono* means 'one' and *poly* means 'many'.) Engineers can make different types of plastics, which can have very different properties, by using different monomers.

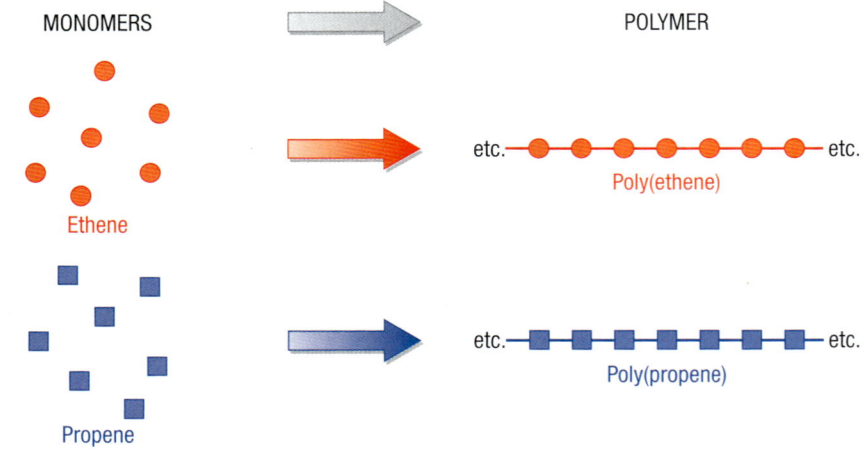

Figure 2 *Polymers are made from many smaller molecules called monomers*

Figure 3 *Polymers produced from compounds derived from crude oil are all around us and are part of our everyday lives*

Ethene (C_2H_4) is the smallest unsaturated hydrocarbon molecule. It can be turned into a polymer known as poly(ethene) or polythene. Poly(ethene) is a really useful plastic as it is easy to shape, strong, and transparent (unless manufacturers add colouring material to it). 'Plastic' carrier bags, some drinks bottles, dustbins, washing-up bowls, and clingfilm are all examples of uses of poly(ethene).

Propene (C_3H_6) is another alkene. Engineers can also make polymers with propene as the monomer. The polymer formed is called poly(propene). It forms a very strong, tough plastic. Manufacturers can use it to make many things, including carpets, milk crates, and ropes.

How do monomers join together?

When alkene molecules join together, the double bond between the carbon atoms in each molecule 'opens up'. It is replaced by single bonds as thousands of molecules join together. The reaction is called **polymerisation**.

Ethene monomers → Poly(ethene)

You can also write this more simply as:

where n is a large number

Many single ethene monomers → Long chain of poly(ethene)

Activity

Modelling polymerisation

Use a molecular model kit to show how ethene molecules polymerise to form poly(ethene).

Make sure you can see how the equation shown above represents the polymerisation reaction you have modelled.

■ Describe what happens to the bonds in the reaction.

■ Think up a model to demonstrate the polymerisation of ethene, using people in your class as monomers.

Then evaluate the ideas of other groups.

Study tip

The C=C double bond in ethene (an alkene) makes it much more reactive than ethane (an alkane).

Summary questions

1 a Define a monomer and a polymer.
  b What is the name for the reaction between a large number of monomers?
  c Write an equation to represent the formation of poly(ethene) from ethene.
  d State **two** uses of poly(ethene).

2 Why is ethene the smallest possible alkene molecule whereas methane is the smallest alkane?

3 a Draw the displayed formula of a propene molecule, showing all its bonds.
  b Draw a diagram to show how propene molecules join together to form poly(propene).
  c State **two** uses of poly(propene).
  d Explain the polymerisation reaction in part **b**.

Key points

■ Plastics are made of polymers.

■ Polymers are large molecules made when many monomers (small, reactive molecules) join together.

■ The reaction between monomers to form a polymer is called polymerisation.

13.3

The properties of polymers

Learning objectives

After this topic, you should know:

■ that the properties of polymers depend on their monomers

■ that changing reaction conditions can modify the polymers made

■ the differences between thermosetting and thermosoftening polymers.

As you know, **polymers** are made from chemicals derived from crude oil. Small molecules called monomers join together to make much bigger molecules called polymers. As the monomers join together they produce a tangled web of very long chain molecules. Poly(ethene) is an example.

The properties of a polymer depend on:

■ the monomers used to make it, and

■ the conditions chosen to carry out the reaction.

Different monomers

Polymer chains can be made from many different monomers. The monomers chosen make a big difference to the properties of the polymer made. Consider the properties of the polymers in the bag in Figure 1 and the electrical socket in Figure 4.

Different reaction conditions

There are two types of poly(ethene). One is called high density (HD) and the other low density (LD) poly(ethene). Both are made from ethene monomers but they are formed under different reaction conditions.

■ Using very high pressures and a trace of oxygen, ethene forms LD poly(ethene). The polymer chains are randomly branched and they cannot pack closely together so LD poly(ethene) has a lower density.

■ Using a catalyst at 50 °C and a slightly raised pressure, ethene makes HD poly(ethene). This is made up of straighter poly(ethene) chains. They can pack more closely together than branched chains so HD poly(ethene) has a higher density. The HD poly(ethene) also has a higher softening temperature and is stronger than LD poly(ethene).

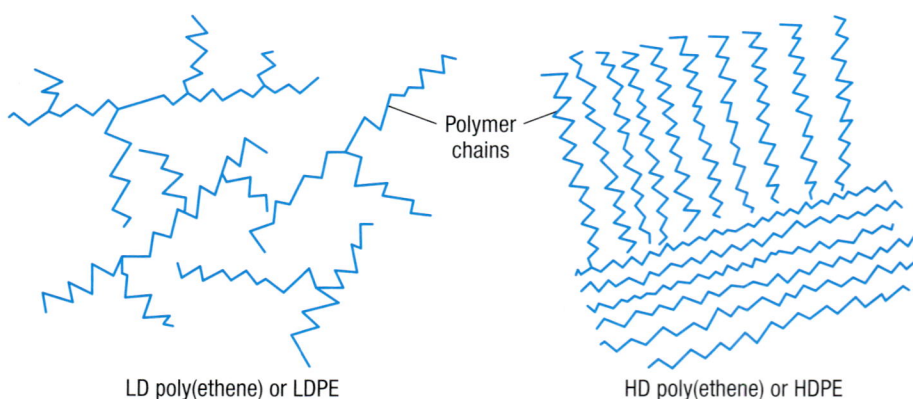

LD poly(ethene) or LDPE   HD poly(ethene) or HDPE

Figure 2 *The branched chains of LD poly(ethene) cannot pack as tightly together as the straighter chains in HD poly(ethene), giving the polymers different properties*

Thermosoftening and thermosetting polymers

It's useful to classify polymers by looking at what happens to them when they are heated. Some will soften quite easily. They will reset when they cool down. These are called **thermosoftening polymers**. They are made up of individual polymer chains that are tangled together.

Figure 1 *The forces between the molecules in poly(ethene) are relatively weak as there are no strong covalent bonds (cross links) between the molecules. This means that this plastic softens fairly easily when heated.*

Other polymers do not melt when heated. These are called **thermosetting polymers**. These have strong covalent bonds forming 'cross links' between their polymer chains (see Figure 3.)

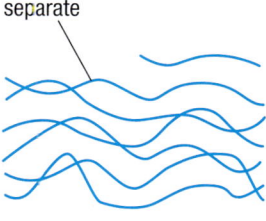

The tangled web of polymer chains are relatively easy to separate

Thermosoftening polymer

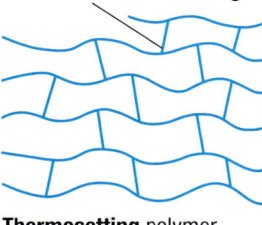

Chains fixed together by strong covalent bonds – this is called **cross linking**

Thermosetting polymer

Figure 3 *Extensive cross linking by covalent bonds between polymer chains makes a thermosetting plastic that is heat-resistant and rigid*

Bonding in and between polymer chains

The atoms in polymer chains are held together by very strong covalent bonds. This is true for all plastics. But the size of the forces *between* polymer molecules in different plastics can be very different.

In thermosoftening polymers the forces between the polymer chains are weak. When the polymer is heated, these weak intermolecular forces are broken. The polymer becomes soft. When the polymer cools down, the intermolecular forces bring the polymer molecules back together. Then the polymer hardens again. This type of polymer can be remoulded.

However, thermosetting polymers are different. Their monomers make covalent bonds between the polymer chains when they are first heated in order to shape them. These covalent bonds are strong, and they stop the polymer from softening. The covalent 'cross links' between chains do not allow them to separate. Even if heated strongly, the polymer will still not soften. Eventually, the polymer will char at high enough temperatures.

Summary questions

1. **a** How are the polymer chains arranged in a thermosoftening polymer?
   b i What is the difference between the structures of a thermosoftening and a thermosetting polymer?
   ii Give two differences in the properties of the two types of polymer.

2. Why do manufacturers use thermosetting rather than thermosoftening polymers to make the handles of pans?

3. Polymer A starts to soften at 100 °C whilst polymer B softens at 50 °C. Polymer C resists heat but eventually starts to char if heated to very high temperatures.
   Explain this using ideas about intermolecular forces.

4. There are two types of poly(ethene), high density (HD) and low density (LD) poly(ethene).
   a What is varied in order to produce the different types of poly(ethene)?
   b Which statement is true?
   HDPE is a thermosetting polymer and LDPE is thermosoftening polymer.
   Both HDPE and LDPE are thermosoftening polymers.
   c Explain the differing densities of HDPE and LDPE in terms of their structures.

Practical

Modifying a polymer

Take some PVA glue . . .

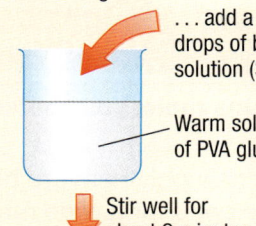

. . . add a few drops of borax solution (3.2%)

Warm solution of PVA glue

Stir well for about 2 minutes

Slime

The glue becomes slimy because the borax makes the long polymer chains in the glue link together to form a jelly-like substance.

- How could you investigate if the properties of slime depend on how much borax you add?

Safety: Wear gloves and eye protection. Borax solution should not exceed 8%.

Figure 4 *Electrical sockets are made out of thermosetting polymers. If the plug or wires get hot, the socket will not soften.*

Key points

- Monomers affect the properties of the polymers that they produce.

- Changing reaction conditions can also change the properties of the polymer that is produced.

- Thermosoftening polymers will soften or melt easily when heated because their intermolecular forces are relatively weak. Thermosetting polymers will not soften because of their 'cross linking' but will eventually char if heated very strongly.

13.4 New and useful polymers

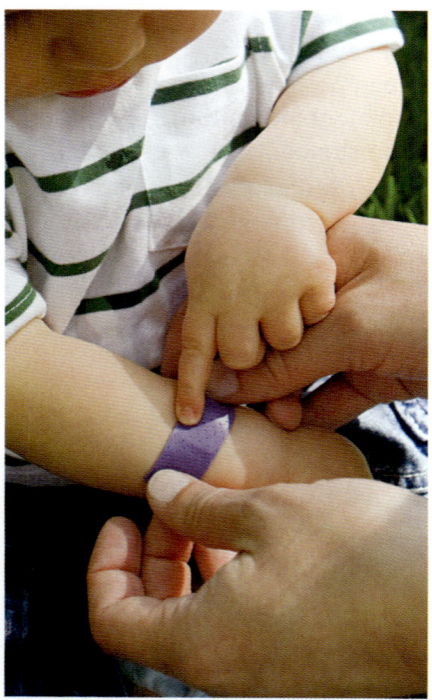

Figure 1 *A sticking plaster is often needed when you cut yourself, but taking the plaster off can be painful too*

Chemists can design new polymers to make materials with special properties to do particular jobs. Medicine is one area where we are beginning to see big benefits from these 'polymers made to order'.

New polymer materials will eventually take over from the current metal fillings for teeth, which contain mercury. Working with the toxic mercury every day is a potential hazard to dental workers.

Other developments include:

■ new softer linings for dentures (false teeth)

■ new packaging materials

■ implants that can slowly release drugs into a patient.

Light-sensitive plasters

You probably know how uncomfortable pulling a plaster off your skin can be. But for some of us taking off a plaster is really painful. Both very old and very young people have quite fragile skin. But now a group of chemists has made a plaster where the 'stickiness' can be switched off before the plaster is removed. The plaster uses a light-sensitive polymer (see Figure 2 below).

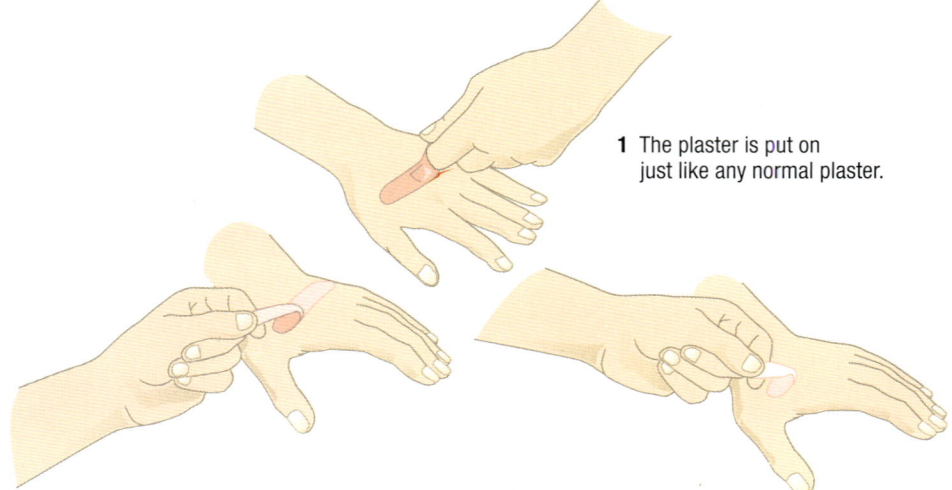

1 The plaster is put on just like any normal plaster.

2 To remove the plaster, the top layer is peeled away from the lower layer which stays stuck to the skin.

3 Once the lower layer is exposed to the light, the adhesive becomes less sticky, making it easy to peel off your skin.

Figure 2 *This plaster uses a light-sensitive polymer*

Hydrogels

Hydrogels are polymer chains with a few cross-linking units between chains. This makes a matrix (framework) that can trap water. These hydrogels can be used as wound dressings. They let the body heal in moist, sterile conditions. This makes them useful for treating burns.

The latest 'soft' contact lenses are also made from hydrogels. To change the properties of hydrogels, scientists can vary the amount of water in the matrix structure of the hydrogel.

Practical

Evaluating plastics

Plan an investigation to compare and evaluate the suitability of different plastics for a particular use.

For example, you might look at treated and untreated fabrics for waterproofing and 'breatheability' (gas permeability), or different types of packaging.

Shape memory polymers

A new 'shape memory polymer' is being developed for surgeons which will make stitching in awkward places easier. When a shape memory polymer is used to stitch a wound loosely, the temperature of the body makes the thread tighten and close the wound, applying just the right amount of force.

This is an example of a **smart polymer**, that is, one that changes in response to changes in its surroundings. In this case a change in temperature causes the polymer to change its shape. Later, after the wound is healed, the polymer is designed to dissolve and is harmlessly absorbed by the body. So, there will be no need to go back to hospital to have the stitches out.

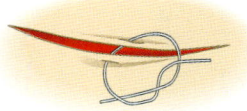  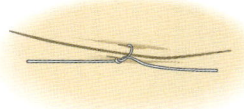

Figure 3 *A shape memory polymer uses the temperature of the body to make the thread tighten and close the wound*

New uses for old polymers

The bottles that people buy fizzy drinks in are a good example of using a plastic because of its properties. These bottles are made out of a plastic called PET.

The polymer it is made from is ideal for making drinks bottles. It produces a plastic that is very strong and tough, and which can be made transparent. The bottles made from this plastic are much lighter than glass bottles. This means that they cost less to transport and are easier for us to carry around.

The PET from recycled bottles is used to make polyester fibres for clothing, such as fleece jackets, and the filling for duvet covers. School uniforms and football shirts are now also made from recycled drinks bottles.

Did you know ... ?

PET is an abbreviation for poly(ethene terephthalate). It takes five 2-litre PET lemonade bottles to make one T-shirt.

Key points

- New polymers are being developed all the time. They are designed to have properties that make them specially suited for certain uses.

- Smart polymers may have their properties changed by light, temperature, or other changes in their surroundings.

- We are now recycling more plastics and finding new uses for them.

Summary questions

1 a What is meant by:
    i   a 'designer polymer'?
    ii a 'smart polymer'?
  b i   Why is PET used to make drinks bottles?
    ii State **two** uses of recycled PET.
    iii Is PET a thermosetting polymer or a thermosoftening polymer? How did you decide?

2 a What is a hydrogel?
  b Give **two** uses of hydrogels.

3 Explain how a light-sensitive plaster works.

4 Explain how stitches made from smart polymers work.

13.5 Plastic waste

After this topic, you should know:
- the problems caused by the disposal of plastics
- what biodegradable polymers are
- how polymers can be made biodegradable.

One of the problems with plastics is what to do with them when you have finished with them. Too much plastic ends up as rubbish in our streets. Even the beaches in the most remote parts of the world can be polluted with the eyesore of plastic waste. Wildlife can also get trapped in the waste or eat the plastics and die.

Not only that, just think of all the plastic packaging that goes in the bin after shopping. Most of it ends up as rubbish in landfill sites. Other landfill rubbish rots away quite quickly as microorganisms in the soil break it down. However, many waste polymers last for hundreds of years before they are broken down completely, taking up valuable space in our landfill sites. So, what was a useful property during the working life of the plastic (its lack of reactivity) becomes a disadvantage in a landfill site.

Figure 1 *Finding space to dump and bury our waste is becoming a big problem*

Biodegradable polymers

Scientists are working to solve the problems of plastic waste. Humans are now making more plastics that do rot away in the soil when dumped. The polymers in these plastics are called **biodegradable**. They can be broken down by microorganisms.

Scientists have found different ways to speed up the decomposition:
- One way uses granules of cornstarch built into a plastic. The microorganisms in soil feed on the starch. This breaks the plastic up into small pieces more quickly.
- Other types of polymer have been developed that are made from plant products.

A polymer called PLA, poly(lactic acid), can be made from cornstarch. The polymer is totally biodegradable. It is used in food packaging. However, it cannot be put in a microwave oven, which limits its use in ready-meal packaging.

Manufacturers can also make plastic carrier bags using PLA. In carrier bags the PLA is mixed with a traditional polymer. This makes sure the bag is strong enough but will still biodegrade a lot more quickly.

Using polymers such as PLA also helps preserve our supplies of crude oil. Remember that crude oil is the raw material for many traditional polymers, such as poly(ethene).

Figure 2 *The breakdown of a biodegradable polymer. PLA can be designed to break down in a few months.*

Disadvantages of biodegradable polymers

The use of a food crop like corn to make polymers can raise the same issues as biofuels. Farmers who sell their crops to be turned into fuel and plastics could cause higher food prices. The lack of basic food supplies could result in starvation in developing countries. Another problem is the destruction of tropical forests to create more farmland. This will destroy the habitats of wildlife and could affect climate change.

Other degradable polymers used for bags will break down in light. However, they will not decompose when buried in a landfill site. One of the best solutions is to reuse the same plastic carrier bags over and over again.

Summary questions

1 Why are waste plastics proving to be a problem for us?

2 Explain what is meant by a 'biodegradable' polymer.

3 **a** Why are plastics whose raw materials are plants becoming more popular?

   b PLA is a biodegradable plastic. What is its monomer?

4 Non-biodegradable polymers such as poly(ethene) can be made to decompose more quickly by mixing them with additives. These enable the polymer chain to be broken down by reacting with oxygen.

   a Why might this be a waste of money if the plastic is buried and compressed under other waste in a landfill site?

   b What could be a disadvantage of using this type of degradable plastic?

   c Explain how cornstarch can play a role in helping to solve the problems of plastic waste.

Practical

Investigating cornstarch

Cornstarch can be fermented to make the starting material for PLA. However, cornstarch itself also has some interesting properties. You can make your own plastic material directly from cornstarch.

■ Investigate how varying the proportions of cornstarch and water affect the product.

Safety: Be aware of allergies and sensitisation.

∞ links

For information on the issues of using biofuels, look back at 12.4 'Alternative fuels'.

Key points

■ Non-biodegradable polymers cause unsightly rubbish, can harm wildlife, and take up space in landfill sites.

■ Biodegradable polymers are decomposed by the action of microorganisms in soil.

■ Making plastics with starch granules in their structure helps the microorganisms break down a polymer.

■ Scientists can make biodegradable polymers from plant material such as cornstarch.

Chapter summary questions

1 Propene is a hydrocarbon molecule containing three carbon atoms and six hydrogen atoms.

  a What is the chemical formula of propene?

  b Draw the displayed formula of propene, showing all its bonds.

  c Is propene a saturated molecule or an unsaturated molecule? Explain your answer.

  d You are given two unlabelled test tubes. One test tube contains propane gas, whilst the other test tube contains propene gas.
  Explain how you could test which tube contains which gas, stating clearly the results obtained in each case.

  e Propene molecules will react together to form long chains.
    i What is this type of reaction called?
    ii Name the product of the reaction.
    iii What is the general name given to the many propene molecules that react together?
    iv Compare the properties of the reactants to those of the product.

2 Give a symbol equation to show the reaction of ethene to make poly(ethene).

3 a Write a word equation and a balanced symbol equation, including state symbols, for the reaction between ethene and steam.

  b If 56 tonnes of ethene reacted with excess steam, and 69 tonnes of ethanol were collected, what is the percentage yield of ethanol?
  (A_r values: H = 1, C = 12, O = 16)

  c Give an advantage and a disadvantage of making ethanol from ethene compared with making it from sugar obtained from plant material.

4 Chemists have developed special waterproof materials made from polymers. It is easy to make plastic material that is waterproof but they tend to make the wearer hot and sweaty. However, there are now polymer materials that have pores 2000 times smaller than a drop of water. However, the tiny pores are 700 times larger than a water molecule.

  a Explain why these materials are described as 'breathable'.

  b Would you describe the 'breathable' polymer material as a smart polymer? Explain your answer.

5 Read the following information then answer the questions.

  New biodegradable polymers are now being developed. They use normal crops as their starting material. They do not use up our decreasing supplies of crude oil. The crop is fermented with microorganisms which make a polymer called PHB, poly(hydroxybutyrate). In America, scientists have used genetic engineering to grow plants which make PHB themselves. They are continuing their research to increase the yield of the polymer in crops such as potatoes.

Look at this cycle for the plastic PHB:

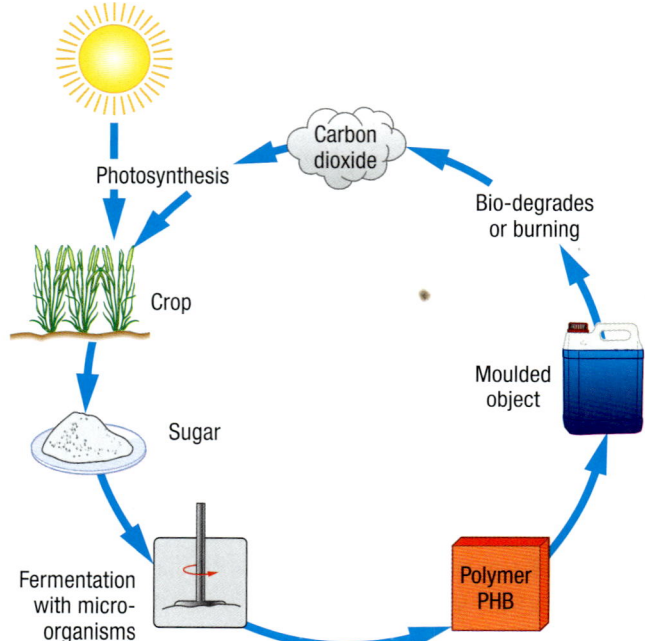

  a Why might people object to the crops that make PHB directly?

  b Does the use of PHB increase the amount of carbon dioxide in the air? Explain your answer.

  c Why is it difficult to say whether PHB made directly by plants will be cheaper than the PHB shown in the cycle?

Practice questions

1 The fractions from crude oil that contain the biggest molecules have few uses. Cracking converts them into more useful substances, including alkenes.

  a Describe the conditions used in **two** different types of cracking. (4)

  b Unlike alkanes, alkenes are *unsaturated*.
    i What feature is present in an alkene but not in an alkane? (1)
    ii Give the displayed formula of the simplest alkene. (1)
    iii Propene has the molecular formula C_3H_6.

      Deduce the molecular formula of the alkene with four carbon atoms in each molecule. (1)

  c Copy and complete this equation for a cracking reaction.

    $C_{12}H_{26} \rightarrow C_8H_{18} +$ ----------- (1)

  d Alkanes and alkenes can be distinguished by a simple chemical test.

    State the reagent used and the result with each type of compound. (3)

  e The traditional way to make ethanol is by fermentation.

    State how it is made using one of the products of cracking. (3)

2 Most alkenes are used to make polymers.

  a State **two** features of polymer molecules that alkenes do not have. (2)

  b Poly(ethene) is made in two forms, depending on the conditions used for polymerisation. These are described as LD and HD forms.
    i State the meanings of the terms 'LD' and 'HD'. (1)
    ii Both forms are described as 'thermosoftening'. What is the meaning of this term? (1)

  c Another type of polymer is described as thermosetting.
    i How does a thermosetting polymer behave when heated? (1)
    ii Describe how the bonding in a thermosetting polymer differs from the bonding in a thermosoftening polymer. (2)

  d Some polymers are 'biodegradable'.

    State the meaning of this term and **one** advantage of these polymers over non-biodegradable polymers. (2)

3 A manufacturer plans to produce objects made from two new polymers. Polymer 1 needs to withstand heat without changing shape. Polymer 2 needs to be able to change shape but does not need to withstand heat.

The diagrams show two types of polymer.

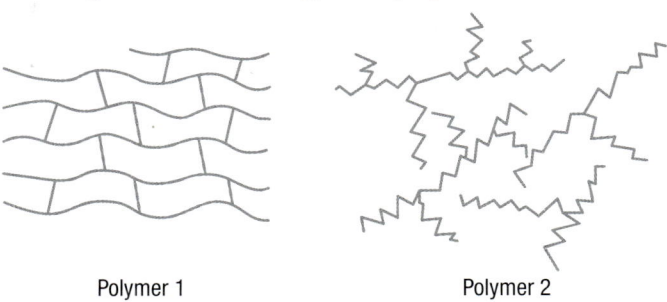

Polymer 1   Polymer 2

Use information from the diagrams and your knowledge and understanding of polymers to identify the types of polymer that could be used as Polymer 1 and Polymer 2. You should support your choices by referring to the properties, structure, and bonding in polymers. (6)

14.1

Structures of alcohols, carboxylic acids, and esters

Learning objectives

After this topic, you should know:

- the names and formulae of the first three alcohols, carboxylic acids, and esters

- how to represent the structures of alcohols, carboxylic acids, and esters.

Did you know ... ?

Not all carbon compounds are classified as organic compounds. The oxides of carbon and compounds containing carbonate and hydrogencarbonate ions are classified as inorganic compounds.

The substances that form the basis of all living things, including ourselves, are organic compounds. Organic molecules all contain carbon atoms. The carbon atoms tend to form the 'backbone' of organic molecules.

You have already met some organic compounds in Chapters 12 and 13. You will probably recall how to draw molecules of alkanes and alkenes (see Figure 1):

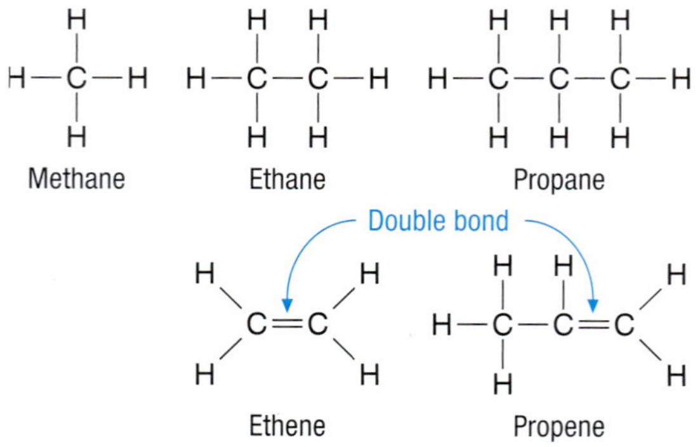

Figure 1 *The structures of the smallest alkane and alkene molecules. The lines between atoms represent covalent bonds. Notice the double bonds in ethene and propene. This type of 2D drawing to show the structure of a molecule is called a 'displayed formula'.*

Both alkanes and alkenes are made of only carbon and hydrogen atoms. However, there are many more 'families' of organic compounds that also contain a few other types of atom.

In this chapter you will learn about some organic compounds made up of carbon, hydrogen, and oxygen. The three 'families' you will look at are called alcohols, carboxylic acids, and esters.

Alcohols

You have also met a member of the alcohol 'family' before. This was ethanol, which can be used as a biofuel. So what is an alcohol?

Imagine removing an H atom from an alkane molecule and replacing it with an O–H group. This would give us an alcohol molecule. The –OH group of atoms is an example of a **functional group**.

A functional group gives a 'family' of organic compounds their characteristic reactions. A 'family' of compounds with the same functional group is called a **homologous series**.

links

For more information on the use of ethanol as a biofuel, look back at 12.4 'Alternative fuels'.

You need to know the first three members of the homologous series of alcohols:

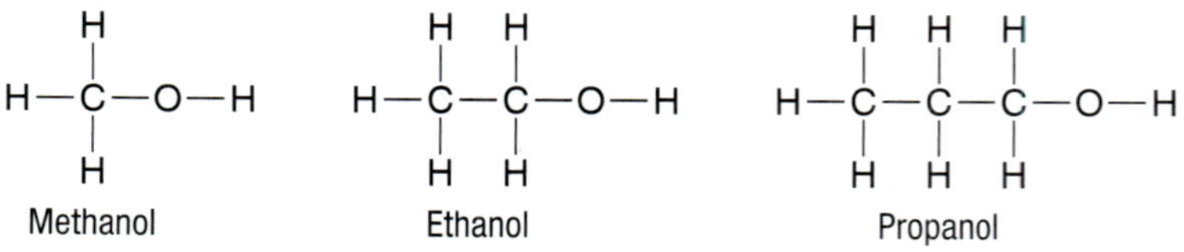

Figure 2 *The displayed formulae and names of the first three alcohol molecules. The alcohols are named from the alkane with the same number of carbon atoms. Just take the 'e' from the end of the alkane's name and replace it with 'ol'.*

Study tip

If you draw displayed formulae, make sure you show all the bonds, including those in the functional group (as lines between atoms) and all the atoms (as their chemical symbols).

If you were asked for the formula of ethanol, you might count the atoms in a molecule and write C_2H_6O. This is correct but chemists can give more information in a **structural formula**. This doesn't show all the bonds, as in a displayed formula, but shows what is bonded to each carbon atom. So for ethanol, a chemist will often show its formula as CH_3CH_2OH (often shortened to C_2H_5OH). Can you see how this relates to the structure of ethanol shown in Figure 2?

Carboxylic acids

You will certainly know of one carboxylic acid. Ethanoic acid is the main acid in vinegar. All carboxylic acids contain the –COOH functional group. Look at the first three members of the homologous series of carboxylic acids below:

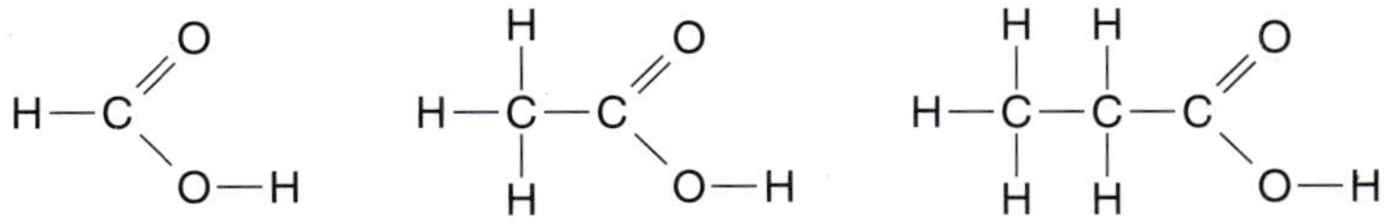

Methanoic acid   Ethanoic acid   Propanoic acid

Figure 3 *The displayed formulae and names of the first three carboxylic acid molecules*

You show the structural formula of each of the carboxylic acids in Figure 3 as $HCOOH$, CH_3COOH, and CH_3CH_2COOH (or C_2H_5COOH).

Esters

Esters are closely related to carboxylic acids. If you replace the H atom in the –COOH group by a hydrocarbon (alkyl) group, such as $-CH_3$ or $-C_2H_5$, you get an ester. Here is the ester called ethyl ethanoate:

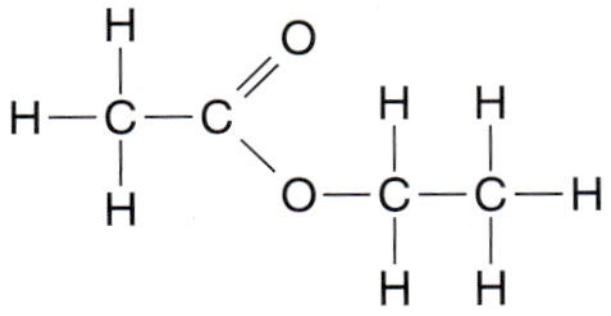

Ethyl ethanoate

Figure 5 *The displayed formulae of the ester ethyl ethanoate*

An ester's structural formula always has the –COO– functional group in it. The structural formula of ethyl ethanoate is $CH_3COOCH_2CH_3$ (or $CH_3COOC_2H_5$).

Figure 4 *Vinegar contains less than 10 per cent ethanoic acid in an aqueous solution but the acid provides the characteristic sharp taste and smell*

∞ links

For information about the homologous series of carboxylic acids and esters, see 14.3 'Carboxylic acids and esters'.

Summary questions

1 Which homologous series do the following compounds belong to?
   a $CH_3COOCH_2CH_2CH_3$
   b $CH_3CH_2CH_2CH_2CH_2CH_2OH$
   c $CH_3CH_2CH_2CH_2COOH$

2 Name the following compounds:
   a $CH_3CH_2CH_2OH$
   b $CH_3COOC_2H_5$
   c $HCOOH$

3 Draw a displayed formula showing all the bonds in:
   a ethanol
   b ethyl ethanoate
   c butanoic acid, which contains four carbon atoms.

Key points

■ The homologous series of alcohols contain the –OH functional group.

■ The homologous series of carboxylic acids contain the –COOH functional group.

■ The homologous series of esters contain the –COO– functional group.

14.2 Properties and uses of alcohols

Learning objectives

After this topic, you should know:

- some properties of alcohols
- the main uses of alcohols
- what is produced when ethanol is fully oxidised.

links

For more information on how ethanol is manufactured by fermentation, look back to 12.4 'Alternative fuels', and from ethene in 13.1 'Cracking hydrocarbons'.

Figure 1 *Alcohols are used as solvents in perfumes*

Alcohols, especially ethanol, are commonly used in everyday products. Ethanol is the main alcohol in alcoholic drinks. It is made by fermenting sugars from plant material and is becoming an important alternative fuel to petrol and diesel. Ethanol can also be made in industry from ethene and steam in the presence of a catalyst.

Alcohols dissolve many of the same substances as water. In fact the alcohols with smaller molecules mix very well with water, giving neutral solutions. The alcohols can also dissolve many other organic compounds. This property makes them useful as solvents. For example, you can remove ink stains from permanent marker pens using methylated spirits.

Methylated spirits ('meths') is mainly ethanol but has the more toxic methanol mixed with it. It also has a purple dye and other substances added to make it unpleasant to drink. Alcohols are also used as solvents in products such as perfumes, aftershaves, and mouthwashes.

Reactions of alcohols

Practical/Demonstration

Comparing the reactions of methanol, ethanol, and propanol

a Ignite and observe the flame in three spirit burners – one containing methanol, one ethanol, and the other propanol.
Compare the three combustion reactions.

b Watch your teacher add a small piece of sodium metal to each of the alcohols.
Compare the reactions. Which gas is given off?

c Watch your teacher boil some of each alcohol with acidified potassium dichromate(VI) solution.
What do you see happen in each reaction?

Safety: Ethanol is highly flammable and harmful. Methanol is highly flammable and toxic. Propanol is highly flammable and irritant. No naked flames. Wear chemical splashproof eye protection.

Combustion

The use of ethanol (and also methanol) as fuels shows that the alcohols are flammable. Ethanol is used in spirit burners and can be used as a biofuel in cars. It burns with a 'clean' blue flame:

$$\text{ethanol} + \text{oxygen} \rightarrow \text{carbon dioxide} + \text{water}$$
$$C_2H_5OH + 3O_2 \rightarrow 2CO_2 + 3H_2O$$

 Did you know ... ?

Ethanol is the main solvent in many perfumes but a key ingredient in some perfumes is octanol. It evaporates more slowly and so holds the perfume on the skin for longer.

Reaction with sodium

The alcohols react in a similar way to water when sodium metal is added. The sodium effervesces (gives off bubbles of gas), producing hydrogen gas, and gets smaller and smaller as it forms a solution of sodium ethoxide in ethanol. The reactions of alcohols are not as vigorous as the reaction between sodium and water. For example, with ethanol:

$$\text{sodium} + \text{ethanol} \rightarrow \text{sodium ethoxide} + \text{hydrogen}$$
$$2Na + 2C_2H_5OH \rightarrow 2C_2H_5ONa + H_2$$

Oxidation

Combustion is one way to oxidise an alcohol. However, when you use chemical oxidising agents, such as potassium dichromate(VI), you get different products. An alcohol is oxidised to a carboxylic acid when boiled with acidified potassium dichromate(VI) solution. So ethanol can be oxidised to ethanoic acid:

$$\text{ethanol} + \text{oxygen atoms from oxidising agent} \rightarrow \text{ethanoic acid} + \text{water}$$
$$C_2H_5OH + 2[O] \rightarrow CH_3COOH + H_2O$$

The same reaction takes place if ethanol is left exposed to air. Microbes in the air produce ethanoic acid from the ethanol. This is why bottles of beer or wine taste and smell like vinegar when they are left open for too long.

Figure 2 *Alcohols are flammable. They produce carbon dioxide and water in their combustion reactions.*

Practical

Oxidation by microbes in air

Add 5 cm³ of ethanol to 50 cm³ of water in a conical flask, mix, and test the pH of the solution formed.

Then mix 5 cm³ of ethanol with 50 cm³ of water in a second conical flask but this time seal the flask with a stopper.

Leave both solutions for a few weeks, swirling occasionally.

- What happens to the pH of the solutions?
- Explain your observations.

Safety: Ethanol is highly flammable and harmful.

Summary questions

1 Give the name and structural formula of the main alcohol in alcoholic drinks.

2 List the main uses of alcohols.

3 a Write a word equation and a balanced symbol equation for the reaction between sodium metal and methanol.

  b Name and give the structural formula of the acidic product formed when methanol is boiled with acidified potassium dichromate(VI), a chemical oxidising agent.

  c Write a word equation and a balanced symbol equation for the complete combustion of methanol.

4 Methanol, ethanol, and propanol are all liquids at room temperature. Plan an investigation to see which alcohol – methanol, ethanol, or propanol – releases most energy per gram when it burns.

Key points

- Alcohols are used as solvents and fuels, and ethanol is the main alcohol in alcoholic drinks.

- Alcohols burn in air, forming carbon dioxide and water.

- With sodium metal, alcohols react to form a solution, and hydrogen gas is given off.

- Ethanol can be oxidised to ethanoic acid, either by chemical oxidising agents or by the action of microbes in the air. Ethanoic acid is the main acid in vinegar.

14.3 Carboxylic acids and esters

Learning objectives

After this topic, you should know:

- how to recognise carboxylic acids from their properties

- some uses of carboxylic acids and esters

- why carboxylic acids are described as weak acids

- how to make esters.

∞ **links**

For more information on the structure of carboxylic acids, look back to 14.1 'Structures of alcohols, carboxylic acids, and esters'.

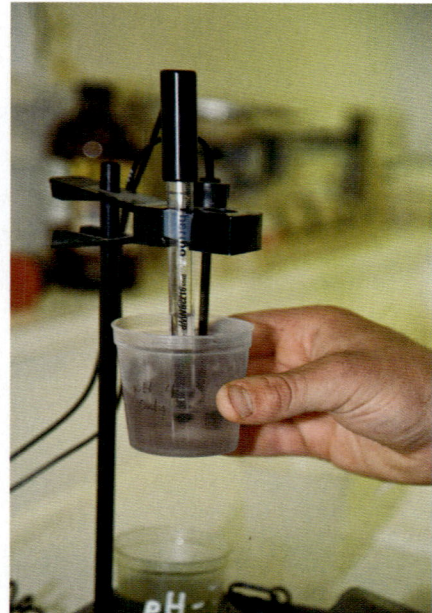

Figure 2 *Testing the pH of a solution using a pH sensor*

∞ **links**

Look back to Chapter 7 to see the typical reactions of acids.

You have already learnt about the structure of carboxylic acids. The most well-known carboxylic acid is ethanoic acid.

Figure 1 *Ethanoic acid, CH_3COOH, is the main acid in vinegar. Its old name was 'acetic acid'. Carboxylic acids are also used to make polyester fibres.*

Carboxylic acids, as their name suggests, form acidic solutions when they dissolve in water. You can look at one of their reactions in the practical on the next page.

Carboxylic acids have the typical reactions of all acids. For example, reaction with a metal carbonate forms a salt, water, and carbon dioxide. In the practical at the top of page 189:

ethanoic acid + sodium carbonate → sodium ethanoate + water + carbon dioxide

$$2CH_3COOH(aq) + Na_2CO_3(s) \rightarrow 2CH_3COONa(aq) + H_2O(l) + CO_2(g)$$

Why are carboxylic acids called 'weak acids'?

You will see in the practical on the next page how CO_2 gas is given off more slowly when a metal carbonate reacts with a carboxylic acid as compared with hydrochloric acid of the same concentration. Carboxylic acids are called **weak** acids, as opposed to **strong** acids such as hydrochloric acid.

The pH of a $0.1\,mol/dm^3$ solution of hydrochloric acid (a strong acid) is 1.0. Yet a $0.1\,mol/dm^3$ solution of ethanoic acid (a weak acid) has a pH of only 2.9. The solution of ethanoic acid is not as acidic even though the two solutions have the same concentration. Why is this?

Acids must dissolve in water before they show their acidic properties. This is because in water all acids ionise (split up). Their molecules split up to form $H^+(aq)$ ions and negative ions. It is the $H^+(aq)$ ions that all acidic solutions have in common. For example, in hydrochloric acid, the HCl molecules all ionise in water:

$$HCl(aq) \xrightarrow{\text{water}} H^+(aq) + Cl^-(aq)$$

We say that strong acids ionise *completely* in solution. However, in weak acids most of the molecules stay as they are. Only some will ionise in their solutions. A position of **equilibrium** is reached in which both molecules and ions are present. So, in ethanoic acid:

$$CH_3COOH(aq) \underset{\text{water}}{\rightleftharpoons} CH_3COO^-(aq) + H^+(aq)$$

ethanoic acid   ethanoate ions   hydrogen ions

Therefore, given two aqueous solutions of equal concentration, the strong acid will have a higher concentration of $H^+(aq)$ ions than the solution of the weak acid. So, a weak acid has a higher pH (and therefore reacts more slowly with a metal carbonate).

Practical

Comparing ethanoic acid and hydrochloric acid

Write down your observations to compare ethanoic acid with hydrochloric acid of the same concentration.

a Take the pH of solutions of both acids.

b Add a little sodium carbonate to solutions of both acids.

Why did you use the same concentrations of each acid in the experiment?

Safety: Wear eye protection.

Making esters

Carboxylic acids also react with alcohols to make esters. Water is also formed in this reversible reaction. A strong acid, usually sulfuric acid, is used as a catalyst. For example:

$$\text{ethanoic acid} + \text{ethanol} \xrightleftharpoons[\text{catalyst}]{\text{sulfuric acid}} \text{ethyl ethanoate} + \text{water}$$

$$CH_3COOH + C_2H_5OH \rightleftharpoons CH_3COOC_2H_5 + H_2O$$

In general:

$$\textbf{carboxylic acid} + \textbf{alcohol} \xrightleftharpoons[\text{catalyst}]{\text{strong acid}} \textbf{ester} + \textbf{water}$$

Here is another example:

ethanoic acid + methanol ⇌ methyl ethanoate + water

$$\text{ethanoic acid} + \text{methanol} \xrightleftharpoons[\text{catalyst}]{\text{sulfuric acid}} \text{methyl ethanoate} + \text{water}$$

$$CH_3COOH + CH_3OH \rightleftharpoons CH_3COOCH_3 + H_2O$$

Another example is:

$$\text{propanoic acid} + \text{ethanol} \xrightleftharpoons[\text{catalyst}]{\text{sulfuric acid}} \text{ethyl propanoate} + \text{water}$$

$$C_2H_5COOH + C_2H_5OH \rightleftharpoons C_2H_5COOC_2H_5 + H_2O$$

The esters formed have distinctive smells. They are volatile (evaporate easily). Many smell sweet and fruity. This makes them ideal to use in perfumes and food flavourings.

Summary questions

1 a Which gas is made when propanoic acid reacts with potassium carbonate?

b Name the salt formed.

c Write a balanced symbol equation for the reaction.

2 a Write a word equation, including the catalyst, to show the reversible reaction between methanoic acid and ethanol.

b Name the ester formed from:
 i ethanoic acid and ethanol
 ii methanoic acid and propanol.

c Esters are volatile compounds. What does this mean?

d Give two uses of esters.

3 Explain in detail why propanoic acid is described as a weak acid.

Demonstration

Making esters

Your teacher will show you how to make different esters using carboxylic acids and alcohols.

After neutralising the acid with sodium hydrogencarbonate, carefully smell the test tubes containing the different esters formed.

■ Write word equations for each reaction.

Key points

■ Solutions of carboxylic acids have a pH value less than 7. Carbonates gently fizz in their acidic solutions, releasing carbon dioxide gas.

■ Aqueous solutions of weak acids, such as carboxylic acids, have a higher pH value than solutions of strong acids with the same concentration.

■ Esters are made by reacting a carboxylic acid and an alcohol together, with a strong acid catalyst.

■ Esters are volatile, fragrant compounds used in flavourings and perfumes.

Chapter summary questions

1 Look at the three organic molecules **A, B,** and **C** below:

A

B

C

Answer these questions about **A, B,** and **C**:

a Which one is a carboxylic acid?

b Which one is an alcohol?

c Which homologous series of organic compounds does **B** belong to?

d Which of the compounds can be represented as the structural formula $CH_3CH_2COOCH_2CH_2CH_2CH_3$?

e Using a structural formula as shown in part **d**, give the structural formulae of the other two compounds.

2 a i Describe what you would **see** happen if a small piece of sodium metal was dropped into a beaker containing some ethanol.

   ii Name the gas given off in this reaction.

   iii Name the sodium compound formed in the reaction.

   iv The sodium in the compound formed is present as ions. Give the formula of a sodium ion.

   v Write a balanced symbol equation for the reaction of sodium with ethanol.

b Using your knowledge of the periodic table and trends in reactivity, suggest how ethanol's reaction with lithium metal would differ from its reaction with sodium.

3 a Draw the displayed formula of methanoic acid, showing all the atoms and bonds.

b **i** Some potassium carbonate powder is dropped into a test tube of a solution of methanoic acid. How would you positively identify the gas given off? Include a labelled diagram of the apparatus you could use.

   ii Write a word equation and balanced symbol equation for the reaction.

c i Which aqueous solution of the following would you expect to have the highest pH value if they all had the same concentration?

   A hydrochloric acid

   B sulfuric acid

   C methanoic acid

   D nitric acid

   E ethanol

  ii Which of the five solutions listed above would have the second highest pH value? Explain your choice.

4 a What is the name of this compound?

b Name the carboxylic acid and alcohol used to make the compound in part **a** and give one use for each.

c Write a word equation and a balanced symbol equation, showing how this compound can be made. Include the name of the catalyst.

d The compound shown in part **a** is volatile. What does this mean?

e Name the carboxylic acid and the alcohol you would use to make methyl propanoate.

5 Describe how you could distinguish between samples of propanol, propanoic acid, and ethyl ethanoate using simple experimental tests. Give the results of any tests suggested.

6 You are given a $0.5\,mol/dm^3$ solution of nitric acid, and a $0.5\,mol/dm^3$ solution of methanoic acid.

a What does 'a $0.5\,mol/dm^3$ solution' mean?

b What can you predict about the relative pH values of the two solutions?

c Explain in detail your answer to part **b**, including equations to show the ionisation of nitric acid and methanoic acid in aqueous solution.

Practice questions

1 Ethanol is often known as alcohol, but is just one member of a homologous series of alcohols. Methanol and propanol are two other members of the same homologous series.

   a State **two** features that members of a homologous series have in common. (2)

   b **i** What is the molecular formula of methanol? (1)
      ii Draw the displayed formula of propanol. (1)

   c A small piece of sodium is added to a beaker containing methanol.

      Describe **two** observations you could make. (2)

   d A sample of propanol burns completely in air.

      Name the products formed. (2)

   e Ethanol burns completely in air in an oxidation reaction.

      Write a balanced symbol equation for this reaction. (2)

   f Ethanol can also be oxidised by microbial action.
      i What is the organic product of this reaction? (1)
      ii What is the common name of the solution containing this organic product? (1)

2 Carboxylic acids are compounds in a homologous series.

Like alcohols they dissolve in water to form colourless solutions.

   a A student was given separate solutions of ethanol and ethanoic acid.

      Describe how these could be distinguished by adding sodium carbonate. Include **one** observation for each compound. (2)

   b The student was given separate solutions of ethanoic acid and nitric acid with the same concentration.

      These solutions can be distinguished by using a pH meter.

      Choose a possible pH value for each solution.
      i The pH value of ethanoic acid is   1 5 7 9 13 (1)
      ii The pH value of nitric acid is   1 5 7 9 13 (1)

   c The student was shown these equations:

$$CH_3COOH \rightleftharpoons CH_3COO^- + H^+$$
$$HNO_3 \rightarrow H^+ + NO_3^-$$

      Explain how the arrows show that ethanoic acid is weaker than nitric acid. (2)

3 Esters are organic compounds formed in the reaction between an alcohol and a carboxylic acid. A reaction of this type is known as esterification.

   a An example of an esterification reaction is shown by this equation:

$$HCOOH + CH_3CH_2CH_2OH \rightleftharpoons HCOOCH_2CH_2CH_3 + H_2O$$

      i Name the two reactants. (2)
      ii Draw the displayed formula of the ester. (1)

   b What is the name of the ester made from ethanol and ethanoic acid? (1)

   c State **two** uses for esters. (2)

Investigations

After this topic, you should know:

- what 'continuous variables' and 'categoric variables' are

- what is meant by 'repeatable evidence', 'reproducible evidence' and 'valid evidence'

- what the link is between the independent and dependent variable

- what a 'hypothesis' and a 'prediction' are

- how to reduce risks in hazardous situations.

Science works for us all day, every day. Working as a scientist you will have knowledge of the world around you and particularly about the subject you are working with. You will observe the world around you. An enquiring mind will then lead you to start asking questions about what you have observed.

Science usually moves forward by slow, steady steps. Each small step is important in its own way. It builds on the body of knowledge that you already have.

Thinking scientifically

Deciding on what to measure

Variables can be one of two different types:

- A **categoric variable** is one that is best described by a label (usually a word). The type of metal is a categoric variable, for example, magnesium or zinc.

- A **continuous variable** is one that you measure, so its value could be any number. Temperature (as measured by a thermometer or temperature sensor) is a continuous variable, for example, 37.6 °C, 45.2 °C. Continuous variables can have values (called a quantity) that can be given by any measurements made (e.g., mass, volume).

When designing your investigation you should always try to measure continuous **data** whenever you can. If there is no way to measure your variable then you have to use a label (categoric variable).

Making your investigation repeatable, reproducible, and valid

When you are designing an investigation you must make sure that you check results are **repeatable** (using the same method), and then plan for somebody else to carry out the experiment, or use different apparatus, to check if results are **reproducible**.

You must also make sure you are measuring the actual thing you want to measure. If you do not, your data cannot be used to answer your original question. This seems very obvious but it is not always quite so easy. You need to make sure that you have controlled as many other variables as you can, so that no-one can say that your investigation is not **valid**.

How might an independent variable be linked to a dependent variable?

The **independent variable** is the one you choose to vary in your investigation.

The **dependent variable** is used to judge the effect of varying the independent variable.

These variables may be linked together. If there is a pattern to be seen (e.g., as one thing gets bigger the other also gets bigger), it may be that:

- changing one has caused the other to change

- the two are related, but one is not necessarily the cause of the other.

Starting an investigation

Observation

As scientists you use observations to ask questions. You can only ask useful questions if you know something about the observed event. You will not have all of the answers, but you know enough to start asking the correct questions.

When you are designing an investigation you have to observe carefully which variables are likely to have an effect.

What is a hypothesis?

A **hypothesis** is an idea based on observation that has some really good science to try to explain it.

When making hypotheses you can be very imaginative with your ideas. However, you should have some scientific reasoning behind those ideas so that they are not totally bizarre.

Remember, your explanation might not be correct, but you think it is. The only way you can check out your hypothesis is to make it into a prediction and then test it by carrying out an investigation.

observation + knowledge → hypothesis → prediction → investigation

Study tip

Observations, backed up by creative thinking and good scientific knowledge can lead into a hypothesis.

Starting to design an investigation

An investigation starts with a question, followed by a **prediction**. You, as the scientist, predict that there is a **relationship** between two variables.

You should think about carrying out a preliminary investigation to find the most suitable range and interval for the independent variable.

Making your investigation safe

Remember that when you design your investigation, you must:
- look for any potential **hazards**
- decide how you will reduce any **risk**.

You will need to write these down in your plan:
- write down your plan
- make a risk assessment
- make a prediction
- draw a blank table ready for the results.

Key points

- Continuous data can give you more information than other types of data.

- You must design investigations that produce repeatable, reproducible, and valid results if you are to be believed.

- Be aware that just because two variables are related, does not mean that there is a causal link.

- Hypotheses can lead to predictions and investigations.

- You must make a risk assessment, make a prediction, and write a plan.

Setting up investigations

Learning objectives

After this topic, you should know:

- what a 'fair test' is
- how to set up a valid investigation
- what a 'control group' is
- how to decide on the variables, range, and intervals
- how to ensure accuracy and precision
- the causes of error and anomalies.

Study tip

Trial runs will tell you a lot about how your investigation might work out. They should get you to ask yourself:

- Do I have the correct conditions?
- Have I chosen a sensible range?
- Have I got enough readings that are close together?
- Will I need to repeat my readings?

Study tip

Just because your results show precision does not mean your results are accurate.

Imagine you carry out an investigation into the energy value of a type of fuel. You get readings of the amount of energy released that are all about the same. This means that your data will have precision, but it doesn't mean that they are necessarily accurate.

Fair testing

A **fair test** is one in which only the independent variable affects the dependent variable. All other variables are controlled in order to carry out a valid investigation. They are known as **control variables**.

This is easy to set up in the laboratory, but almost impossible in fieldwork. Investigations in the environment are not that simple and easy to control. There are complex variables that are changing constantly.

So how can you set up the fieldwork investigations? The best you can do is to make sure that all of the many variables change in much the same way, except for the one you are investigating. Then at least the plants you might be monitoring for the effects of pollution are getting the same weather, even if it is constantly changing.

If you are investigating two variables in a large population then you will need to do a survey. Again, it is impossible to control all of the variables. Imagine scientists were investigating the effect of a new drug on diabetes. They would have to choose people of the same age and same family history to test. The larger the sample size tested, the more valid the results will be.

Control groups are used in these investigations to try to make sure that you are measuring the variable that you intend to measure. When investigating the effects of a new drug, the control group will be given a placebo. The control group think they are taking a drug but the placebo does not contain the drug. This way you can control the variable of 'thinking that the drug is working' and separate out the effect of the actual drug.

Designing an investigation

Accuracy

Your investigation must provide **accurate** data. Accurate data is essential if your results are going to have any meaning.

How do you know if you have accurate data?

It is very difficult to be certain. Accurate results are very close to the true value. It is not always possible to know what that true value is.

- Sometimes you can calculate a theoretical value and check it against the experimental evidence. Close agreement between these two values could indicate accurate data.
- You can draw a graph of your results and see how close each result is to the line of best fit.
- Try repeating your measurements with a different instrument and see if you get the same readings.

How do you get accurate data?

- Using instruments that measure accurately will help.
- The more carefully you use the measuring instruments, the more accuracy you will get.

Precision

Your investigation must provide data with sufficient **precision**. If it doesn't then you will not be able to make a valid conclusion.

How do you get precise and repeatable data?

- You have to repeat your tests as often as necessary to improve repeatability.
- You have to repeat your tests in exactly the same way each time.
- Use measuring instruments that have the appropriate scale divisions needed for a particular investigation. Smaller scale divisions have better **resolution**.

Making measurements

Using instruments

You cannot expect perfect results. When you choose an instrument you need to know that it will give you the accuracy that you want, that is, it will give you a true reading.

When you choose an instrument you need to decide how precise you need to be. Some instruments have smaller scale divisions than others. Instruments that measure the same thing can have different sensitivities. The resolution of an instrument refers to the smallest change in a value that can be detected. Choosing the wrong scale can cause you to miss important data or make silly conclusions.

You also need to be able to use an instrument properly.

Errors

Even when an instrument is used correctly, the results can still show differences. Results may differ because of a **random error**. This is most likely to be due to a poor measurement being made. It could be due to not carrying out the method consistently. The effects of random errors are reduced by taking sets of repeat readings and calculating their mean values.

The **error** may be a **systematic error**. This means that the method was carried out consistently but an error was being repeated. Measurements will be consistently high or consistently low as a result of systematic error.

Anomalies

Anomalies are results that are clearly out of line. They are not those that are due to the natural variation that you get from any measurement. These should be looked at carefully. There might be a very interesting reason why they are so different. If they are simply due to a random error then they should be ignored.

If anomalies can be identified whilst you are doing an investigation, then it is best to repeat that part of the investigation. If you find anomalies after you have finished collecting the data for an investigation, then they must be discarded.

?? Did you know …?

Imagine measuring the temperature after a set time when a fuel is used to heat a fixed volume of water.

Two students repeated this experiment, four times each. Their results are marked on the thermometer scales below:

- A **precise** set of results is grouped closely together.
- An accurate set of results will have a mean (average) close to the true value.

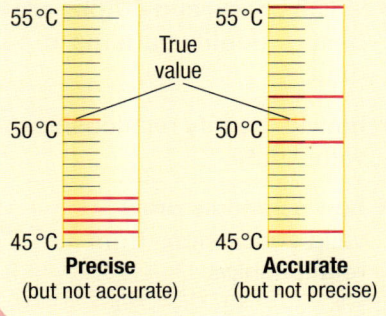

Precise (but not accurate)　　**Accurate** (but not precise)

Key points

- Care must be taken to ensure fair testing.
- You can use a trial run to make sure that you choose the best values for your variables.
- Careful use of the correct equipment can improve accuracy.
- If you repeat your results carefully you can improve precision.
- Results will nearly always vary. Better instruments give more accurate results.
- Resolution in an instrument is the smallest change that it can detect.
- Human error can produce random and systematic errors.
- You must examine anomalies.

Using data

After this topic, you should know:

- what is meant by the 'range' and the 'mean' of a set of data
- how data should be displayed
- which charts and graphs are best to identify patterns in data
- how to identify relationships within data
- how scientists draw valid conclusions from relationships
- how to evaluate the reproducibility of an investigation.

Presenting data

Tables

Tables are really good for getting your results down quickly and clearly. You should design your table before you start your investigation.

The range of the data

Pick out the maximum and the minimum values and you have the **range**. You should always quote these two numbers when asked for a range. For example, the range is between … (the lowest value) and … (the highest value) and don't forget to include the units!

The mean of the data

To find the **mean** of a set of data, add up all of the measurements and divide by how many there are.

Bar charts

If you have a categoric independent variable and a continuous dependent variable then you should use a **bar chart**.

Line graphs

If you have a continuous independent variable and a continuous dependent variable then use a **line graph**.

Scatter graphs

These are used in much the same way as a line graph, but you might not expect to be able to draw such a clear line of best fit. For example, to find out if the melting point of an element is related to its density you might draw a scatter graph of your results.

Using data to draw conclusions

Identifying patterns and relationships

Now you have a bar chart or a graph of your results you can begin looking for patterns in your results. You must have an open mind at this point.

Firstly, there could still be some anomalous results. You might not have picked these out earlier. How do you spot an anomaly? It must be a significant distance away from the pattern, not just within normal variation.

A line of best fit will help to identify any anomalies at this stage. Ask yourself – do the anomalies represent something important or were they just a mistake?

Secondly, remember a line of best fit can be a straight line or it can be a curve – you have to decide from your results.

The line of best fit will also lead you into thinking what the relationship is between your two variables. You need to consider whether your graph shows a linear relationship. This simply means can you be confident about drawing a straight line of best fit on your graph? If the answer is yes, then does this line have a positive or negative gradient?

A **directly proportional** relationship is shown by a straight line with a positive gradient that goes through the origin (0, 0).

Your results might also show a curved line of best fit. These can be predictable, complex, or very complex!

Drawing conclusions

Your graphs are designed to show the relationship between your two chosen variables. You need to consider what that relationship means for your conclusion. You must also take into account the repeatability, reproducibility, and the validity of the data you are considering.

You will continue to have an open mind about your conclusion.

You will have made a prediction. This could be supported by your results, it might not be supported, or it could be partly supported. It might suggest some other hypothesis to you.

You must be willing to think carefully about your results. Remember it is quite rare for a set of results to completely support a prediction and be completely repeatable.

Look for possible links between variables. It may be that:
- changing one has caused the other to change
- the two are related, but one is not necessarily the cause of the other.

You must decide which is the most likely. Remember a positive relationship does not always mean a causal link between the two variables.

Your conclusion must go no further than the **evidence** that you have. Any patterns you spot are only strictly valid in the range of values you tested. Further tests are needed to check whether the pattern continues beyond this range.

The purpose of the prediction was to test a hypothesis. The hypothesis can:
- be supported,
- be refuted, or
- lead to another hypothesis.

You have to decide which it is on the evidence available.

Evaluation

If you are still uncertain about a conclusion, it might be down to the repeatability, the reproducibility, and the validity of the results. You could check reproducibility by:
- looking for other similar work on the Internet or from others in your class
- getting somebody else to redo your investigation
- trying an alternative method to see if you get the same results.

Key points

- The range states the maximum and the minimum value.
- The mean is the sum of the values divided by how many values there are.
- Tables are best used during an investigation to record results.
- Bar charts are used when you have a categoric independent variable and a continuous dependent variable.
- Line graphs are used to display data that are continuous.
- Drawing lines of best fit helps us to study the relationship between variables. The possible relationships are linear, positive and negative; directly proportional; predictable and complex curves.
- Conclusions must go no further than the data available.
- The reproducibility of data can be checked by looking at other similar work done by others, perhaps on the Internet. It can also be checked by using a different method or by others checking your method.

Calcium carbonate reacts with dilute hydrochloric acid. The equation for the reaction is:

$$CaCO_3(s) + 2HCl(aq) \rightarrow CaCl_2(aq) + H_2O(l) + CO_2(g)$$

A student investigated how changing the concentration of dilute hydrochloric acid altered how quickly carbon dioxide gas was made.

The gas was collected over water and its volume measured every 20 seconds.

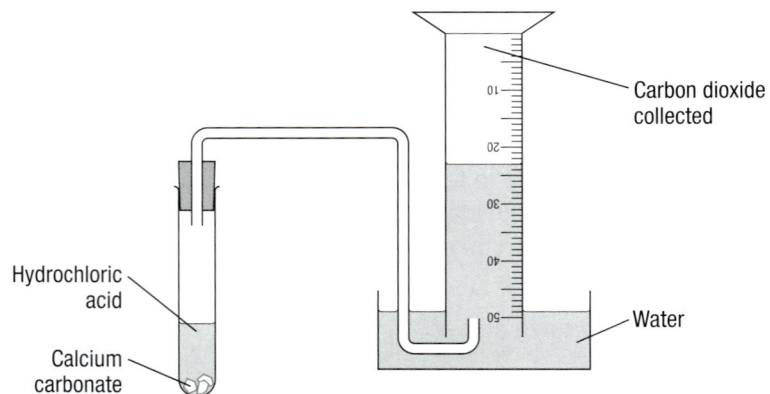

Hydrochloric acid
Calcium carbonate
Carbon dioxide collected
Water

1 (a) The diagram shows the apparatus during one experiment.

What volume of carbon dioxide gas has been collected?

23 cm³

The volume is correct and units have been given. Level has been read to bottom of the meniscus.

(1)

The graph shows the results the student obtained.

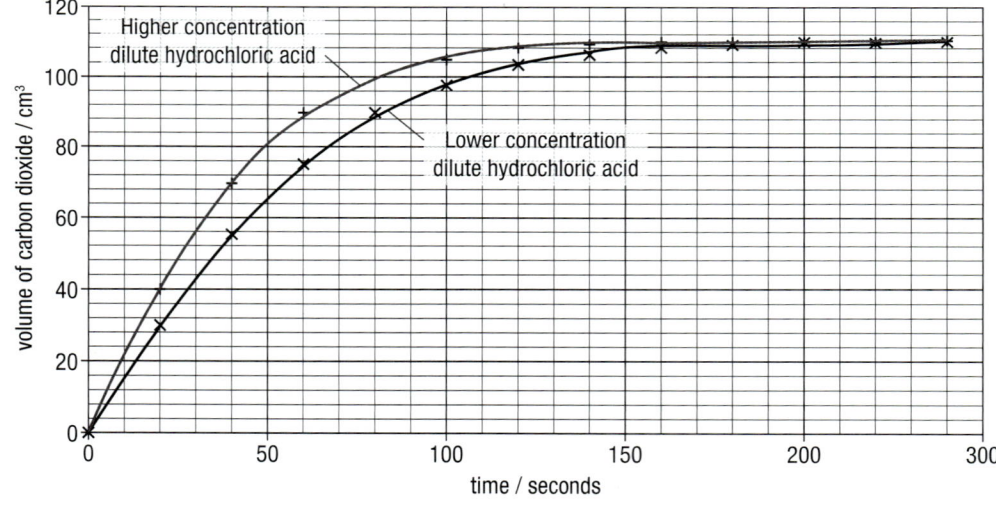

There are two marks available for the answer, so there will normally be two things that need to be said to gain full marks. This candidate has spotted that whilst the final volume of gas collected is the same (the graphs level off at the same height) the line for the more concentrated acid is steeper – this means that at any time before the reaction is complete, more gas has been collected when the more concentrated acid is used.

1 (b) Describe the effect of changing the concentration of the dilute hydrochloric acid on the volume of carbon dioxide collected during the experiment.

When the concentration of the acid is higher the gas is collected more quickly but the same final volume of gas is collected.

(2)

Due to the fact that carbon dioxide is slightly soluble in water, a small amount of it will dissolve in the water in the water bath. This will reduce the volume of gas that can be collected in the measuring cylinder.

1 (c) Carbon dioxide gas is slightly soluble in water. What effect will this have on the volume of gas collected?

It will reduce the volume of gas collected.

(1)

1 (d) The student carried out her investigation so that it was a fair test. Select one variable that must be kept constant for the experiment to be a fair test and explain why it must be kept constant.

The surface area of the calcium carbonate must be kept constant. If the surface area was higher, then there would be more collisions of hydrogen ions with calcium carbonate per second. This would make the reaction faster.

(3)

> A variable which must be controlled has been identified (other variables that must be controlled include the temperature of the acid). This scores one mark. The remaining two marks are awarded for explaining why changing the selected variable would change the results.

Another student investigated the mass of calcium carbonate used on the final volume of carbon dioxide gas collected. The method she followed was:

1. Place 25.0 cm³ of dilute hydrochloric acid in a boiling tube.

2. Add a known mass of calcium carbonate to the dilute acid in the boiling tube.

3. Connect the boiling tube to a gas syringe as shown in the diagram.

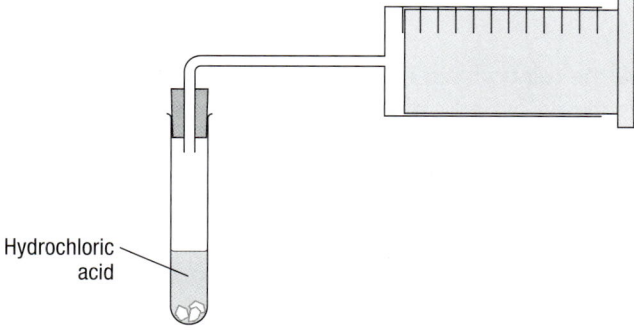

Hydrochloric acid

The graph shows the results she obtained.

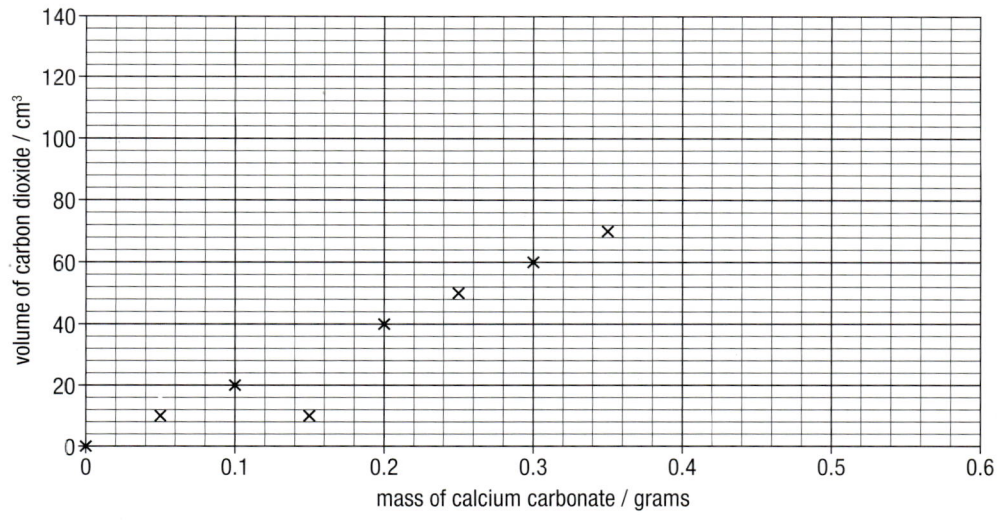

1 (e) Name an item of laboratory apparatus most suitable for measuring 25.0 cm³ of hydrochloric acid.

A volumetric pipette could be used.

(1)

> A volumetric pipette is the best apparatus to measure a fixed volume such as 25.0 cm³. A burette could also be used.

The line drawn is straight (it has been done with the aid of a rule) and it ignores the anomalous point.

An alternative way to say this is 'they are directly proportional'. A straight line through the origin shows a directly proportional relationship. If the line is straight but does not pass through the origin then you can say it is a linear relationship (but it is not directly proportional).

1 (f) Draw a line of best fit through the points on the graph.

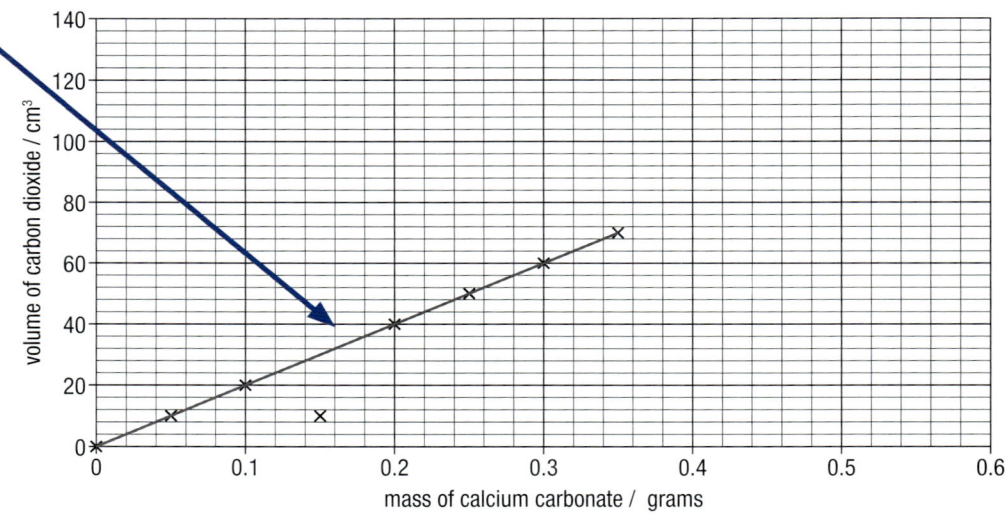

(1)

1 (g) Describe the relationship between the mass of calcium carbonate used and the volume of carbon dioxide gas collected.

As the mass of calcium carbonate is doubled, the volume of gas collected also doubles.

(2)

1 (h) There is an anomalous point on the graph. Circle the anomalous point and explain what could have happened to cause the anomalous point.

The point that is some distance from the line has been correctly identified.
The explanation gives a correct reason why the volume of gas collected could have been too small and explains why the reason given would result in too small a volume. Other possible explanations include the mass of calcium carbonate being too small so less carbon dioxide would be made.

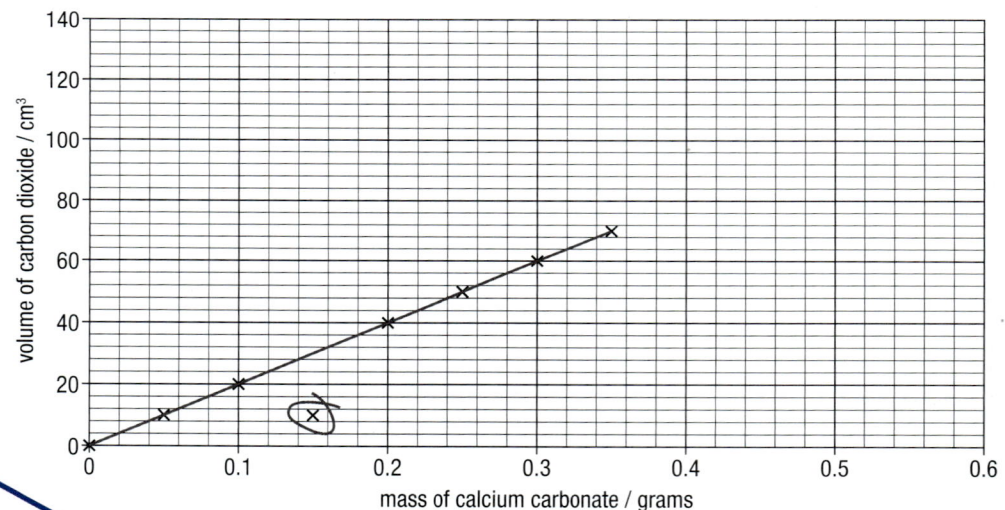

The bung may not have been replaced correctly. This would have allowed some of the gas to escape, and so less was collected.

(3)

Repeating the experiment does not necessarily make the results more accurate, but it does let us check to see if the results are precise. Calculating a mean using the repeated results will reduce any random errors and so the mean values should be more accurate.

1 (i) What should the student have done to check the precision of her results and improve the accuracy of the data she has plotted?

The experiment should have been repeated at least three times and then a mean value calculated for the volume of gas collected for each mass of calcium carbonate.

(2)

1 (j) Use the graph to predict the volume of gas that would be collected if 0.55 g of calcium carbonate were used in the experiment. Show all of your working.

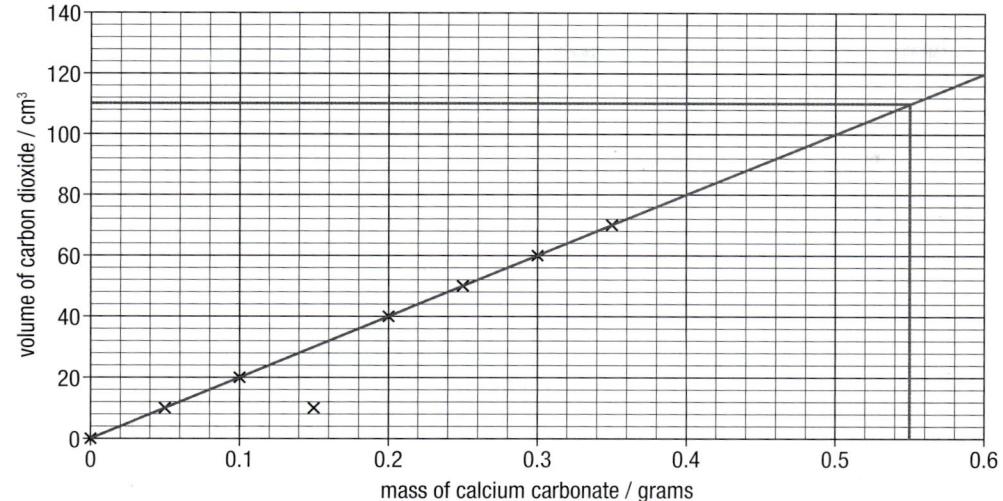

The volume of gas would be 110 cm³.

(2)

This scores both marks because the candidate has extrapolated the graph line and then read off the correct value from the graph. The value has been read from the graph accurately.

The student then repeated the experiment using 0.55 g of calcium carbonate. She obtained a much smaller volume of gas than she expected. Her teacher suggested that this was because she had not used enough acid to react with all the calcium carbonate.

1 (k) Explain how she could check whether her teacher's suggestion was correct.

She could repeat the experiment using a larger volume of acid. If the teacher is correct then the volume of gas will increase.

(2)

This scores both marks as the candidate has said what should be done and the result that should be obtained if the teacher is correct.

The student calculated the volume of gas, at 25 °C, she should have made in some of the runs. The table shows the volumes she collected and how much should have been made.

Mass calcium carbonate/g	Volume carbon dioxide collected/cm³	Theoretical volume of carbon dioxide/cm³
0.10	20	24
0.20	40	48
0.30	60	72

All of the gas volumes are too small. This means there is a problem with the procedure being used.

1 (l) What type of error do her results show?

It is a systematic error.

(1)

1 (m) Explain how an error of this type could have occurred in each run.

As soon as the calcium carbonate has been added to the acid carbon dioxide gas is made. Some of this gas escapes before the bung is placed in the boiling tube connected to the gas syringe.

(2)

This answer scores both marks as it gives a problem that would cause a systematic error and explains how that error makes the gas volume too small. Another correct answer would be if the gas was being collected at a temperature below 25 °C as at a lower temperatures the gas will have a smaller volume.

Glossary

A

Accurate A measurement is considered accurate if it is judged to be close to the true value.

Acid A sour substance which can attack metal, clothing, or skin. When dissolved in water, its solution has a pH value less than 7. Acids are proton (H^+ ion) donors.

Activation energy The minimum energy needed for a reaction to take place.

Alkali metal Elements in Group 1 of the periodic table, e.g., lithium (Li), sodium (Na), potassium (K).

Alkali Its solution has a pH value more than 7.

Alkane Saturated hydrocarbon with the general formula C_nH_{2n+2}, e.g., methane, ethane, and propane.

Alkene Unsaturated hydrocarbon which contains a carbon–carbon double bond. Its general formula is C_nH_{2n}, e.g., ethene C_2H_4.

Alloy A mixture of two or more elements, at least one of which is a metal.

Aluminium A low density, corrosion-resistant metal used in many alloys, including those used in the aircraft industry.

Anhydrous Describes a substance that does not contain water.

Anode The positive electrode in electrolysis.

Anomalies Results that do not match the pattern seen in the other data collected or are well outside the range of other repeat readings. They should be retested and if necessary discarded.

Aqueous solution The mixture made by adding a soluble substance to water.

Atmosphere The relatively thin layer of gases that surround planet Earth.

Atom The smallest part of an element that can still be recognised as that element.

Atomic number The number of protons (which equals the number of electrons) in an atom. It is sometimes called the proton number.

B

Balanced (equation) A symbol equation in which there are equal numbers of each type of atom on either side of the equation.

Bar chart A chart with rectangular bars with lengths proportional to the values that they represent. The bars should be of equal width. Also called a bar graph.

Base The oxide, hydroxide, or carbonate of a metal that will react with an acid, forming a salt as one of the products. (If a base dissolves in water it is called an alkali.) Bases are proton (H^+ ion) acceptors.

Biodegradable Materials that can be broken down by microorganisms.

Biodiesel Fuel for cars made from plant oils.

Biofuel Fuel made from animal or plant products.

Bond dissociation energy The energy required to break a specific chemical bond.

Brine A solution of sodium chloride in water.

Burette A long glass tube with a tap at one end and markings to show volumes of liquid. It is used to add precisely known volumes of liquids to a solution in a conical flask below it.

C

Carbon monoxide A toxic gas whose formula is CO.

Carbon steel Alloy of iron containing controlled, small amounts of carbon.

Catalyst A substance that speeds up a chemical reaction but remains chemically unchanged itself at the end of the reaction.

Catalytic converter Fitted to exhausts of vehicles to reduce pollutants released.

Categoric variable See Variable – categoric.

Cathode The negative electrode in electrolysis.

Chromatography The process whereby small amounts of dissolved substances are separated by running a solvent along a material such as absorbent paper.

Climate change The change in global weather patterns that could be caused by excess levels of greenhouse gases in the atmosphere.

Closed system A system in which no matter or energy enters or leaves.

Collision theory An explanation of chemical reactions in terms of reacting particles colliding with sufficient energy for a reaction to take place.

Compound A substance made when two or more elements are chemically bonded together. For example, water (H_2O) is a compound made from hydrogen and oxygen.

Contact process The industrial process for the manufacture of sulfuric acid, H_2SO_4.

Continuous variable See Variable – continuous.

Control group If an experiment is to determine the effect of changing a single variable, a control is often set up in which the independent variable is not changed, therefore enabling a comparison to be made. If the investigation is of the survey type a control group is usually established to serve the same purpose.

Control variable See Variable – control.

Covalent bond The bond between two atoms that share one or more pairs of electrons.

Covalent bonding The attraction between two atoms that share one or more pairs of electrons.

Cracking The reaction used in the oil industry to break down large hydrocarbons into smaller, more useful ones. This occurs when the hydrocarbon vapour is either passed over a hot catalyst or mixed with steam and heated to a high temperature.

D

Data Information, either qualitative or quantitative, that has been collected.

Delocalised electron Bonding electron that is no longer associated with any one particular atom.

Dependent variable See Variable – dependent.

Diffusion The automatic mixing of liquids and gases as a result of the random motion of their particles.

Directly proportional A relationship that, when drawn on a line graph, shows a positive linear relationship that passes through the origin.

Displacement reaction A reaction in which a more reactive element takes the place of a less reactive element in one of its compounds or in solution.

Distillation Separation of a liquid from a mixture by evaporation followed by condensation.

Dot and cross diagram A drawing to show only the arrangement of the outer shell electrons of the atoms or ions in a substance.

Double bond A covalent bond made by the sharing of two pairs of electrons.

E

Electrolysis The breakdown of a substance containing ions into elements by electricity.

Electrolyte A liquid, containing free-moving ions, which is broken down by electricity in the process of electrolysis.

Electron A tiny particle with a negative charge. Electrons orbit the nucleus in atoms or ions.

Electronic structure A set of numbers to show the arrangement of electrons in their shells (or energy levels), e.g., the electronic structure of a potassium atom is 2,8,8,1.

Electroplating The process of depositing a thin layer of metal on an object during electrolysis.

Element A substance made up of only one type of atom. An element cannot be broken down chemically into any simpler substance.

Empirical formula The simplest ratio of atoms of each element in a compound.

End point The point in a titration where the reaction is complete and titration should stop.

Endothermic A reaction that *takes in* energy from the surroundings.

Energy level see Shell.

Equilibrium The point in a reversible reaction in which the forward and backward rates of reaction are the same. Therefore, the amounts of substances present in the reacting mixture remain constant.

Error Sometimes called an uncertainty.

Error – random Causes readings to be spread about the true value, due to results varying in an unpredictable way from one measurement to the next. Random errors are present when any measurement is made, and cannot be corrected. The effect of random errors can be reduced by making more measurements and calculating a new mean.

Error – systematic Causes readings to be spread about some value other than the true value, due to results differing from the true value by a consistent amount each time a measurement is made. Sources of systematic error can include the environment, methods of observation, or instruments used. Systematic errors cannot be dealt with by simple repeats. If a systematic error is suspected, the data collection should be repeated using a different technique or a different set of equipment, and the results compared.

Ethene An alkene with the formula C_2H_4.

Evidence Data which has been shown to be valid.

Exothermic A reaction that *gives out* energy to the surroundings.

F

Fair test A fair test is one in which only the independent variable has been allowed to affect the dependent variable.

Fermentation The reaction in which the enzymes in yeast turn glucose into ethanol and carbon dioxide.

Flammable Easily ignited and capable of burning rapidly.

Fraction Groups of hydrocarbons separated from crude oil. Each fraction contains molecules with a similar number of carbon atoms.

Fractional distillation A way to separate liquids from a mixture of liquids by boiling off the substances at different temperatures, then condensing and collecting the liquids.

Fuel cell An electrical cell in which the energy released in the oxidation of a fuel is used to generate electricity. Typical fuels used are hydrogen or methane gases.

Fullerene Form of the element carbon that can exist as large cage-like structures, based on hexagonal rings of carbon atoms.

Functional group An atom or group of atoms that give organic compounds their characteristic reactions.

G

Gas A state of matter.

Giant covalent structure A huge 3-D network of covalently-bonded atoms (e.g., the giant lattice of carbon atoms in diamond or graphite).

Giant lattice A huge 3-D network of atoms or ions (e.g., the giant ionic lattice in sodium chloride).

Giant structure See Giant lattice.

Global dimming The reflection of sunlight by tiny solid particles in the air.

Group All the elements in the columns (labelled 1 to 7 and 0) down the periodic table.

H

Half equation An equation that describes reduction (gain of electrons) or oxidation (loss of electrons), such as the reactions that take place at the electrodes during electrolysis. For example: $Na^+ + e^- \rightarrow Na$.

Halogens The elements found in Group 7 of the periodic table.

Hazard Something (e.g., an object, a property of a substance, or an activity) that can cause harm.

High-alloy steel Expensive alloy of iron mixed with relatively large proportions of other metals, e.g., stainless steel which contains nickel and chromium along with the iron.

Homologous series A group of related organic compounds that have the same functional group, e.g., the molecules of the homologous series of alcohols all contain the –OH group.

Hydrated Describes a substance that contains water in its crystals, e.g., hydrated copper sulfate.

Hydration A reaction in which water (H_2O) is chemically added to a compound.

Hydrocarbon A compound containing only hydrogen and carbon.

Hypothesis A proposal intended to explain certain facts or observations.

I

Incomplete combustion When a fuel burns in insufficient oxygen, producing carbon monoxide as a toxic product or carbon (soot) particles.

Independent variable See Variable – independent.

Inert Unreactive.

Intermolecular forces The attraction between the individual molecules in a covalently-bonded substance.

Interval The quantity between readings, for example, a set of 11 readings equally spaced over a distance of 1 m would have an interval of 10 cm.

Ion A charged particle produced by the loss or gain of electrons.

Ionic bond The electrostatic force of attraction between positively and negatively charged ions.

Ionic bonding The electrostatic force of attraction between positively and negatively charged ions.

Ionic equation An equation that shows only those ions or atoms that change in a chemical reaction.

Isotope Atoms that have the same number of protons but a different number of neutrons, i.e., they have the same atomic number but different mass numbers.

L

Limewater The common name for calcium hydroxide solution.

Line graph Used when both variables are continuous. The line should normally be a line of best fit, and may be straight or a smooth curve.

Liquid A state of matter.

Low-alloy steel Alloy of iron containing small amounts (1 to 5 per cent) of other metals.

M

Macromolecule Giant covalent structure.

Mass number The number of protons plus neutrons in the nucleus of an atom.

Mean The arithmetical average of a series of numbers.

Metal ore A rock in which there is sufficient metal compound (or occasionally metal) to make it economically viable to extract the metal.

Mixture When some elements or compounds are mixed together and intermingle but do not react together (i.e., no new substance is made). A mixture is not a pure substance.

Mole The amount of substance in the relative atomic or formula mass of a substance in grams.

Molecular formula The chemical formula that shows the actual numbers of atoms of each element in a particular molecule (e.g., C_2H_4).

Molecule A group of atoms bonded together, e.g., PCl_5.

Monomers Small reactive molecules that react together in repeating sequences to form a very large molecule (a polymer).

N

Nanoscience The study of very tiny particles or structures between 1 and 100 nanometres in size, where 1 nanometre = 10^{-9} metres.

Neutral A solution with a pH value of 7 which is neither acidic nor alkaline. Alternatively, something that carries no overall electrical charge, that is, neither positively nor negatively charged.

Neutralisation The chemical reaction of an acid with a base in which they cancel each other out, forming a salt and water. If the base is a carbonate or hydrogen carbonate, carbon dioxide is also produced in the reaction.

Neutron A dense particle found in the nucleus of an atom. It is electrically neutral, carrying no charge.

Nitrogen oxide Gaseous pollutant given off from motor vehicles. It is a cause of acid rain.

Noble gases The very unreactive gases found in Group 0 of the periodic table.

Non-renewable Something which cannot be replaced once it is used up.

Nucleus (of an atom) The very small and dense central part of an atom which contains protons and neutrons.

O

Oxidation The reaction when oxygen is added to a substance (or when electrons are lost).

Oxidised A reaction where oxygen is added to a substance (or when electrons are lost from a substance).

P

Particulate Small solid particle, e.g., carbon (soot), given off from motor vehicles as a result of incomplete combustion of its fuel.

Percentage yield The actual mass of product collected in a reaction divided by the maximum mass that could have been formed in theory, multiplied by 100.

Periodic table An arrangement of the elements in the order of their atomic numbers, forming groups and periods.

pH scale A number which shows how strongly acidic or alkaline a solution is. Acids have a pH value of less than 7 (pH 0 is strongly

acidic). Alkalis have a pH value above 7 (pH 14 is strongly alkaline). A neutral liquid has a pH value of 7.

Pipette A glass tube used to measure accurate volumes of liquids.

Polymer A substance made from very large molecules made up of many repeating units, e.g., poly(ethene).

Polymerisation The reaction of monomers to make a polymer.

Precipitate An insoluble solid formed by a reaction taking place in solution.

Precise A precise measurement is one in which there is very little spread about the mean value. Precision depends only on the extent of random errors – it gives no indication of how close the results are to the true value.

Precision A precise set of repeat readings will be closely grouped together.

Prediction A forecast or statement about the way something will happen in the future. In science it is not just a simple guess, because it is based on some prior knowledge or on a hypothesis.

Product A substance made as a result of a chemical reaction.

Propene An alkene with the formula C_3H_6.

Proton A tiny positive particle found inside the nucleus of an atom.

R

Random error See Error – random.

Range The maximum and minimum values of the independent or dependent variables. It is important in ensuring that any pattern is detected.

Reactant A substance you start with before a chemical reaction takes place.

Reactivity series A list of elements in order of their reactivity. The

most reactive element is put at the top of the list.

Reduction A reaction in which oxygen is removed (or electrons are gained).

Relationship The link between the variables that were investigated. These relationships may be: causal, i.e., changing x is the reason why y changes; by association, i.e., both x and y change at the same time, but the changes may both be caused by a third variable changing; by chance occurrence.

Relative atomic mass, A_r The average mass of the atoms of an element compared with carbon-12 (which is given a mass of exactly 12). The average mass must take into account the proportions of the naturally occurring isotopes of the element.

Relative formula mass, M_r The total of the relative atomic masses, added up in the ratio shown in the chemical formula of a substance.

Repeatable A measurement is repeatable if the original experimenter repeats the investigation using the same method and equipment and obtains the same results.

Reproducible A measurement is reproducible if the investigation is repeated by another person, or by using different equipment or techniques, and the same results are obtained.

Resolution This is the smallest change in the quantity being measured (input) of a measuring instrument that gives a perceptible change in the reading.

Reversible reaction A reaction in which the products can re-form the reactants.

R_f (retention factor) A measurement from chromatography. It is the distance a spot of substance has been carried above the baseline divided by the distance of the solvent front.

Risk The likelihood that a hazard will actually cause harm. We can reduce risk by identifying the hazard and doing something to protect against that hazard.

Rusting The corrosion of iron.

S

Salt A compound formed when some or all of the hydrogen in an acid is replaced by a metal (or by an ammonium ion), for example, potassium nitrate, KNO_3 (from nitric acid).

Saturated hydrocarbon A hydrocarbon with only single bonds between its carbon atoms. This means that it contains as many hydrogen atoms as possible in each molecule.

Shell (or energy level) An area in an atom, around its nucleus, where electrons are found.

Smart polymer Polymers that change in response to changes in their environment.

Solid A state of matter.

Stainless steel A chromium-nickel alloy of steel which does not rust.

State symbol The abbreviations used in balanced symbol equations to show if reactants and products are solid (s), liquid (l), gas (g) or dissolved in water (aq).

State of matter Tells us whether a substance is a solid, a liquid, or a gas.

Steel An alloy of iron with small amounts of carbon or other metals, such as nickel and chromium, added.

Sulfur dioxide A toxic gas whose formula is SO_2. It causes acid rain.

Symbol equation A balanced chemical equation showing the formula of each reactant and product in the reaction, e.g., $H_2 + Cl_2 \rightarrow 2HCl$

Systematic error See Error – systematic.

T

Thermal decomposition The breakdown of a compound by heating it.

Thermosetting polymer Polymer that can form extensive cross-linking between chains, resulting in rigid materials which are heat-resistant.

Thermosoftening polymer Polymer that forms plastics which can be softened by heating, then remoulded into different shapes as they cool down and set.

Titration A method for measuring the volumes of two solutions that react together.

Transition element Element from the central block of the periodic table. It has typical metallic properties and forms coloured compounds.

U

Universal indicator A mixture of indicators that can change through a range of colours depending on the pH of a solution. Its colour is matched to a pH number using a pH scale. It shows how strongly acidic or alkaline liquids and solutions are.

Unsaturated hydrocarbon A hydrocarbon whose molecules contains at least one carbon–carbon double bond.

V

Valid Suitability of the investigative procedure to answer the question being asked.

Variable Physical, chemical, or biological quantity or characteristic.

Variable – categoric Categoric variables have values that are labels. For example, names of plants or types of material.

Variable – continuous Can have values (called a quantity) that can be given by measurement (e.g., light intensity, flow rate, temperature).

Variable – control A variable which may, in addition to the independent variable, affect the outcome of the investigation and therefore has to be kept constant or at least monitored.

Variable – dependent The variable for which the value is measured for each and every change in the independent variable.

Variable – independent The variable for which values are changed or selected by the investigator.

Viscosity The resistance of a liquid to flowing or pouring; a liquid's 'thickness'.

W

Word equation A way of describing what happens in a chemical reaction by showing the names of all reactants and the products they form.

Y

Yield See Percentage yield.

Answers

1 Atomic structure

1.1

1 E.g.

	General properties	Average distance between particles	Arrangement of particles	Movement of particles
Solid	Fixed shape; incompressible	Particles are touching	Regular pattern	Vibrate on the spot
Liquid	No fixed shape; can flow; very difficult to compress	Most particles are touching	Irregular, random	Slip and slide over and around each other
Gas	No fixed shape; spreads out to fill its container; easily compressed	Large distances	Irregular, random	Can move very quickly. In a random manner, between collisions

2 As the particles of gas are cooled down their average speed decreases until at its condensation point the particles become much closer together. They form a liquid in which the particles are moving randomly, slipping over and around each other. As the liquid is cooled, the average speed of the particles decreases until at its freezing point the particles stop moving around randomly and remain in fixed positions, vibrating on the spot. As the solid is cooled the average rate of the vibrations slows down.

3 a freeze/solidify   d boil/evaporate
   b condense   e sublime
   c melt

4 As the temperature is increased, the particles in a gas gain energy and their average speed is increased and they take up more space if the pressure remains constant. Therefore, the density decreases as the same mass of gas occupies a larger volume.
   By compressing a gas the pressure is increased as particles collide with the walls of their container more frequently. So, the same mass of gas occupies a smaller volume and the density of the gas increases.

5 The strength of the attractive forces between particles varies in different substances so those with stronger forces of attraction will have higher melting points than those with weaker forces of attraction.

6 Plan a fair test varying one factor, e.g., temperature of the water or surface area of the paper towel, keeping all other variables constant. Monitor the rate of evaporation by measuring the mass of the wet paper towel on the electronic balance at regular time intervals.

1.2

1 a The food particles released from the hot food diffuse through the air outside the shop and passers-by detect these with their sense of smell.
   b The particles of water move around faster in hot water than in cold water so they collide into the sugar particles in the solid more frequently and with more energy. This breaks down the regular arrangement of sugar particles in the solid and spreads them out more quickly as the sugar solution is formed.
   c Hot water releases water particles into the air as its most energetic particles are moving fast enough to escape from their neighbouring water particles at the surface. The glass in the window is colder than the air so any water particles diffusing through the air to the window will cool down and slow down; then the vapour will condense back to liquid on the glass.
   d Petrol is a volatile liquid so particles can evaporate from its surface quite easily. It is also flammable. In a petrol station there are likely to be particles from petrol diffusing through the air which could be ignited by any naked flame or a lighted cigarette.

2 a The movement of one substance through another due to the random motion of the particles in liquids or gases.
   b The average speed of the particles in a gas is greater than that in a liquid so the mixing together of particles happens more quickly in gases.

3 The jittery movement of the pollen grains must be caused by collisions with the particles in liquid water which are too small to be seen under the microscope.

4 As hydrogen bromide is heavier than hydrogen chloride the white ring of smoke would form nearer the end of the tube releasing particles (molecules) of hydrogen bromide.

ammonia + hydrogen bromide → ammonium bromide
$NH_3(g)$ + $HBr(g)$ → $NH_4Br(s)$

1.3

1 a C Atoms are solid spheres that cannot be split into simpler particles.
   b soda and lime
2 a The electron
   b A diffuse cloud of positive charge with tiny negatively charged electrons spread throughout.
3 Rutherford said that the positive charge in an atom was concentrated into a very small volume at the centre of the atom (in the nucleus) and that the electrons were orbiting the centre of the atom/nucleus.
4 He found that energy emitted from electron transitions could only have certain fixed energies, so he refined the 'orbiting electrons' in Rutherford's nuclear model to 'orbiting electrons in energy levels (or shells) at fixed distances from the nucleus'.

1.4

1 a

Metals	Non-metals
Barium	Phosphorus
Vanadium	Krypton
Mercury	
Potassium	
Uranium	

   b Non-metal because it does not have the physical properties of a metal, e.g., it does not conduct electricity.
2 A mixture of elements can be separated by physical means because atoms from the different elements are not chemically bonded to each other whereas they are if they have reacted together.
3 One diagram showing an element in which all the atoms present must of the same type and another showing a compound containing more than one type of atom chemically combined.
4 A tiny central nucleus surrounded by orbiting electrons.
5 a Na from Latin *natrium*   c Pb from Latin *plumbum*
   b Au from Latin *aurum*   d K from Latin *kalium*

1.5

1

Sub-atomic particle	Location	Relative charge
Proton	In the nucleus	+1
Neutron	In the nucleus	0
Electron	Orbiting the nucleus	−1

2 a It increases by 1.
   b It increases by 8.
   c Metals – Li, Be, Na, Mg, Al
      Non-metals – H, He, C, N, O, F, Ne, P, S, Cl, Ar
      Metalloids – B, Si
3 Atoms contain an equal number of protons (carrying a relative charge of +1) and electrons (carrying a relative charge of −1) so the opposite charges cancel each other out.
4 Cobalt, Co, atomic number = 27, mass number = 59.
5 a 7 protons, 7 electrons, and 7 neutrons.
   b 17 protons, 17 electrons, and 18 neutrons.
   c 47 protons, 47 electrons, and 61 neutrons.
   d 92 protons, 92 electrons, and 143 neutrons.

1.6

1 a The first or innermost shell.
   b 2 in the first then 8 in the second.
2 a 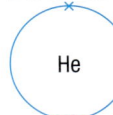   b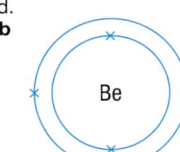

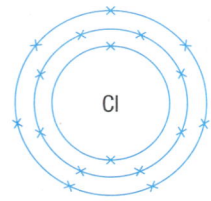

c (Cl)   d (Ar)

3   **a**   2,8,8,1   **b**   1 electron
4   phosphorus, P
5   **a**   Because they all have one electron (the same number of electrons) in the highest energy level (outermost shell) of their atoms.
    b   lithium   + oxygen →   lithium oxide
       sodium   + oxygen →   sodium oxide
       potassium + oxygen → potassium oxide

1.7

1

Sub-atomic particle	Relative mass	Relative charge
Proton	1	+1
Neutron	1	0
Electron	Very small $\left(\frac{1}{2000}\right)$	−1

2   **a**   5 protons and 6 neutrons   **d**   17 protons and 20 neutrons
    b   7 protons and 7 neutrons   **e**   53 protons and 74 neutrons
    c   12 protons and 12 neutrons
3   Atoms of the same element with different numbers of neutrons/Atoms with the same number of protons but different numbers of neutrons/ Atoms with the same atomic number but different mass numbers
4   **a**   Density
    b   They have the same electronic structures so have the same number of electrons in their highest energy level/outermost shell.
5   **a**   0.00024 g/cm³   **c**   Radioactive
    b   Highly flammable

Answers to end of chapter summary questions

1   **a**   The change from a solid directly to a gas.
    b

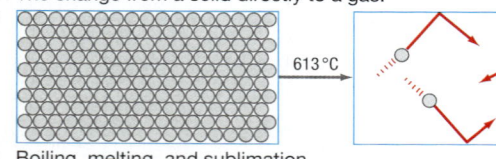

613 °C

    c   Boiling, melting, and sublimation
2   **a**   Particles (molecules) of hydrogen sulfide and sulfur dioxide diffuse through the particles (molecules) of air in the long tube. Where they meet, they collide and react with each other to form the yellow powder.
    b   Diffusion
    c   Because the particles (molecules) of sulfur dioxide travel down the tube/diffuse more slowly than the particles (molecules) of hydrogen sulfide (showing sulfur dioxide molecules have a greater mass than hydrogen sulfide molecules).
    d   A long tube showing a jagged, random path
    e   The yellow powder would appear sooner because the molecules of gas move faster at higher temperatures.
3   **a**   Bromine – liquid, Caesium – solid, Fluorine – gas, Strontium – solid, Xenon – gas
    b   Caesium
    c   The two gases have much lower densities because they have large spaces between their particles, unlike the solids and liquid whose particles are touching (so many more particles fit into the same volume compared with the gases).
    d   Bromine – Br, Caesium – Cs, Fluorine – F, Strontium – Sr, Xenon – Xe
    e   Bromine – non-metal, Caesium – metal, Fluorine – non-metal, Strontium – metal, Xenon – non-metal
4   **a**   **i**   Protons and neutrons
       ii   2 electrons in the first shell and 8 in the second
    b   **i**   It is neutral because the number of protons (+) and electrons (–) is the same in an atom so the charges cancel out.
       ii   Atomic number is the number of protons.
          Mass number is the number of protons plus neutrons.
5   **a**   **i**   Non-metal   **v**   Noble gases, Group 0
       ii   Metals   **vi**   Helium, Argon, Krypton, Xenon, Radon
       iii   10 protons   **vii**   2,8
       iv   10 neutrons   **viii**   They are very stable arrangements of electrons.
    b   **i**   88
       ii   2 electrons because it is in Group 2

---

    iii   Metal
    iv   2,8,8,2
    v   Medical use of radioactive isotopes – e.g., tracing the movement of materials inside the body; treating cancer/chemotherapy Industrial use – e.g., nuclear fuel rods to generate electricity/ gauging thickness of material in manufacturing/finding faults in materials

Answers to end of chapter practice questions

1   Solid   *(1 mark)*
    Gas   *(1 mark)*
    Liquid   *(1 mark)*
2   **a**   Element – all atoms are the same.   *(1 mark)*
       Compound – two different atoms/elements combined together.   *(1 mark)*
    b   Ammonia is a compound because only one type of molecule, each containing two types of atom.   *(1 mark)*
       Air is a mixture because more than one type of molecule.   *(1 mark)*
    c   **i**   Condensation/liquefaction   *(1 mark)*
       ii   Evaporation/boiling   *(1 mark)*
       iii   Sublimation   *(1 mark)*
3   **a**   Blue   *(1 mark)*
       Darker at bottom of beaker/lighter at top of beaker   *(1 mark)*
    b   Becomes smaller/disappears   *(1 mark)*
    c   Dissolving/solution   *(1 mark)*
       Diffusion   *(1 mark)*
4   **a**   White solid/smoke   *(1 mark)*
       Closer to ammonia end   *(1 mark)*
    b   g   g   s   *(1 mark)*
    c   They travel further than hydrogen chloride molecules.   *(1 mark)*
5   **a**   Atoms of same element with same numbers of protons but different numbers of neutrons.   *(1 mark)*
    b   Same number of electrons (in outer shell)   *(1 mark)*
    c   **i**   10   *(1 mark)*
       ii   10   *(1 mark)*
       iii   10   *(1 mark)*
       iv   22   *(1 mark)*
    d $^{10}_{20}\text{Ne}$ *(1 mark)*
6   **a**   2   *(1 mark)*
    b   8   *(1 mark)*
    c   3   *(1 mark)*
    d   5   *(1 mark)*
    e   16   *(1 mark)*

2 Structure and bonding

2.1

1   **a**   Covalent bond
    b   Ionic bond
2   The atoms of Group 1 lose their single electron in the outermost shell (highest energy level) by transferring it to a Group 7 atom, which has seven electrons in the outermost shell (highest energy level). The atoms become ions which have the stable electronic arrangements of a noble gas.
3   **a**   2,8 for the Al^{3+} ion (3 electrons lost)
    b   2,8 for the $F^−$ ion (1 electron gained)
    c   2,8,8 for the K^+ ion (1 electron lost)
    d   2,8 for $O^{2−}$ (2 electrons gained)

2.2

1

Atomic number	Atom	Electronic structure of atom	Ion	Electronic structure of ion
9	F	2,7	$F^−$	2,8
3	Li	2,1	Li^+	2
16	S	2,8,6	$S^{2−}$	2,8,8
20	Ca	2,8,8,2	Ca^{2+}	2,8,8

2   Metal atoms lose electrons when they form ions (therefore their ions have more positively-charged protons than negatively-charged electrons) resulting in positively-charged ions. Non-metal atoms gain electrons when they form ions (therefore their ions have more negatively-charged electrons than positively-charged protons) resulting in negatively-charged ions.
3   **a**   Groups 1, 2, and 3; the charge on their ions = (group number)+
    b   Groups 5, 6, and 7; the charge on their ions = (8 – group number)–
4   **a**   In KBr the 1– charge on the Br⁻ ion is cancelled out by the 1+ charge on K^+ ion whereas two K^+ ions are needed to cancel out the 2– charge on an $O^{2−}$ ion.
    b   In MgO the 2– charge on the $O^{2−}$ ion is cancelled out by the 2+ charge on Mg^{2+} ion whereas two Cl⁻ ions are needed to cancel out the 2+ charge on an Mg^{2+} ion.

5 a

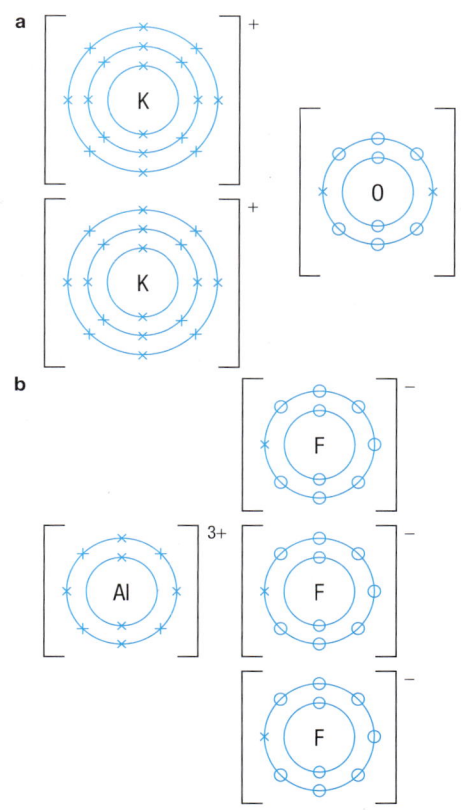

b

3 a

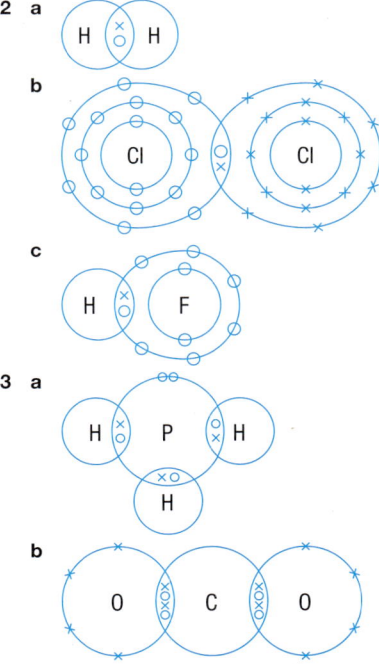

2.3

1 Hydrogen iodide, sulfur dioxide, and nitrogen(III) chloride because these contain only non-metallic elements.

2 a
b
c

3 a
b

4 Hydrogen attains the electronic structure of helium.
Chlorine attains the electronic structure of argon.

5 The electrostatic attraction between the bonding pair of negatively charged electrons for each of the positively charged nuclei on either side of the pair hold the atoms together.

2.4

1 The particles are arranged in regular patterns.

2 a Because the metal atoms donate their outer shell electrons into a 'sea' of electrons so have more protons (+) than electrons (–).
b Electrons that are free-moving within a structure and are no longer associated with any particular atom.

3 Magnesium atoms have two electrons in their outer shell (highest energy level). These are donated into the 'sea' of delocalised electrons which are free to move around. The electrostatic attraction of the negatively charged delocalised electrons for the positively charged Mg^{2+} ions bonds the ions in the giant metallic lattice.

4 The 'glue' model can help us imagine the positively charged metal ions being stuck in position in the giant lattice but a glue itself becomes solid when it sets so is not helpful in visualising the fluid movement of the delocalised electrons throughout the structure.

2.5

1 There are strong electrostatic forces of attraction between the oppositely charged ions in the giant lattice, acting in all directions, which result in a lot of energy being required to separate the ions in the process of melting.

2 The ions in the solid are fixed in position and are only free to move (and carry their charge to electrodes) when they are molten or dissolved in water.

3 There is a higher concentration of ions from the dissolved salts in seawater than there are in freshwater.

4 Positive electrode – chloride ions, bromide ions, oxide ions, iodide ions
Negative electrode – lithium ions, calcium ions, sodium ions, zinc ions, barium ions

5 Aluminium oxide because it contains 3+ ions as opposed to the 1+ ions in sodium oxide so you can expect stronger electrostatic forces of attraction between the metal ions and the oxide ions (the smaller size of the aluminium ions also contributes towards this).

2.6

1 Forces of attraction between molecules

2 a Oxygen
b Chlorine
c Hydrogen sulfide; hydrogen; sulfur

3 a It has a giant covalent structure with strong covalent bonds holding the atoms strongly into the regular giant lattices.
b Nitrogen gas has a very strong triple covalent bond between its nitrogen atoms within each N_2 molecule but only very weak forces of attraction between the N_2 molecules/weak intermolecular forces between N_2 molecules. Therefore molecules easily separated from each other.

4 Good electrical insulator/does not conduct electricity as there is no overall charge on its molecules.

5 There are only weak intermolecular forces between HCl molecules whereas in the giant ionic lattice of sodium chloride, very strong ionic bonds are present, which form strong electrostatic forces of attraction in all directions.

2.7

1 a Diamond, graphite, fullerenes
b Allotropes

2 Very high melting point and boiling point, hard, insoluble in water, electrical insulator

3 a Delivering drugs to parts of the body where they are needed
b The shape of the C_{60} molecule was similar to a building in Montreal designed by the architect Buckminster Fuller in 1967.

4 The layers of graphite atoms have weak forces between them so can slip and slide over each other to act as a lubricant.

5 The carbon atoms in graphite's layers are arranged in hexagons, with each carbon atom forming three strong covalent bonds to its nearest neighbours. As carbon atoms have four electrons in their outer shell available for bonding, this leaves one spare outer electron on each carbon atom. This electron is free to move along the layers of carbon atoms. So these free-moving delocalised electrons can drift along the layers enabling graphite to conduct electricity. In diamond all the outer shell electrons are involved in covalent bonding so there are no free electrons.

2.8

1 a Because the layers of atoms (positively charged ions) can slide over each other relatively easily
b i Malleable
ii Ductile

2 a The differently sized atoms inserted into the regular arrangement of metal atoms (positively charged ions) is disrupted, making it more difficult for the layers to slip past each other.
b A mixture of metals that once deformed will revert to its original shape at a certain temperature

3 Nitinol can apply a force to the teeth once it is warmed up in the mouth as it changes to its original shape whereas a traditional metal must be adjusted manually to change the force applied to the teeth.

4 They conduct electricity because the 'sea' of delocalised electrons can drift through the metal's giant structure because they are free-moving.
They conduct heat because the free-moving delocalised electrons gain energy when heated and can move around more quickly within the metal, spreading the energy through the metal quickly.

2.9

1 The study of particles between 1 and 200 nm in size

2 As an antibactericide in fridges and in sprays in operating theatres

3 a They have a large surface area to volume ratio.
  b Explosions caused by sparks/particles escaping into the air or dangers of being breathed in.
4 a In sun-screens to block ultraviolet light and in face creams to deliver active ingredients deeper beneath the surface of the skin
  b Delivering drugs to the tumour itself inside the body or being absorbed by tumours and then absorbing energy from lasers to damage the tumour by affecting its proteins
5 Student's views expressed in a balanced argument.

Answers to end of chapter summary questions

1 a 1   **d** 6
  b 2   **e** 4
  c 7
2 a

Giant covalent	Giant ionic	Simple molecules
silicon dioxide	magnesium oxide	ammonia
graphite	lithium chloride	hydrogen bromide

  b Ammonia, and hydrogen bromide
  c Graphite because although it has a giant covalent structure it does conduct electricity
  d

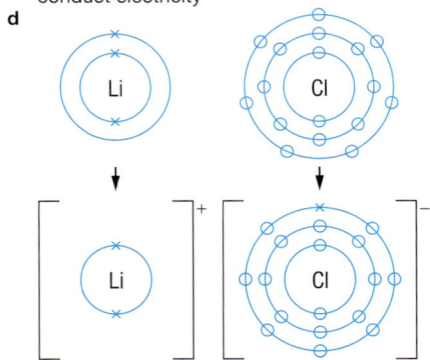

3 a Hydrogen iodide, chlorine(VII) oxide, phosphorus(V) fluoride
  b They contain only non-metallic elements.
  c Ionic bonding
  d i HI   **ii** $CaBr_2$
4

	chloride, Cl^-	oxide, O^{2-}	sulfate, SO_4^{2-}	phosphate(V), PO_4^{3-}
potassium, K^+	KCl	K_2O	K_2SO_4	K_3PO_4
magnesium, Mg^{2+}	$MgCl_2$	MgO	$MgSO_4$	$Mg_3(PO_4)_2$
iron(III), Fe^{3+}	$FeCl_3$	Fe_2O_3	$Fe_2(SO_4)_3$	$FePO_4$

5 a    **b**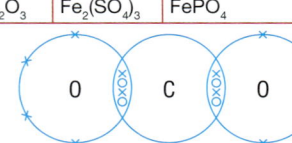

Answers to end of chapter practice questions

1 a Electrons transfer   *(1 mark)*
    From lithium to oxygen   *(1 mark)*
    Each lithium atoms loses one electron, each oxygen atom gains two electrons.   *(1 mark)*
  b Li^+ *(1 mark)*
    O^{2-} *(1 mark)*
  c Ions attract each other strongly.   *(1 mark)*
    Many bonds need to be broken.   *(1 mark)*
  d Ions cannot move when solid.   *(1 mark)*
    Ions move to the electrodes when molten.   *(1 mark)*
2 a Diagram showing two electrons between H and F   *(1 mark)*
    Six more electrons in outer shell of F   *(1 mark)*
  b Covalent   *(1 mark)*
  c Attractions between molecules are weak.   *(1 mark)*
    Easily overcome/need little energy to break   *(1 mark)*
  d H–Br   *(1 mark)*
3 a Covalent bonds   *(1 mark)*
    Many/strong bonds   *(1 mark)*
    Hard to overcome/need a lot of energy to break   *(1 mark)*
  b No ions   *(1 mark)*
    No electrons free to move   *(1 mark)*
4 a Carbon   *(1 mark)*
    In hexagonal rings   *(1 mark)*
  b In drug delivery   *(1 mark)*

c 1–200 nanometres   *(1 mark)*
d High surface area to volume ratio   *(1 mark)*
5 a i Atoms/ions in layers   *(1 mark)*
       Layers slide   *(1 mark)*
  ii Delocalised/free electrons   *(1 mark)*
       Move   *(1 mark)*
b i Alloy   *(1 mark)*
  ii Atoms of other metals are different in size   *(1 mark)*
       Make it harder for layers of copper atoms to slide   *(1 mark)*

3 The periodic table

3.1

1 a Repeated at regular intervals
  b i A vertical column of elements
    ii A horizontal row of elements
2 There are many more metals.
3 They have the same number of electrons in the highest energy level (outermost shell).
4 a 2   **b** 3   **c** 1   **d** 2
  e 8   **f** 2   **g** 8   **h** 7
5 Because of their very stable electron arrangements

3.2

1 To keep them out of contact with air/oxygen and water vapour
2 The melting point decreases with increasing atomic number/going down the group.
3 They all have one electron in their outermost shell (highest energy level) which they lose when forming ions.
4 The metal would explode on contact with water as hydrogen gas is liberated very quickly. The alkali metals get more reactive going down the group so caesium, being near the bottom of Group 1, is extremely reactive. (Alkaline caesium hydroxide would also be formed in the reaction.)

caesium + water → caesium hydroxide + hydrogen
$2Cs(s) + 2H_2O(l) → 2CsOH(aq) + H_2(g)$

5 a $2Cs(s) + I_2(g) → 2CsI(s)$ **b** $2Cs(s) + Br_2(g) → 2CsBr(s)$

3.3

1 a They are good conductors of electricity and heat, are hard and strong, have high densities, and have high melting points.
  b Mercury has a lower than expected melting point.
2 a Iron(II) chloride
  b Chromium(III) oxide
3 Copper(I) oxide, Cu_2O and copper(II) oxide, CuO
4 Vanadium(V) oxide, V_2O_5

3.4

1 a Their melting points increase.
  b Their reactivity decreases.
2 a Fluorine – gas; Chlorine – gas; Bromine – liquid; Iodine – solid
  b The halogens
3 a

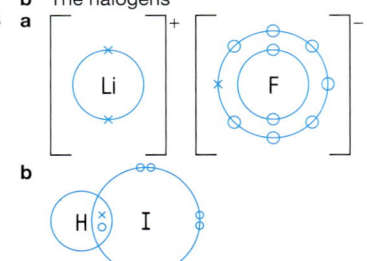

  b

4 a Lithium + bromine → lithium bromide
  b Bromine water + potassium iodide → potassium bromide + iodine solution
5 a $2Na(s) + I_2(g) → 2NaI(s)$
  b $Cl_2(aq) + 2NaI(aq) → 2NaCl(aq) + I_2(aq)$

3.5

1 Helium – despite the fact that it has the smallest positive charge on its nucleus, its outer electrons are nearest to the nucleus and have no inner shells of electrons shielding them from the nuclear charge.
2 The electron in the outermost shell in potassium is attracted less strongly to the positive nucleus than in lithium. The attraction is weaker in potassium because the distance between the outer electron and the nucleus is greater. As well as that, the outer electron in potassium experiences a greater shielding effect from more inner shells of electrons, making it easier for potassium to lose its outer electron than it is for lithium.
3 Going down the group, the outermost shell gets further away from the attractive force of the nucleus so it is harder for a bromine atom to gain an electron. The outer shell in bromine will also be shielded by more

inner electrons, again reducing the attraction of the nucleus for an electron so bromine is less reactive than fluorine.

4 a Ca is more reactive than Mg – the Group 2 elements get more reactive going down the group, as their atoms react by losing their two outer shell electrons, which get easier to lose because of the increasing distance between the nucleus and the outer electrons and the greater shielding effect in Ca compared with Mg.

  b Group 6 non-metals get less reactive going down the group, as their atoms can react by gaining two electrons, and this is more difficult for sulfur atoms than for oxygen atoms, as sulfur's outer shell is further from the attractive force of its nucleus and is shielded by more inner electrons.

Answers to end of chapter summary questions

1 a i Group 7
    ii Group 0
  b i Transition elements
    ii Group 1/alkali metals
    iii Group 1
    iv Between Groups 2 and 3
    iii Group 0/noble gases
    iv Group 7/halogens

2 a Seven electrons as it is in Group 7.
  b Solid
  c i Ionic
    ii White
    iii NaAt
    iv $2Na + At_2 \rightarrow 2NaAt$
    v Yes, because chlorine is more reactive than astatine so would displace it from solution as the element.

3 a i Soft
    ii Good electrical conductivity
    iii Low melting point
  b i 1+
    ii

Rubidium compound	Chemical formula
Rubidium iodide	RbI
Rubidium fluoride	RbF
Rubidium hydroxide	RbOH

    iii White; they will all dissolve in water.
  c i $2Rb(s) + 2H_2O(l) \rightarrow 2RbOH(aq) + H_2(g)$
    ii $2Rb(s) + Cl_2(g) \rightarrow 2RbCl(s)$
  d Rubidium will react more vigorously than potassium because it can lose its outer electron more easily as it is further from the nucleus than potassium's outer electron and it is shielded from the nuclear charge by one more inner shell of electrons than potassium.

4 a i   ii
  b

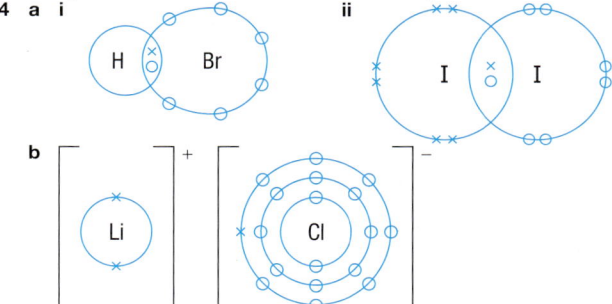

5 a i Copper(II) sulfate is blue and sodium sulfate is white.
    ii Copper can form more than one type of ion (2+ or 1+ ion) whereas sodium can only form a 1+ ion.
  b Copper will have a much higher melting point, greater hardness, and higher density than sodium.
  c Sodium is much more reactive than copper. For example, sodium reacts vigorously with water, giving off hydrogen gas and forming an alkaline solution of sodium hydroxide whereas copper does not react with water.

Answers to end of chapter practice questions

1 a Melting point/boiling point/density (any two of) *(2 marks)*
  b i Fizzing/bubbles/effervescence
     Metal floats or moves on the surface
     Metal gets smaller or disappears
     (any two) *(2 marks)*
    ii Potassium melts. *(1 mark)*
     Flame *(1 mark)*
    iii Blue *(1 mark)*
     Alkali present/OH^- ions formed. *(1 mark)*
    iv $2K + 2H_2O \rightarrow 2KOH + H_2$
     Four correct formulae *(1 mark)*
     Balancing *(1 mark)*
  c Potassium loses an electron further away from nucleus. *(1 mark)*
    Lost electron less strongly attracted to nucleus. *(1 mark)*
    Lost electron repelled/shielded by more inner electrons. *(1 mark)*

2 a Fe^{2+} *(1 mark)*
    Fe^{3+} *(1 mark)*
  b $Fe(OH)_2$ is green. *(1 mark)*
    $Fe(OH)_3$ is brown. *(1 mark)*
  c i Valency/charge on the ion *(1 mark)*
    ii Catalysts *(1 mark)*

3 a $Ca + Cl_2 \rightarrow CaCl_2$
    Three correct formulae *(1 mark)*
    Balancing *(1 mark)*
  b i No reaction *(1 mark)*
    ii Sodium chloride + bromine *(1 mark)*
    iii Potassium bromide + iodine *(1 mark)*
  c Electron gained by fluorine gets closer to nucleus. *(1 mark)*
    Gained electron more strongly attracted to nucleus. *(1 mark)*
    Gained electron repelled/shielded by fewer inner electrons. *(1 mark)*

4 Metals

4.1

1 a Good conductors of heat and electricity; high melting points, strong; high density
  b The central block between Groups 2 and 3
  c Aluminium

2 a It is very unreactive and is malleable.
  b To make gold more hard-wearing

3 The silver and gold would be too expensive to use (except in small amounts in specialist uses).

4 They are 'light-weight' alloys due to the low density of aluminium, so more passengers and/or cargo can be carried on the aircraft; and the alloying makes them strong enough to withstand the high stresses put on an aeroplane in flight.

5 The copper alloys are harder-wearing so won't be worn away as quickly as softer coins made from pure copper. In alloying copper, differently-sized atoms of other metals are introduced into the regular layers of copper atoms, distorting the layers and making it more difficult for them to slide over each other.

4.2

1 a It contains too much carbon.
  b Cast iron contains about 4% carbon whereas pure iron contains only iron. This makes cast iron very hard and brittle as opposed to pure iron which is relatively soft and malleable.
  c For example, man-hole covers

2 a Haematite (iron ore/iron(III) oxide); coke (carbon); limestone (calcium carbonate); air
  b Carbon monoxide
    $CO_2(g) + C(s) \rightarrow 2CO(g)$
  c Iron(III) oxide + carbon $\rightarrow$ iron + carbon dioxide
    $2Fe_2O_3(s) + 3C(s) \rightarrow 4Fe(l) + 3CO_2(g)$
  d As a gas, carbon monoxide can diffuse throughout the blast furnace and come into contact with lots of iron(III) oxide, as opposed to solid carbon which can only react with iron(III) oxide it is directly in contact with.

3

Type of steel	Main useful properties
Low carbon steel	Relatively soft and malleable
High carbon steel	Hard/strong
Chromium–nickel steel	Corrosion resistant

4 a They will not rust so are easier to sterilise and stay sharp.
  b i Low carbon steel as it can be stamped into shapes
    ii High carbon steel as it is very strong and hard

4.3

1 a lithium + water $\rightarrow$ lithium hydroxide + hydrogen
    $2Li(s) + 2H_2O(l) \rightarrow 2LiOH(aq) + H_2(g)$
  b zinc + steam $\rightarrow$ zinc oxide + hydrogen
    $Zn(s) + H_2O(g) \rightarrow ZnO(s) + H_2(g)$

2 a Gas given off/fizzing; magnesium gets smaller and smaller then disappears; heat energy released/gets hot.
  b metal + an acid $\rightarrow$ a salt + hydrogen
  c magnesium + sulfuric acid $\rightarrow$ magnesium sulfate + hydrogen
    $Mg(s) + H_2SO_4(aq) \rightarrow MgSO_4(aq) + H_2(g)$

3 a Because they are so unreactive they don't react with water at all.
  b Because they are so reactive they will react with water vapour (and oxygen) in the air.
  c Zinc will react with acidic foods whereas tin will not at room temperature.

4 Aluminium is protected by a tough/impervious layer of aluminium oxide.

4.4

1 a No reaction

b $Zn(s) + CuSO_4(aq) \rightarrow ZnSO_4(aq) + Cu(s)$

c $Mg(s) + FeCl_2(aq) \rightarrow MgCl_2(aq) + Fe(s)$

2 a Tungsten is below hydrogen in the reactivity series.

b $WO_3 + 3H_2 \rightarrow W + 3H_2O$

3 a $Zn(s) + Fe^{2+}(aq) \rightarrow Zn^{2+}(aq) + Fe(s)$

b Zinc atoms lose two electrons to form zinc(II) ions. Zinc atoms have been oxidised as oxidation is the loss of electrons. Iron(II) ions are reduced as they form iron atoms when they gain two electrons from zinc (and reduction is the gain of electrons).

4.5

1 a metal carbonate + acid → a salt + water + carbon dioxide

b metal carbonate $\xrightarrow{\text{heat}}$ metal oxide + carbon dioxide

2 sodium carbonate + hydrochloric acid → sodium chloride + water + carbon dioxide

$Na_2CO_3(s) + 2HCl(aq) \rightarrow 2NaCl(aq) + H_2O(l) + CO_2(g)$

3 $CuCO_3(s) \rightarrow CuO(s) + CO_2(g)$

4 $CuCO_3(s) + 2HNO_3(aq) \rightarrow Cu(NO_3)_2(aq) + H_2O(l) + CO_2(g)$

5 Limewater is a solution of alkaline calcium hydroxide. Carbon dioxide is a weakly acidic gas so it reacts with the alkaline limewater, forming a suspension of solid particles of insoluble calcium carbonate which precipitates out of solution making it cloudy.

The reaction is:

calcium hydroxide + carbon dioxide → calcium carbonate + water
(limewater)   (an insoluble white precipitate)

$Ca(OH)_2(aq) + CO_2(g) \rightarrow CaCO_3(s) + H_2O(l)$

Answers to end of chapter summary questions

1 a Vigorous reaction, metal changes to metal oxide.

X + steam → X oxide + hydrogen

$X(s) + H_2O(g) \rightarrow XO(s) + H_2(g)$

b Fizzes readily, X gets smaller and eventually dissolves away completely into the acid.

X + sulfuric acid → X sulfate + hydrogen

$X(s) + H_2SO_4(aq) \rightarrow XSO_4(aq) + H_2(g)$

c Because we do not know how much less reactive than calcium and more reactive than magnesium X is, so bubbles of hydrogen may be given off too slowly to be seen immediately or they might be liberated slowly on contact with water.

d i You would expect the more reactive metal X to displace the less reactive copper from solution – so the solution will lose its blue colour, X will dissolve into solution, and copper metal will be deposited as a brown solid.

$X(s) + Cu(NO_3)_2(aq) \rightarrow X(NO_3)_2(aq) + Cu(s)$

ii $X(s) + Cu^{2+}(aq) \rightarrow X^{2+}(aq) + Cu(s)$

iii $X(s) \rightarrow X^{2+}(aq) + 2e^-$ X atoms lose two electrons so they are oxidised

$Cu^{2+}(aq) + 2e^- \rightarrow Cu(s)$ Cu^{2+} ions gain two electrons so they are reduced.

e i Just above copper

ii No change/no reaction as Y is less reactive than magnesium so cannot displace it from a solution of one of its compounds.

2 a zinc + copper(II) oxide → zinc oxide + copper

$Zn(s) + CuO(s) \rightarrow ZnO(s) + Cu(s)$

b Iron + zinc nitrate will not react.

c Iron + magnesium oxide will not react.

d Magnesium + copper(II) sulfate will react.

$Mg(s) + CuSO_4(aq) \rightarrow MgSO_4(aq) + Cu(s)$

3 a i aluminium + iron(III) oxide → aluminium oxide + iron

$2Al + Fe_2O_3 \rightarrow Al_2O_3 + 2Fe$

ii Displacement/redox reaction

b Aluminium and copper(II) oxide would react more vigorously than aluminium and iron(III) oxide.

c No change/no reaction

d You would expect it to react in air, water, and acid more readily than it does so you would not expect it to be used outdoors or for drinks cans.

4 a Gold

b Chromium

c Aluminium

d Copper

e Magnesium

Answers to end of chapter practice questions

1 a i Concentration/temperature/volume (any two of) *(2 marks)*

ii Surface area *(1 mark)*

b i Magnesium *(1 mark)*

ii X has the same observation. *(1 mark)*

c Timer *(1 mark)*

Gas syringe/measuring cylinder/boiling tube *(1 mark)*

2 a $Zn + CuSO_4 \rightarrow ZnSO_4 + Cu$

Reactants *(1 mark)*

Products *(1 mark)*

b $Fe + Cu^{2+} \rightarrow Fe^{2+} + Cu$ *(1 mark)*

c i Reduction and oxidation occur. *(1 mark)*

ii Iron is oxidised *(1 mark)*

Because it loses electrons *(1 mark)*

5 Electrolysis

5.1

1 a The breakdown of a compound by electricity

b Electrolyte

c Ionic

2

	Cathode (–)	Anode (+)
a	zinc	iodine
b	lithium	bromine
c	iron	fluorine
d	sodium	oxygen
e	potassium	chlorine

3 $2NaCl(l) \rightarrow 2Na(l) + Cl_2(g)$

4 a $CuCl_2(aq)$ and $AgNO_3(aq)$

b Copper and silver are below hydrogen in the reactivity series.

5 In the solid, the ions are in fixed positions in the giant ionic lattice; but when molten or in aqueous solution the ions are free to move around within the liquid and can carry their charge to the oppositely charged electrode.

5.2

1 a i They lose electrons.   **ii** Oxidation

b i They gain electrons.   **ii** Reduction

2

	Cathode	Anode
a	lithium	oxygen
b	copper	chlorine
c	hydrogen	oxygen

3 a $2Cl^- \rightarrow Cl_2 + 2e^-$

b $2Br^- \rightarrow Br_2 + 2e^-$

c $Mg^{2+} + 2e^- \rightarrow Mg$

d $Al^{3+} + 3e^- \rightarrow Al$

e $K^+ + e^- \rightarrow K$

f $2H^+ + 2e^- \rightarrow H_2$

g $2O^{2-} \rightarrow O_2 + 4e^-$

h $4OH^- \rightarrow O_2 + 2H_2O + 4e^-$

5.3

1 a Because in solid aluminium oxide the ions are fixed in position in the giant lattice so can only move to electrodes if molten when they are free to move around within the liquid.

b To lower the melting point of aluminium oxide and save energy.

2 Oxygen is produced at the hot carbon anodes and reacts with the carbon to produce carbon dioxide gas, in effect gradually burning away the anodes.

3 a Cathode: $Al^{3+}(l) + 3e^- \rightarrow Al(l)$

Anode: $2O^{2-}(l) \rightarrow O_2(g) + 4e^-$

b Oxide ions are oxidised as they lose electrons and aluminium ions are reduced as they gain electrons.

4 a Lots of energy is needed to purify the aluminium oxide which involves heating to decompose aluminium hydroxide formed in the process; melting the aluminium oxide/cryolite mixture; electrical energy in the electrolytic cells.

b Aluminium in nature is combined/bonded to other elements in compounds that are difficult to break down. It was not until scientists could use electrical cells to pass electricity through substances that aluminium could be extracted from one of its molten compounds by electrolysis.

5.4

1 a Chlorine, hydrogen, and sodium hydroxide solution

b Chlorine is used to make bleach and plastics;

hydrogen is used to make margarine;

sodium hydroxide is used to make bleach, paper, and soap.

2 a At anode: $2Cl^-(aq) \rightarrow Cl_2(g) + 2e^-$

b At cathode: $2H^+(aq) + 2e^- \rightarrow H_2(g)$

3 a Molten sodium chloride will produce sodium metal and chlorine, whereas sodium chloride solution produces chlorine, hydrogen, and sodium hydroxide solution.

b The difference at the cathode in aqueous solution is caused by the presence of $H^+(aq)$ ions from the ionisation of water, so hydrogen is produced and not sodium as with molten sodium chloride.

Sodium hydroxide solution is formed from the excess aqueous sodium ions and hydroxide ions left in solution when aqueous hydrogen and chloride ions are discharged.

5.5

1 Nickel, chromium, silver, gold, tin, zinc, copper (any four)
2 To protect from corrosion; to improve appearance; to increase durability; to save money by using a thin layer of a precious metal
3 **a** At the anode/positive electrode
   b Aqueous zinc ions, $Zn^{2+}(aq)$
4
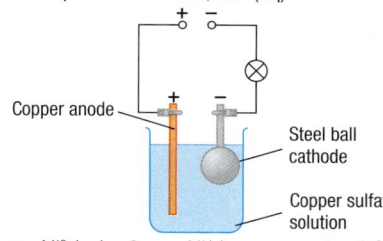

Copper anode
Steel ball cathode
Copper sulfate solution

5 **a** $Ni^{2+}(aq) + 2e^- \rightarrow Ni(s)$ **b** $Cr^{3+}(aq) + 3e^- \rightarrow Cr(s)$

5.6

1

	At the cathode	At the anode
With copper electrodes	copper electrode becomes thicker/gains mass	copper electrode becomes thinner/loses mass
With carbon electrodes	copper deposited on surface of carbon electrode	bubbles of (oxygen) gas given off

2 With copper electrodes, the copper anode undergoes a chemical reaction turning into copper(II) ions so is known as 'active', whereas the carbon electrodes just carry electrons to and from the electrolyte and do not undergo any chemical change themselves so are known as 'inert'(meaning unreactive).
3 **a** At the cathode: $Cu^{2+}(aq) + 2e^- \rightarrow Cu(s)$
      At the anode: $Cu(s) \rightarrow Cu^{2+}(aq) + 2e^-$
   b Reduction occurs at the cathode as copper(II) ions **gain** electrons. Oxidation occurs at the anode as copper atoms **lose** electrons.

Answers to end of chapter summary questions

1 **a** B   **d** A
   b A   **e** B
   c B   **f** A
2 **a** Anode: iodide, fluoride, oxide, bromide
      Cathode: potassium, calcium, magnesium, aluminium
   b i $Mg^{2+} + 2e^- \rightarrow Mg$ **ii** $2Br^- \rightarrow Br_2 + 2e^-$ or $2Br^- - 2e^- \rightarrow Br_2$
3 **a** **A** = chlorine, **B** = hydrogen, and **C** = sodium hydroxide solution
   b For example, making bleach and plastics, sterilising water
   c 11–14
   d Lighted spill burns with a 'pop'.
   e Anode: $2Cl^-(aq) \rightarrow Cl_2(g) + 2e^-$ or $2Cl^-(aq) - 2e^- \rightarrow Cl_2(g)$
      Cathode: $2H^+(aq) + 2e^- \rightarrow H_2(g)$
4 **a** $2H_2O(l) \rightarrow 2H_2(g) + O_2(g)$
   b $H_2O(l) \rightleftharpoons H^+(aq) + OH^-(aq)$
   c Anode: $4OH^-(aq) \rightarrow 2H_2O(l) + O_2(g) + 4e^-$
      Cathode: $2H^+(aq) + 2e^- \rightarrow H_2(g)$
   d Electrical supply/battery/cell/power pack
5 **a** $Li^+ + e^- \rightarrow Li$
   b $Sr^{2+} + 2e^- \rightarrow Sr$
   c $2F^- \rightarrow F_2 + 2e^-$ or $2F^- - 2e^- \rightarrow F_2$
   d $2O^{2-} \rightarrow O_2 + 4e^-$ or $2O^{2-} - 4e^- \rightarrow O_2$
6 Description to include:
   Diagram of circuit with tin anode, iron cathode, and tin(II) nitrate solution
   At anode: oxidation i.e., loss of electrons; $Sn \rightarrow Sn^{2+} + 2e^-$ or $Sn - 2e^- \rightarrow Sn^{2+}$
   At cathode: reduction i.e., gain of electrons; $Sn^{2+} + 2e^- \rightarrow Sn$
7 **a** Beaker A – copper deposited on cathode (negative electrode);
      copper(II) ions released from anode (positive electrode).
      Beaker B – copper deposited on cathode (negative electrode); oxygen gas given off at anode (positive electrode).
   b Beaker A – at the cathode: $Cu^{2+} + 2e^- \rightarrow Cu$;
      at the anode: $Cu \rightarrow Cu^{2+} + 2e^-$
      Beaker B – at the cathode: $Cu^{2+} + 2e^- \rightarrow Cu$;
      at the anode: $4OH^- \rightarrow H_2O + O_2 + 4e^-$
   c Beaker A – when the copper anode is worn away completely and all the copper(II) ions in solution are converted to copper at the cathode
      Beaker B – when all the copper(II) ions have been removed from solution/deposited as copper atoms at the cathode

Answers to end of chapter practice questions

1 **a** Bromine   (1 mark)
      $2Br^- \rightarrow Br_2 + 2e^-$
      Both formulae correct   (1 mark)
      Balancing   (1 mark)
   b Lead   (1 mark)
      $Pb^{2+} + 2e^- \rightarrow Pb$
      Both formulae correct   (1 mark)
      Balancing   (1 mark)
   c Electrons move.   (1 mark)
   d Ions cannot move when solid/ions move when molten.   (1 mark)
   e Ions gain electrons.   (1 mark)
2 **a** H^+ (1 mark)
   b $4OH^- \rightarrow 2H_2O + O_2 + 4e^-$
      All formulae correct   (1 mark)
      Balancing   (1 mark)
   c Sulfate/$SO_4{}^{2-}$ (1 mark)
   d It increases.   (1 mark)
   e It takes twice as many electrons to discharge the same volume of oxygen compared with hydrogen/same number of electrons discharges twice the volume of H^+ ions as OH^- ions, e.g., 2 moles of electrons give 1 mole of H_2 but only 0.5 mole of O_2. (1 mark)
3 **a** Hydrogen less reactive/sodium more reactive   (1 mark)
   b Greater concentration of Cl^- ions   (1 mark)
   c Sodium hydroxide   (1 mark)
   d i Fuel/making margarine   (1 mark)
     ii Bleach/making plastics   (1 mark)
4 **a i** Nickel for the anode and copper spoon for the cathode (1 mark)
     ii Nickel chloride/nitrate/sulfate   (1 mark)
     iii $Ni \rightarrow Ni^{2+} + 2e^-$ (1 mark)
   b Appearance/durability   (1 mark)
   c Graphite conducts electricity/plastic does not conduct electricity (1 mark)
5 **0 marks** No relevant content
   Level 1 (1–2 marks) There is a brief description of the electrolysis of aluminium oxide.
   Level 2 (3–4 marks) There is some description of the electrolysis of aluminium oxide.
   Level 3 (5–6 marks) There is a clear, balanced, and detailed description of the electrolysis of aluminium oxide.
   Examples of chemistry points made in the response:
   • aluminium oxide is melted/made liquid
   • cryolite is used as a solvent/to decrease the amount of energy or electricity used
   • carbon/graphite used as the positive electrode
   • aluminium ions are attracted to the negative electrode
   • at the negative electrode aluminium is formed or aluminium ions gain electrons
   • oxide ions are attracted to the positive electrode
   • oxygen is formed at the positive electrode or oxide ions lose electrons
   • the oxygen reacts with carbon to make carbon dioxide or carbon dioxide formed at positive electrode
   • $Al^{3+} + 3e^- \rightarrow Al$
   • $2O^{2-} \rightarrow O_2 + 4e^-$

6 Chemical Analysis

6.1

1 A mixture is made up of two or more elements or compounds which are not chemically combined together.
2 The relative proportions of each substance/component of the mixture can be in any ratio (it is not fixed as in a compound).
3 **a**

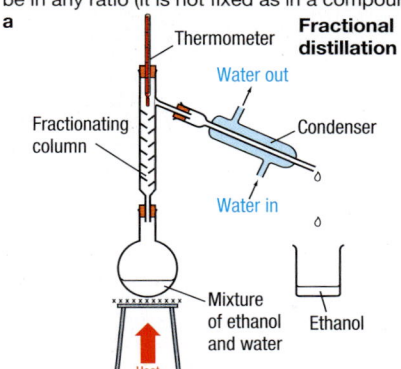

Thermometer   **Fractional distillation**
Water out
Condenser
Fractionating column
Water in
Mixture of ethanol and water
Ethanol
Heat

b At 78 °C/the boiling point of ethanol, much of the water vapour rising up the fractionating column would condense and return to the heated flask instead of entering the condenser to be collected with the ethanol.

4 Method A – Add water to the mixture, stir, and filter. Sulfur residue will be left on the filter paper. Wash with distilled water and leave sulfur to dry. Evaporate water from the filtrate of sodium nitrate solution on a water bath until the point of crystallization. Then leave the saturated solution to crystallise.
Method B – Add xylene to the mixture, stir, and filter. Sodium nitrate residue will be left on the filter paper. Wash with xylene and leave sodium nitrate to dry. Evaporate xylene from the filtrate of sulfur solution using water bath (electrically heated/no naked flame) in a fume cupboard to crystallise.

6.2

1 Allow the paper to dry and measure the distance from the pencil line to the solvent front and to each dye left on the paper. Divide the distance each dye moved by the distance of the solvent front to get the R_f values, then match the values against tables of data to identify the dyes.

2 0.77

3 The solubility of Z in the solvent is greater than the solubility of Y in the same solvent. To identify Y and Z, their R_f values are matched against values of known substances in the 50–50 water–ethanol solvent at 20 °C in a database or databook.

4 Because changing the solvent and the temperature both affect the attraction of the compound to the solvent (mobile phase) and paper (stationary phase)/its solubility in the solvent/its distribution between the mobile phase and the stationary phase, so the distance the spot travels along the paper will differ under different conditions, resulting in different R_f values.

6.3

1 The hydrogen reacts with the oxygen gas in the air to form steam/water, transferring energy to the surroundings in an exothermic reaction, warming them up and making the popping sound.

2 a magnesium carbonate + hydrochloric acid → magnesium chloride + water + carbon dioxide

b E.g., place the mouths of the two angled tubes together and wait a while, giving time for the carbon dioxide to build up in the reaction tube before spilling over and sinking down into the tube with the limewater. Stopper this tube and shake the limewater. Alternatively, place a dropping pipette into the reaction tube above the acid and withdraw the carbon dioxide building up in the tube. Squirt this into the limewater. Repeat a few times.

3 Collect the gas given off above the anode in a small test tube and insert a glowing splint to see if it re-lights. If it does, oxygen must be given off. Then hold a damp piece of blue litmus paper above the solution near the anode/in test tube of gas collected and see if it turns white/is bleached. If it does, then chlorine gas is given off.

4 a i $Mg(s) + H_2SO_4(aq) \rightarrow MgSO_4(aq) + H_2(g)$
ii $Mg(s) + 2H^+(aq) \rightarrow Mg^{2+}(aq) + H_2(g)$

b The magnesium is oxidised, as its atoms each lose two electrons; and the hydrogen ions are reduced, as they each receive an electron (before two bond together to make hydrogen molecules).

5 For example – Use zinc/magnesium + dilute acid to obtain different proportions of hydrogen and air by collecting the gas over water, starting with an inverted test tube full of water in order to collect 100% hydrogen (make sure the first bubbles are allowed to escape, as this will be displaced air from the apparatus). Stopper the tube and label it 100%. Then vary the proportions of water in the collection test tube. Invert the test tube each time and collect a range of different proportions of hydrogen to air mixtures. Measure the volume of water in a full tube using a measuring cylinder. Then measure and mark different proportions of that total volume on the side of four more different tubes to give the rough proportion of hydrogen in the mixture. Then test each tube in turn with a lighted splint positioned at a set distance from a sound meter. Record the maximum sound level for each mixture.

6.4

1 a White

b Add sodium hydroxide solution – all will form a white precipitate but the aluminium hydroxide precipitate formed would dissolve in excess sodium hydroxide solution (Al^{3+} identified). Take the original magnesium and calcium compounds and carry out a flame test – the calcium ions will produce a brick red flame (Ca^{2+} identified) and the magnesium ions will not colour the flame (so Mg^{2+} identified).

c $Mg^{2+}(aq) + 2OH^-(aq) \rightarrow Mg(OH)_2(s)$

2 a Dip a wire loop in concentrated hydrochloric acid and heat to clean it → dip the loop into conc. acid again before dipping in the metal compound to be tested → hold the loop in the edge of a roaring Bunsen flame → match the colour of the flame to the known flame test colours of metal ions.

b (Apple) green flame

3 a brown precipitate
b nothing observed (no change)
c nothing observed (no change)
d nothing observed (no change)
e yellow
f K^+
g Ca^{2+}
h Al^{3+}
i Fe^{2+}
j Li^+

6.5

1 Dilute hydrochloric acid, limewater, dilute nitric acid, silver nitrate solution, barium chloride solution

2 Potassium iodide (potassium ions have a lilac flame test; and silver iodide, a pale yellow precipitate, is formed with silver nitrate solution.)

3 a magnesium chloride + silver nitrate → magnesium nitrate + silver chloride
$MgCl_2(aq) + 2AgNO_3(aq) \rightarrow Mg(NO_3)_2(aq) + 2AgCl(s)$

b potassium carbonate + hydrochloric acid → potassium chloride + water + carbon dioxide
$K_2CO_3(s) + 2HCl(aq) \rightarrow 2KCl(aq) + H_2O(l) + CO_2(g)$

c aluminium sulfate + barium chloride → aluminium chloride + barium sulfate
$Al_2(SO_4)_3(aq) + 3BaCl_2(aq) \rightarrow 2AlCl_3(aq) + 3BaSO_4(s)$

4 a $Ag^+(aq) + Br^-(aq) \rightarrow AgBr(s)$
b $Ba^{2+}(aq) + SO_4^{2-}(aq) \rightarrow BaSO_4(s)$
c $2H^+(aq) + CO_3^{2-}(s) \rightarrow CO_2(g) + H_2O(l)$

5 To dissolve the compound and to remove any carbonate ions as they would also form a precipitate with the silver ions, interfering with the test. Also, any other acids added would produce precipitates with silver nitrate solution, e.g., sulfuric acid would make a precipitate of silver sulfate.

Answers to end of chapter summary questions

1 a Nothing observed/dissolves.
b White precipitate forms.
c White precipitate forms.
d (Brick) red flame
e Sodium iodide
f Copper(II) carbonate

2 Lithium sulfate

3 a Lilac flame test to identify potassium ions
Add nitric acid, then silver nitrate solution – a cream precipitate will identify bromide ions.
b KBr
c Ionic

4 a Compounds have a fixed composition whereas mixtures do not (their proportions vary). Compounds cannot be easily separated into their elements/needing chemical reactions but mixtures can be separated by physical means, as there are no chemical bonds between different substances in a mixture.
b i Crystallisation
ii Filtration
iii (Simple) distillation
iv (Fractional) distillation
c $Pb(NO_3)_2(aq) + 2NaI(aq) \rightarrow PbI_2(s) + 2NaNO_3(aq)$

5 a i Chlorine
ii Hydrogen
iii Carbon dioxide
iv Ammonia
v Oxygen
b i Hydrogen, oxygen, carbon dioxide
ii Ammonia and chlorine are soluble in water.
c i Hydrogen = 2; oxygen = 32; chlorine = 71; ammonia = 17; carbon dioxide = 44
ii Hydrogen and ammonia

6 a R_f of X = 0.6
R_f of Y = 0.2
b X is more soluble/has stronger attraction to ethanol than Y.
c Temperature

Answers to end of chapter practice questions

1 a i Lilac *(1 mark)*
ii K^+ *(1 mark)*
b i White precipitate *(1 mark)*
ii Aluminium hydroxide *(1 mark)*
c Too much added *(1 mark)*
So white precipitate dissolved *(1 mark)*
d Add dilute hydrochloric acid *(1 mark)*
And barium chloride solution *(1 mark)*
White precipitate forms *(1 mark)*

2 a i Carbon dioxide *(1 mark)*
ii Limewater *(1 mark)*
turns cloudy *(1 mark)*
b i Silver chloride *(1 mark)*
ii $Ag^+ + Cl^- \rightarrow AgCl$
Reactants *(1 mark)*
Products *(1 mark)*
c Sodium ions do not form a precipitate. *(1 mark)*
The two iron ions form a precipitate. *(1 mark)*

The colour is not green or brown/colour is a mixture of green and brown. *(1 mark)*

3 a Distillation *(1 mark)*
Difference in boiling point *(1 mark)*
b Filtration *(1 mark)*
Difference in solubility *(1 mark)*
c (Paper) chromatography *(1 mark)*
Difference in solubility in solvent/R_f values *(1 mark)*

4 **0 marks** No relevant content
Level 1 (1–2 marks) There is a brief description of how to do the chromatography experiment.
Level 2 (3–4 marks) There is some description of how to do the chromatography experiment.
Level 3 (5–6 marks) There is a clear, balanced, and detailed description of how to do the chromatography experiment.
Examples of chemistry points made in the response:
- use chromatography paper
- draw pencil line near the bottom
- place drops of the drink and colours on the line
- pour some water into a large beaker or tank
- place the paper in the tank
- with the spots of colour above the water level
- fit the tank with a lid
- leave until the water reaches almost to the top
- leave to dry

7 Acids, bases, and salts

7.1
1 a An alkali is a base that dissolves in water.
b They dissolve in water to produce hydroxide ions.
c $KOH(s) \xrightarrow{\text{water}} K^+(aq) + OH^-(aq)$
2 a $H^+(aq)$ ions
b $HI(g) \xrightarrow{\text{water}} H^+(aq) + I^-(aq)$
3 Distilled water pH 7; sodium hydroxide solution pH above 7, e.g., 14; ethanoic acid pH below 7, e.g., 4
4 A pH sensor and datalogger would be more accurate as the matching of colours against a pH chart by eye is subjective and difficult to judge. A pH sensor is more likely to produce more repeatable measurements over a narrower range than using universal indicator paper.

7.2
1 a acid + a base → a salt + water
b acid + a metal → a salt + hydrogen
2 Copper sulfate solution is heated in an evaporating dish on a water bath. Some of the water is evaporated off from the copper sulfate solution until the point of crystallisation when crystals appear at the edge of the solution. The copper sulfate solution is then left at room temperature for the remaining water to evaporate off slowly, leaving crystals in the dish.
3 a Copper metal does not react with dilute acid.
b Potassium metal is so reactive it will explode in dilute acid.
4 a $Mg(s) + 2HCl(aq) \rightarrow MgCl_2(aq) + H_2$
b $Li_2O(s) + H_2SO_4(aq) \rightarrow Li_2SO_4(aq) + H_2O(l)$

7.3
1 a acid + alkali → a salt + water   b $H^+(aq) + OH^-(aq) \rightarrow H_2O(l)$
2 a Ammonium nitrate
b It is used as a fertiliser to provide plants with nitrogen.
c $NH_4OH + HNO_3 \rightarrow NH_4NO_3 + H_2O$ or $NH_3 + HNO_3 \rightarrow NH_4NO_3$
3 $LiOH(aq) + HCl(aq) \rightarrow LiCl(aq) + H_2O(l)$
Add a known volume of dilute hydrochloric acid (e.g., from a burette) to a measured volume of lithium hydroxide solution which has a few drops of an indicator added (e.g., methyl orange) until the indicator just changes colour. Note how much acid is needed to neutralise a known volume of lithium hydroxide solution. Repeat until consistent volumes of acid are used. Then repeat without the indicator and evaporate off water from the lithium chloride solution by heating until the point of crystallisation is reached. Then leave at room temperature for the rest of the water to evaporate off.
4 a Add solutions of lead nitrate and, e.g., sodium chloride, filter, wash the precipitate of lead chloride collected on the filter paper with distilled water and leave in a warm oven or leave to dry at room temperature.
b For example: $Pb(NO_3)_2(aq) + 2NaCl(aq) \rightarrow PbCl_2(s) + 2NaNO_3(aq)$
c Ionic bonding
5 Add an alkali (e.g., sodium hydroxide or lime) and filter off the precipitate of chromium(III) hydroxide formed.

Answers to end of chapter summary questions

1 a i Nickel(II) oxide
ii $NiO(s) + H_2SO_4(aq) \rightarrow NiSO_4(aq) + H_2O(l)$
iii Neutralisation
b Add excess nickel(II) oxide to warm sulfuric acid and filter off the excess solid. Then heat the nickel(II) sulfate solution in an evaporating dish on a water bath until the point of crystallisation is reached. Stop heating and allow the remaining water to evaporate from the solution at room temperature to leave the crystals of nickel(II) sulfate.
2 a $2LiOH(aq) + H_2SO_4(aq) \rightarrow Li_2SO_4(aq) + 2H_2O(l)$
b $Fe_2O_3(s) + 6HNO_3(aq) \rightarrow 2Fe(NO_3)_3(aq) + 3H_2O(l)$
c $Zn(s) + 2HCl(aq) \rightarrow ZnCl_2(aq) + H_2$
d $Ba(NO_3)_2(aq) + K_2SO_4(aq) \rightarrow BaSO_4(s) + 2KNO_3(aq)$
3 a $CaCO_3$
b i $CaCO_3(s) + 2HCl(aq) \rightarrow CaCl_2(aq) + H_2O(l) + CO_2(g)$
ii

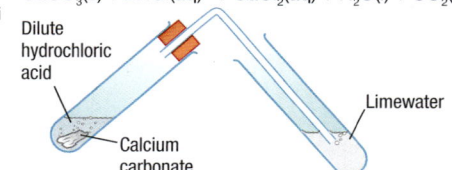

iii calcium hydroxide + carbon dioxide → calcium carbonate + water
$Ca(OH)_2(aq) + CO_2(g) \rightarrow CaCO_3(s) + H_2O(l)$
c i calcium carbonate → calcium oxide + carbon dioxide
$CaCO_3(s) \rightarrow CaO(s) + CO_2(g)$
ii Thermal decomposition
iii 5.5 tonnes
iv Making cement; glass; extracting iron from its ore; building material; raising pH of acidic soil or lakes

Answers to end of chapter practice questions

1 a Solution 1 pH = 2
Solution 2 pH = 7
Solution 3 pH = 12
b Universal indicator *(1 mark)*
It has more colours/it indicates pH values. *(1 mark)*
c Solution 4 *(1 mark)*
2 a i Magnesium chloride *(1 mark)*
Hydrogen *(1 mark)*
ii Zinc nitrate *(1 mark)*
Water *(1 mark)*
iii Copper(II) sulfate *(1 mark)*
Water + carbon dioxide *(1 mark)*
b Bubbles *(1 mark)*
c To make sure all the acid reacts *(1 mark)*
d i Magnesium chloride *(1 mark)*
ii Zinc oxide *(1 mark)*
iii Blue *(1 mark)*
e Evaporate some of the water (on a water bath). *(1 mark)*
Cool the liquid/heat to the point of crystallisation. *(1 mark)*
Filter off the crystals/then leave at room temperature. *(1 mark)*
Dry them on filter paper/until all water has evaporated. *(1 mark)*
3 a $MgCl_2(aq) + 2NaF(aq) \rightarrow MgF_2(s) + 2NaCl(aq)$
All four formulae *(1 mark)*
Balancing *(1 mark)*
All four state symbols *(1 mark)*
b i Lead nitrate *(1 mark)*
Sodium sulfate/other soluble sulfate or sulfuric acid *(1 mark)*
ii To remove soluble reactant. *(1 mark)*

8 Quantitative chemistry

8.1
1 a Because no atoms are created or destroyed in a chemical reaction, equal numbers and types of atoms must appear on both sides of the equation.
b $H_2 + Cl_2 \rightarrow 2HCl$
2 a 17.6 g
b magnesium carbonate → magnesium oxide + carbon dioxide
$MgCO_3(s) \rightarrow MgO(s) + CO_2(g)$
3 a $2KNO_3 \rightarrow 2KNO_2 + O_2$ c $4Fe + 3O_2 \rightarrow 2Fe_2O_3$
b $4Li + O_2 \rightarrow 2Li_2O$
4 $2Na(s) + 2H_2O(l) \rightarrow 2NaOH(aq) + H_2(g)$

8.2
1 a The average relative mass of an atom of an element, taking into account the proportions of different isotopes naturally occurring in that element, compared on a scale in which an atom of carbon-12 has a value of exactly 12

b Because of the averaging of the relative masses of different isotopes in a naturally occurring sample

2 a 256   **c** 159.5
  b 62   **d** 180

3 a 8 times
  b i 4 moles   **ii** 0.005 moles
  c i 0.3 moles   **ii** 500 000 moles

4 a 13.5 g   **c** 0.1 g
  b 5000 g or 5 kg   **d** 74.4 g

8.3

1 5.9%   **4** PCl_3
2 21.2%   **5** Al_2O_3
3 SO_3

8.4

1 2 molecules of HCl or 2 moles of HCl molecules
2 4.0 g
3 a $2H_2O_2(aq) \rightarrow 2H_2O(l) + O_2(g)$
  b 3.4 g
4 a $Ca(s) + 2H_2O(l) \rightarrow Ca(OH)_2(aq) + H_2(g)$
  b 2.0 g

8.5

1 To conserve the Earth's resources and reduce waste and pollution
2 The reaction may be reversible (as products form they react to re-form the reactants again); some reactants may react to give unexpected products; some of the product may be lost in handling or left behind in the apparatus; the reactants may not be pure; there may be losses in separating the product that you want from the reaction mixture.
3 80 g
4 21% (21.2%)
5 a $2NaHCO_3 \rightarrow Na_2CO_3 + H_2O + CO_2$
  b 86.8%

8.6

1 Precise
2 a Meniscus
  b i (Volumetric) pipette and burette
    ii Read to the bottom of the meniscus, with the eye level with that point.
3 a Measure a known volume of sodium hydroxide solution into a conical flask using a pipette, and add a few drops of an acid/base indicator. Then pour the dilute nitric acid into a burette, recording the reading on the burette. Then slowly add small amounts of acid to the flask, swirling it to make sure that the two solutions are mixed. Continue until the indicator in the flask changes colour. Repeat the titration until you get two identical results.
  b $NaOH(aq) + HNO_3(aq) \rightarrow NaNO_3(aq) + H_2O(l)$
4 a For example, methyl orange, phenolphthalein
  b Two distinctly different colours in acidic and alkaline conditions in order to make the end point clear to the experimenter

8.7

1 $KOH(aq) + HNO_3(aq) \rightarrow KNO_3(aq) + H_2O(l)$
2 0.004 (4×10^{-3}) moles
3 0.004 (4×10^{-3}) moles
4 0.32 mol/dm³ and 20.16 g/dm³

8.8

1 The volume of gas occupied by 1 mole of gas at room temperature and pressure (24 dm³ or 24 000 cm³)
2 a i 1.5 mol   **ii** 417 mol
  b i 216 dm³ (or 216 000 cm³)   **ii** 7.2 dm³ (or 7200 cm³)
  c 0.064 g
3 300 dm³ (or 300 000 cm³)
4 0.048 dm³ (or 48 cm³)

Answers to end of chapter summary questions

1 a 34 g   **e** 106 g
  b 64 g   **f** 342 g
  c 28 g   **g** 118 g
  d 40 g
2 a 0.25 moles   **c** 0.05 moles
  b 0.001 moles
3 a 80%   **b** 0.76 g
4 $AlBr_3$
5 a $2Na + Cl_2 \rightarrow 2NaCl$
  b $4Al + 3O_2 \rightarrow 2Al_2O_3$
  c $2Al(OH)_3 \rightarrow Al_2O_3 + 3H_2O$
  d $2Ba(NO_3)_2 \rightarrow 2BaO + 4NO_2 + O_2$
  e $2C_4H_{10} + 13O_2 \rightarrow 8CO_2 + 10H_2O$
6 95.7%
7 a $C_2H_4(g) + H_2O(g) \rightleftharpoons C_2H_5OH(g)$
  b 75%

8 a 1 mole
  b 196 kg
  c 96%
  d Any two valid reasons, e.g., the 2nd stage is a reversible reaction/does not go to completion; reactants/products might be lost in the process at any stage; the sulfur might be impure.
9 a $KOH(aq) + HCl(aq) \rightarrow KCl(aq) + H_2O(l)$
  b i 0.01 moles   **ii** 0.01 moles
  c i 0.8 mol/dm³   **ii** 44.8 g/dm³

Answers to end of chapter practice questions

1 a 32   *(1 mark)*
  b 74   *(1 mark)*
  c M_r of NaOH = 40   *(1 mark)*
    number of moles = $\frac{20}{40}$ = 0.5   *(1 mark)*
  d M_r of NH_4NO_3 = 80   *(1 mark)*
    mass = 2×80 = 160 g   *(1 mark)*
2 a $2HBr + H_2SO_4 \rightarrow 2H_2O + Br_2 + SO_2$ *(1 mark)*
  b M_r of CH_3Br = 95   *(1 mark)*
    Percentage by mass = $\frac{80}{95}$ = 84%   *(1 mark)*
  c % of oxygen = 31.8%   *(1 mark)*

	Na	Br	O	
	15.2	53.0	31.8	*(1 mark)*
	23	80	16	
ratio	0.66	0.66	1.99	*(1 mark)*

    Empirical formula = $NaBrO_3$ *(1 mark)*
3 a Fire/explosion   *(1 mark)*
  b Water   *(1 mark)*
    Hydrogen reacting with oxygen in oxide   *(1 mark)*
  c Weigh and heat in hydrogen again   *(1 mark)*
    Until mass is constant.   *(1 mark)*
  d i 0.64 g   *(1 mark)*
    ii $\frac{0.64}{16}$ = 0.04 mol   *(1 mark)*
    iii $\frac{5.08}{63.5}$ = 0.08 mol   *(1 mark)*
    iv Cu:O ratio is 2:1   *(1 mark)*
    Empirical formula = Cu_2O *(1 mark)*
4 a $H^+ + OH^- \rightarrow H_2O$ *(1 mark)*
  b i Potassium hydroxide   *(1 mark)*
    Smaller volume used   *(1 mark)*
    ii Nitric acid – burette   *(1 mark)*
    Potassium hydroxide – (volumetric) pipette   *(1 mark)*
  c i 26.10 and 26.20   *(1 mark)*
    ii (26.10 + 26.20) ÷2   *(1 mark)*
    = 26.15 cm³   *(1 mark)*
  d (0.120 × 25.00) ÷ 28.50   *(1 mark)*
    = 0.105 mol/dm³   *(1 mark)*
5 a Amount of carbon monoxide = $\frac{56}{28}$ = 2 mol   *(1 mark)*
    mass of hydrogen = 4×2 = 8 g   *(1 mark)*
  b M_r of methanol = 32   *(1 mark)*
    mass of methanol = 2×32 = 64 g   *(1 mark)*
  c % yield = $\frac{16 \times 100}{64}$ *(1 mark)*
    = 25%   *(1 mark)*

9 Rates of reaction

9.1

1 a calcium carbonate + hydrochloric acid → calcium chloride + water + carbon dioxide
    $CaCO_3(s) + 2HCl(aq) \rightarrow CaCl_2(aq) + H_2O(l) + CO_2(g)$
  b All the acid had been used up in the reaction.
  c The line will be horizontal.
2 a i

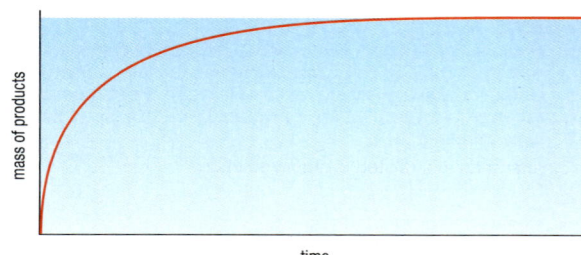

ii

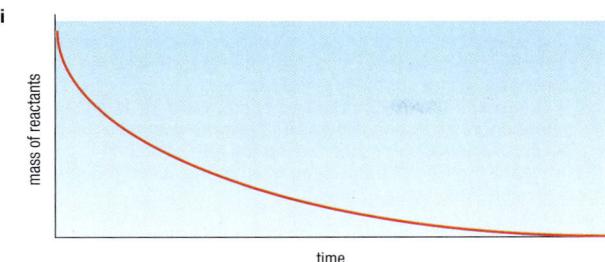

b   The slope of the line tells us the rate of the reaction at that particular time.

9.2

1   Temperature, concentration, surface area, pressure (if gases involved), catalyst (some reactions affected by light)

2   a   Only iron atoms on the surface can react

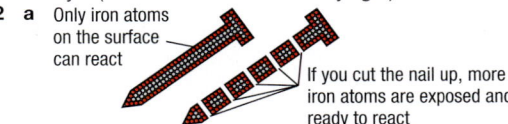

If you cut the nail up, more iron atoms are exposed and ready to react

   b   To control all the other variables/keep the other variables constant in order to make it a fair test

3   The 'activation energy' is the minimum energy needed for reactant particles to react so any collisions with insufficient energy will not react so having no effect on the rate of reaction; only those colliding particles with the activation energy will react.

4   Well-chewed food will be in smaller pieces so will have a greater surface area. Therefore it will have more food molecules exposed to attack by the acid particles in your stomach than unchewed food.

9.3

1   a   Because the frequency of collisions increases and more particles collide with energy greater than the activation energy

   b   By a factor of 2/doubles/twice the rate

2   a   carbon dioxide

   b   It gets shorter.

   c   The reactant hydrogencarbonate and acid particles move around more quickly so they collide more often and there is a higher proportion of collisions with energy greater than the activation energy needed for a reaction to take place/for effective collisions to occur.

3   The chemical reactions that take place in the cooking process will occur more quickly as the water under pressure can be heated to a higher temperature than 100 °C.

9.4

1   a   The lowest (yellow) line/highest concentration because its slope is steepest.

   b   To stop any acid spray escaping whilst allowing the carbon dioxide gas to escape/a rubber bung would trap the carbon dioxide gas given off until the bung blew off under the increasing pressure as the reaction proceeded.

   c   The electronic balance responds to small changes in mass.

2

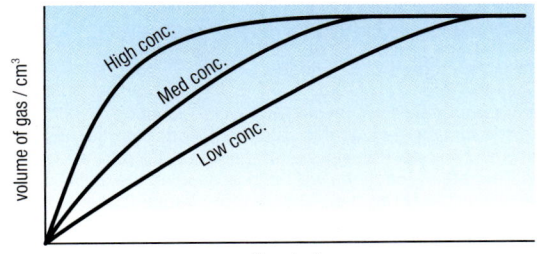

3   The higher the concentration of a gas, the more particles there are in a given volume and each particle of gas will contribute to the overall pressure exerted by the gas. If you double the concentration of a gas, you double the number of particles in a given volume. The particles are moving randomly at high speeds so the force per unit area caused by their collisions with the walls of the container will also double.

4   There are many more collisions between the particles of acid in the undiluted cleaner and the limescale (calcium carbonate) in a given time resulting in a faster reaction than with diluted cleaner. This is because the number of acid particles in a given volume is far higher in the undiluted cleaner than in a diluted cleaning solution.

9.5

1   It remains chemically unchanged.

2   To make them as effective as possible by increasing their surface area as much as possible

3   The reactants are used up as they turn into products but catalysts remain chemically unchanged at the end of a reaction. So catalysts can go on to catalyse further conversion of reactants to products (eventually they do become 'poisoned' by impurities and need to be replaced, but not often).

4   For example – Manufacture of ammonia/Haber process:
nitrogen + hydrogen ⇌ ammonia (uses an iron catalyst)
Manufacture of sulfuric acid/Contact process:
sulfur dioxide + oxygen ⇌ sulfur trioxide (uses a vanadium(v) oxide catalyst)
Manufacture of nitric acid:
ammonia + oxygen ⇌ nitrogen(ii) oxide (nitrogen monoxide) + water (uses platinum/rhodium as a catalyst)
Manufacture of margarine:
unsaturated oil + hydrogen → hydrogenated fat (uses nickel as a catalyst)

5   They speed up the production process, making it more economical; save energy; reduce pollution from, and save limited resources of, fossil fuels.

Answers to end of chapter summary questions

1   a   For example – Measuring the mass of the reaction mixture over time/measuring the volume of gas given off in a gas syringe (or measuring cylinder over water) over time.

   b   Use more concentrated acid; increase the temperature of the acid; use a larger surface area of magnesium (e.g., cut the magnesium ribbon up into smaller pieces).

   c   i   The student could repeat the reaction at different temperatures, keeping the other variables constant i.e., concentration and volume of acid, mass and surface area of magnesium ribbon.

      ii   The higher the temperature of acid, the greater the rate of reaction/faster the reaction.

      iii   As the temperature of the acid is increased the particles of acid (hydrogen ions) in solution gain more energy and move around more quickly (on average). Therefore they will collide with particles (atoms) of the magnesium ribbon more frequently. Not only that, a higher proportion of the collisions will have energy greater than the activation energy so more collisions in any given time will be successful/result in a reaction.

2   a   Carbon dioxide; limewater turns cloudy.

   b   i   calcium carbonate + nitric acid → calcium nitrate + water + carbon dioxide

      ii $CaCO_3(s) + 2HNO_3(aq) \rightarrow Ca(NO_3)_2(aq) + H_2O(l) + CO_2(g)$

   c   They used two solutions of nitric acid with different concentrations, keeping all other variables constant, in a conical flask. This was placed on an electronic balance, marble chips added, with a cotton wool plug in the mouth of the flask and the loss in mass (which is equivalent to the mass of gas produced) was monitored for 2.5 minutes, taking readings every 30 seconds.

   d   Student graphs (both lines drawn as smooth curves and labelled)

   e   The rate of reaction in Investigation B is half that in Investigation A.

   f   The final mass of gas produced in Investigation B is half of that produced in Investigation A.

   g   The initial concentration of the acid in Investigation A is twice/double the concentration of acid in Investigation B.

   h   In Investigation A there are twice as many acid particles ($H^+(aq)$ ions) as in Investigation B in the same volume of acid. So, there will be twice the number of collisions in a given time between the acid particles in solution and particles at the surface of the marble chips. Therefore the rate of reaction in Investigation A is twice the rate in Investigation B.

3   a   They kept the temperature, concentration of acid, volume of acid, and mass of zinc constant and only varied the surface area of the zinc in each different test.

   b   Hydrogen; a lighted spill pops.

   c $Zn(s) + H_2SO_4(aq) \rightarrow ZnSO_4(aq) + H_2(g)$

   d   Line 1

   e   The largest pieces of zinc

   f   The line would be the steepest, rising to the same level as the other three lines, levelling off before them.

   g   The smaller the pieces of zinc, the larger its surface area and the more zinc particles (atoms) are exposed to react with the acid particles ($H^+(aq)$ ions) moving around randomly in the solution. Therefore the frequency of collision between reactant particles increases, causing an increase in the rate of reaction.

4   a   A substance that increases the rate of a reaction but remains chemically unchanged at the end of the reaction

   b $2H_2O_2(aq) \rightarrow 2H_2O(l) + O_2(g)$

   c   For example – Measure the time to collect a fixed volume of oxygen gas using the same mass of potential catalyst (with the same surface area) in the same concentration of hydrogen peroxide solution at the same temperature in each test.

   d   Filter off the insoluble metal oxides after the reaction, wash with distilled water, and dry in a warm oven/leave to dry/dab with filter paper to dry.

Answers to end of chapter practice questions

1 a Acid concentration *(1 mark)*
  Temperature *(1 mark)*
  b Curve is steeper. *(1 mark)*
  Curve becomes horizontal before the others. *(1 mark)*
  c Greater surface area *(1 mark)*
  d i Student 1 *(1 mark)*
    Bigger (%) error in recording short times *(1 mark)*
    ii 1 ÷ 11 *(1 mark)*
    = 0.09 *(1 mark)*
    iii Rate increases as concentration increases *(1 mark)*
    In direct proportion *(1 mark)*
    iv Concentration doubled/twice as many (acid) particles *(1 mark)*
    More collisions between (acid) particles *(1 mark)*
    In a given time/more frequent *(1 mark)*
2 a S/sulfur *(1 mark)*
  Does not dissolve/appears as a precipitate *(1 mark)*
  b i They move more quickly. *(1 mark)*
    ii Collisions are more energetic. *(1 mark)*
3 a A substance that speeds up a chemical reaction *(1 mark)*
  But is not used up during the reaction *(1 mark)*
  b They reduce costs/they last a long time/they help to make a lot of products in a shorter time. *(1 mark)*
  c Greater surface area *(1 mark)*

10 Extent of reaction

10.1

1 A reaction in which reactants form products but the products also react to form the reactants.
2 The mixture of water and phenolphthalein would appear colourless to start with, then would turn pink-purple in the alkaline sodium hydroxide solution, and finally turn back to colourless in the excess acid.
3 HPhe ⇌ $H^+(aq)$ + $Phe^-(aq)$
  colourless   pink-purple
4 For example, toothbrushes, mugs, strip thermometers
5 White anhydrous copper sulfate powder turns blue in the presence of water:

$$CuSO_4 + 5H_2O \rightleftharpoons CuSO_4 \cdot 5H_2O$$
  white   blue

Blue cobalt chloride paper turns pink in the presence of water:

$$CoCl_2 \cdot 2H_2O + 4H_2O \rightleftharpoons CoCl_2 \cdot 6H_2O$$
  blue   pink

10.2

1 The rate of the reverse reaction is equal to the rate of the forward reaction.
2 They remain constant.
3 Because although the concentrations of reactants and products are unchanged overall in the reaction mixture, the forward and reverse reactions are still taking place but at the same rate.
4 Pump more chlorine gas into the mixture as this will increase the rate of the forward reaction and more iodine trichloride will be produced.

10.3

1 Decreases products
2 Increases the amount of hydrogen gas formed
3 a As pressure increases the colour of a mixture lightens as the forward reaction producing more N_2O_4 results in fewer gas molecules.
  b The colour of a mixture darkens as the reverse reaction producing more NO_2 is endothermic.
4 No effect, as the number of gas molecules is the same on either side of the balanced equation.

10.4

1 Nitrogen gas – from air
  Hydrogen gas – from natural gas
2 a $N_2(g) + 3H_2(g) \rightleftharpoons 2NH_3(g)$
  b 450 °C, 200 atmospheres, iron catalyst
3 a Cooled, condensed, and collected as a liquid
  b They are recycled back in to the reaction mixture.
4

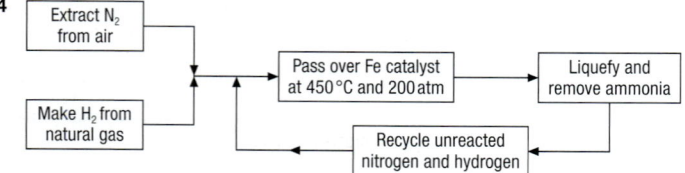

10.5

1 a Increases the yield of ammonia as there are fewer molecules of gas on the right-hand side of the balanced equation so increasing the pressure will favour the forward reaction.
  b Increases the rate of production of ammonia as at higher pressures there are more gas molecules in the same volume so collisions will occur more frequently, increasing the rate of reaction.
2 To increase the rate of production of ammonia (even though both forward and reverse rates are increased)
3 a 31%
  b 10%
  c 94%
  d 82%
  e These conditions are a compromise so that the yield of ammonia is lower than would be ideal but the rate of reaction is reasonable at this temperature and pressure (with the aid of the iron catalyst).

10.6

1 Stage 1   sulfur   +   oxygen   ⇌   sulfur dioxide
           $S(s/l)$ + $O_2(g)$ ⇌ $SO_2(g)$
  Stage 2   sulfur dioxide + oxygen   ⇌   sulfur trioxide
           $2SO_2(g)$ + $O_2(g)$ ⇌ $2SO_3(g)$
  Stage 3   sulfur trioxide + water   ⇌   sulfuric acid
           $SO_3(g)$ + $H_2O(l)$ ⇌ $H_2SO_4(l)$
2 a 450 °C, atmospheric pressure (or just above)/1 atosphere/100 k Pa, vanadium(v) oxide catalyst/V_2O_5
  b Several layers in the reaction vessel; supported on porous 'daisy-shaped' pellets to maximise the surface area for gases to react on
  c Temperature of 450 °C – this is chosen to maintain a reasonable rate of formation of sulfur trioxide, despite the fact that as the temperature is raised the yield of sulfur trioxide decreases. Atmospheric pressure (or just above to push the gases through the pipes and reaction vessels) is chosen as a very high yield is achieved anyway in the reaction vessel so it not worth spending more money compressing the gases. The vanadium(v) catalyst speeds up the rate of formation of sulfur trioxide although it does not affect its yield.
3 Start with: $\frac{1000}{32}$ moles of S. If all of it is converted to sulfuric acid, H_2SO_4 you also have $\left(\frac{1000}{32}\right) \times 2$ moles of H and $\left(\frac{1000}{32}\right) \times 4$ moles of O.

So the mass of H_2SO_4 per kilogram = $[\left(\frac{2000}{32}\right) \times 1]$g of H
$+ [\left(\frac{1000}{32}\right) \times 32]$g of S
$+ [\left(\frac{4000}{32}\right) \times 16]$g of O
= 3062.5 g
= 3.06 kg (to 3 significant figures)

Answers to end of chapter summary questions

1 a There is solid, liquid, and gas present.
  b More ICl_3 is made as the position of equilibrium shifts to favour the forward reaction which produces no molecules of gas, thereby reducing the pressure.
  c The position of equilibrium would shift to the left in order to replace lost chlorine gas.
2 a $A(g) + B(g) \rightleftharpoons 2C(g)$
  b i No effect   ii Increase the rate
  c i Increase the amount of C formed
    ii Increase the rate
3 a Increase in temperature will favour the reverse reaction, making more sulfur dioxide and oxygen, as this reaction is endothermic, absorbing energy and thereby decreasing the temperature. Therefore, yield will be lower. The rate of reactions will be increased as the gas molecules will move around faster so there will be more frequent collisions, with more being successful in producing a reaction as more have energy greater than the activation energy.
  b Increase in pressure will favour the forward reaction as this produces fewer molecules of gas (two molecules of sulfur trioxide on the right-hand side of the equation as opposed to three molecules of sulfur dioxide and oxygen on the left), thereby increasing the yield. The rate of reaction will be increased as there will be more gas molecules in a given volume, producing more frequent collisions.
  c Use of a catalyst has no effect on the position of equilibrium but will increase the rate of both forward and reverse reactions by the same amount as it lowers the activation energy so a higher proportion of the gas molecules have sufficient energy to collide successfully and react.
4 The pressure could be decreased as the left-hand side of the equation has fewer (9) molecules of gas than the right-hand side (10), so would tend to decrease the pressure.
  The temperature could be decreased as the forward reaction is exothermic so releases energy thereby increasing the temperature.
  The platinum catalyst will not affect the yield as it increases the rate of both forward and reverse reactions equally.

5 a i Increases the yield of ammonia
  ii Decreases the yield of ammonia
  iii Looking at pressure, choose one temperature and follow the pattern down a column as the pressure increases.
     Looking at temperature, choose one pressure and follow the pattern across a row as the temperature increases.
 b The pressure is not raised any higher for economic reasons – making a plant to withstand higher pressures costs more, generating higher pressures takes more energy and increases safety risks.
   The temperature cannot be decreased, although this would improve the yield of ammonia, as the rate of reaction would be too slow.
 c No effect on the yield of ammonia

6

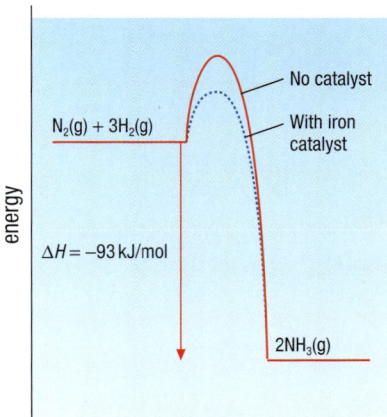

Answers to end of chapter practice questions

1 a The rates of the forward and reverse reactions are the same. *(1 mark)*
   The concentrations of the reactants and products do not change. *(1 mark)*
 b i It becomes brown/darker. *(1 mark)*
   ii The reaction is endothermic. *(1 mark)*
     The equilibrium shifts to the right (as the temperature increases). *(1 mark)*
 c The colour becomes pale yellow/paler. *(1 mark)*
   There are more gas molecules on the right. *(1 mark)*
   The equilibrium shifts to the left (as the pressure increases). *(1 mark)*
2 a i Air *(1 mark)*
   ii Natural gas/water *(1 mark)*
 b Iron catalyst *(1 mark)*
   450 °C *(1 mark)*
   200 atmospheres *(1 mark)*
 c Nitrogen and hydrogen *(1 mark)*
 d Cool/lower temperatures *(1 mark)*
 e $N_2 + 3H_2 \rightarrow 2NH_3$
   All symbols correct *(1 mark)*
   Balancing *(1 mark)*
3 a The temperature decrease raises the yield *(1 mark)*
   Because the reaction is exothermic *(1 mark)*
   So the equilibrium shifts to the right *(1 mark)*
   The pressure increase raises the yield *(1 mark)*
   Because there are fewer molecules of gas on the right *(1 mark)*
   So the equilibrium shifts to the right *(1 mark)*
 b It increases it. *(1 mark)*
 c It increases the rates of the forward and reverse reactions *(1 mark)*
   By the same amount *(1 mark)*
4 a i Ammonia *(1 mark)*
     Hydrogen chloride *(1 mark)*
   ii Ammonium chloride *(1 mark)*
   iii $NH_3 + HCl \rightarrow NH_4Cl$ *(1 mark)*
 b Top of tube too hot for ammonium chloride
   to form/solid forms outside the tube. *(1 mark)*
5. a The Contact process *(1 mark)*
 b 512 tonnes *(2 marks)*
 c i $2SO_2 + O_2 \rightleftharpoons 2SO_3$ *(2 marks)*
   ii Vanadium(v) oxide *(1 mark)*
 d 784 tonnes *(2 marks)*

11 Energy changes in chemical reactions

11.1
1 a Exothermic
 b Endothermic
 c For example, oxidation, combustion, neutralisation, respiration
 d For example, any thermal decomposition, photosynthesis
2 The beaker would feel cold as the process of dissolving absorbed energy from its surroundings which included the beaker and your hand holding it.
3 The enthalpy of the reactants was greater/higher than the products, with the difference released as energy which raised the temperature of the surroundings. ΔH would have a negative value.
4 a $MgCO_3(s) \rightarrow MgO(s) + CO_2(g)$
 b +117 kJ/mol

11.2
1 a It uses the energy released in the oxidation of iron, forming hydrated iron(III) oxide in an exothermic reaction. Sodium chloride (common salt) is used as a catalyst.
 b A supersaturated solution is made to crystallise by pressing a small metal disc. The crystals spread throughout the solution, releasing energy. The crystals are re-dissolved in hot water, ready to use again.
 c The disposable hand warmer lasts longer when activated than the reusable one. However, it can only be used once. Opposite applies to reusable hand warmers.
 d Self-heating cans
2 a To treat injuries with cold packs; to chill drinks in cans
 b Dissolving ammonium nitrate in water
 c i NH_4NO_3 ii For example, fertiliser
3 a Calcium oxide
 b $CaO(s) + H_2O(l) \rightarrow Ca(OH)_2(aq)$
 c It would form a harmful alkaline solution.

11.3
1 a Their size/numerical value/magnitude is the same but they have opposite signs.
 b Water
 c It reacts with water vapour in the air.
 d Energy is released.
2 a $W + X \rightleftharpoons Y + Z$
 b It will absorb 50 kJ (+50 kJ/mol).
3 a 4 moles
 b $CoCl_2 \cdot 2H_2O + 4H_2O \rightleftharpoons CoCl_2 \cdot 6H_2O$
 c Warm them up.

11.4
1 a A = 6720 J (6.72 kJ); B = 3360 J (3.36 kJ); C = 5670 J (5.67 kJ)
 b A = 28 000 J (28 kJ); B = 18 667 J (18.67 kJ); C = 21 000 J (21 kJ)
 c A > C > B
 d A = 1344 kJ/mol; B = 784 kJ/mol; C = 1218 kJ/mol
 e A > C > B
 f They are likely to be much too low in magnitude because of large energy losses to the surroundings in the simple calorimeter.

11.5
1 Because it is a good thermal insulator it will reduce the rate of energy transferred to or from the surroundings.
2 a $NaOH(aq) + HCl(aq) \rightarrow NaCl(aq) + H_2O(l)$
 b Neutralisation
 c 2.0 mol/dm³
 d 1155 J
 e −23.1 kJ/mol

11.6
1 a

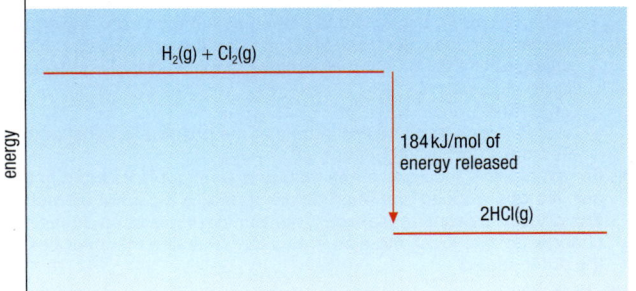

b

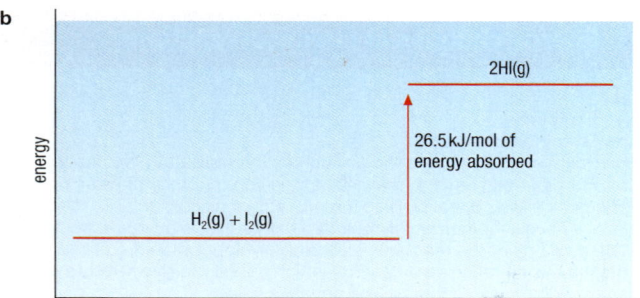

2 a *and* **b**

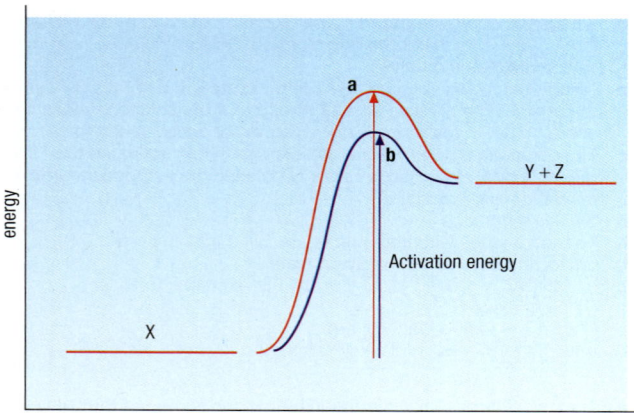

c It reduces the activation energy/energy needed for colliding particles to react so increasing the proportion of collisions in a sample with sufficient energy to cause a reaction and therefore increases the rate of reaction.

3 a Energy must be supplied from the surroundings and absorbed by the molecule whose bonds are being broken to overcome the attraction between atoms. So the separated atoms formed will have a higher energy content than the original molecule.

b

$$H \text{---} C \text{---} H \;+\; \begin{matrix} O \text{---} O \\ O \text{---} O \end{matrix} \;\longrightarrow\; O{=}C{=}O \;+\; \begin{matrix} H \\ H \end{matrix}{>}O \;\; \begin{matrix} H \\ H \end{matrix}{>}O$$

c Bonds broken: 4 C–H; 2 O=O
Bonds made: 2 C=O; 4 O–H

11.7

1 Endothermic
2 The energy required to break a specific bond.
3 a $H_2 + Cl_2 \rightarrow 2HCl$ $\Delta H = -185\,kJ/mol$
   b $2H_2 + O_2 \rightarrow 2H_2O$ $\Delta H = -486\,kJ/mol$

11.8

1 There has to be a difference in reactivity between two different metals to produce a voltage.
2 a Student diagram of the electrical cell set-up with a lamp in the external circuit.
   b Iron
   c Zinc acts as the negative terminal and a source of electrons into the external circuit, because it is more reactive than iron, having a greater tendency to lose electrons and form positive ions. So zinc atoms are changed to Zn^{2+} ions at the negative terminal, donating two electrons to iron(II) ions:
$Zn \rightarrow Zn^{2+} + 2e^-$
and Fe^{2+} ions turn into iron atoms at the positive terminal of the cell:
$Fe^{2+}(aq) + 2e^- \rightarrow Fe(s)$
3 For example – Once one of the reactants runs out/is used up in the dry cell, it stops working and must be discarded as it cannot be recharged. The dry cell is prone to leakage if the zinc casing is used up (as it changes to zinc ions) and dissolves away releasing the inner paste from the outer casing.

11.9

1 combustion, fuel, oxygen, water, carbon dioxide, carbon
2 Ensure that the electricity used is generated using renewable energy sources, such as from wind turbines or hydroelectric plants.

3 a The only product of combustion is water:
$2H_2 + O_2 \rightarrow 2H_2O$
   b Storage – hydrogen is a gas so it is difficult to store/needs to be compressed.
Safety – any leakages can form an explosive mixture with oxygen in the air.

Answers to end of chapter summary questions

1 a and **b**

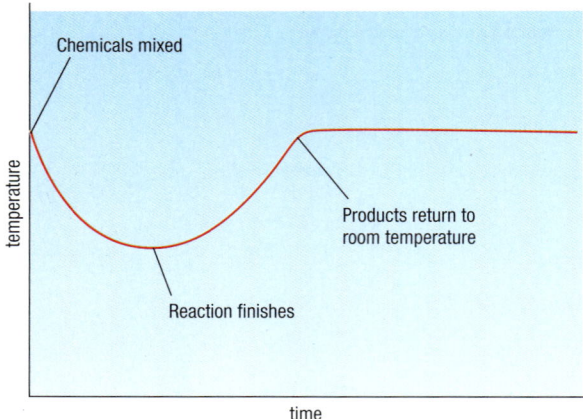

c 504 J
2 a

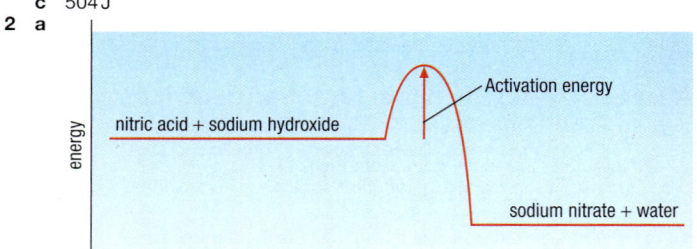

b

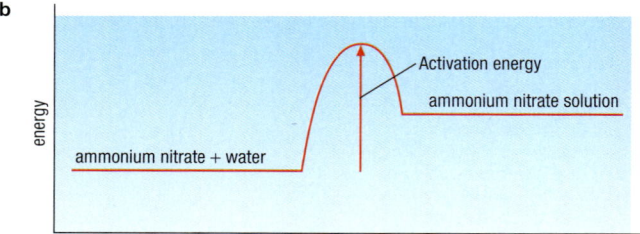

c

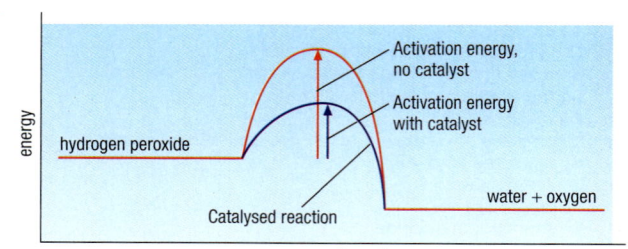

3 a $C_{12}H_{22}O_{11} + 12O_2 \rightarrow 12CO_2 + 11H_2O$
   b To break the bonds in reactants/to supply the activation energy
   c The process of making new bonds in carbon dioxide and water releases energy
   d 85 kJ
4 a −210 kJ/mol
   b The energy absorbed in breaking the bonds in hydrogen peroxide is less than the energy released when the new bonds in water and oxygen are made.
5 a 27.3 kJ
   b 2730 kJ/mol

Answers to end of chapter practice questions

1 a i Exothermic reactions give out heat/produce temperature rise.
    Endothermic reactions take in heat/produce temperature fall. *(2 marks)*
  ii Enthalpy change *(1 mark)*
  iii Negative sign means exothermic/positive sign means endothermic. *(1 mark)*
 b i Energy of reactants is higher than energy of products. *(1 mark)*
  ii Activation energy *(1 mark)*
  iii Catalyst used (in B) *(1 mark)*
  iv C *(1 mark)*
 c –31 kJ/mol *(1 mark)*
2 a i 12.5 °C *(1 mark)*
  ii Student 2 *(1 mark)*
 b i $100 \times 4.18 \times 14.5 = 6061\,J$
    Temperature change *(1 mark)*
    Substitution *(1 mark)*
    Correct answer *(1 mark)*
  ii $6200 \div 0.40 = 15\,500\,J/gram = 15.5\,kJ/gram$
    Substitution *(1 mark)*
    Conversion from J to kJ *(1 mark)*
    Correct answer *(1 mark)*
 c Heat lost (to atmosphere) *(1 mark)*
3 a $(4 \times 412) + (2 \times 496)$ *(1 mark)*
    $= 2640$ *(1 mark)*
 b $(2 \times 743) + (4 \times 463)$ *(1 mark)*
    $= 3338$ *(1 mark)*
 c –698 kJ/mol *(1 mark)*

12 Carbon compounds as fuels

12.1

1 a A mixture of hydrocarbons
 b We rely on fractions from crude oil for most of our fuels, as well as many other products made from them.
 c Because it is a mixture of different substances with different boiling points
2 As a mixture, crude oil contains a wide range of hydrocarbons. Its components all have differing properties, such as boiling points, so it is better to separate the crude oil into fractions which can have different specific uses.
3 a $C_nH_{(2n+2)}$
 b hexane, C_6H_{14}; heptane, C_7H_{16}; octane, C_8H_{18}; nonane, C_9H_{20}; decane, $C_{10}H_{22}$
4 a

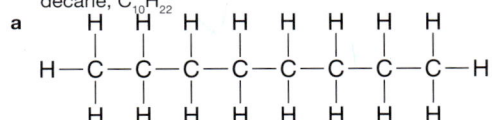

 b 46
 c 15
 d Because they are compounds of hydrogen and carbon atoms only which have the maximum number of hydrogen atoms possible in their molecules/contain only single covalent C–C bonds.

12.2

1 a i The larger the hydrocarbon, the higher the boiling point
  ii The larger the hydrocarbon, the lower the volatility
  iii The larger the hydrocarbon, the higher the viscosity
 b Short-chain hydrocarbon molecules – it will burn with a clean flame/not smoky.
2 Student table that effectively shows the patterns in the boiling points (gets higher), volatility (gets lower), viscosity (gets more viscous), and flammability (gets less flammable) as the size of the hydrocarbon molecules increase.
3 Hot crude oil is heated and enters a fractionating column near the bottom as a vapour. The temperature decreases going up the column. The gases condense when they reach the temperature of their boiling points so different fractions are collected as liquids at different levels in a continuous process. The hydrocarbons with the smallest molecules have the lowest boiling points and are collected at the top of the column where the temperature is lower. Nearer the bottom of the fractionating column, the fractions with the higher boiling points are collected.

12.3

1 a Carbon dioxide and water
 b Sulfur
2 a Acid rain
 b Oxides of nitrogen

3 a From oxidation of sulfur impurities when fuels burn
 b When nitrogen in the air is oxidised in the very high temperatures inside a vehicle engine
 c When fuel, such as diesel, undergoes incomplete combustion
4 a $CH_4(g) + 2O_2(g) \rightarrow CO_2(g) + 2H_2O(l)$
 b $2CH_4(g) + 3O_2(g) \rightarrow 2CO(g) + 4H_2O(l)$
 c Carbon monoxide is a highly toxic gas. As it is colourless and odourless, people cannot detect it and so become poisoned.

12.4

1 Because ethanol can be made from plants which actually absorb carbon dioxide as they photosynthesise which offsets the carbon dioxide released when it is burnt, whereas burning petrol is releasing carbon dioxide from hydrocarbons trapped beneath the surface for millions of years.
2 The energy comes from the Sun, which is used by the plants, such as oilseed rape, in the process of photosynthesis as the biofuel crop is growing.
3 a Ethanol produced by fermenting glucose from sugar cane (or sugar beet) is a renewable source of energy as the crop can be grown and harvested each year. However, ethanol made from ethene which is obtained by cracking products from crude oil is using up a non-renewable fossil fuel.
 b If farmers switch from producing food crops to growing crops for biofuels, there will be less land available to grow food or raise animals. This could result in food shortages and higher food prices.

Answers to end of chapter summary questions

1 a i $C_nH_{(2n+2)}$ ii $C_{18}H_{38}$ iii C_8H_{18}
 b Student graph
 c The higher the number of carbon atoms (the larger the alkane molecule), the higher its boiling point.
 d Actual boiling point of pentane is 36 °C – see student graph for accuracy.
2 a i A compound made up of hydrogen and carbon atoms only
  ii They have the maximum number of hydrogen atoms possible in their molecules/contain only single covalent bonds between carbon atoms.
 b i Propane, C_3H_8
  ii Atoms of carbon and hydrogen
  iii Single covalent bonds
3 a B, because a fractionating column gets cooler nearer the top and the vapour of B will condense at a lower temperature than A.
 b B as it is made up of smaller alkane molecules than A because it has a lower boiling point – and smaller hydrocarbons are more flammable.
 c i A is more viscous than B.
  ii B is more volatile than A.
4 a Climate change
 b i Sulfur dioxide
  ii Remove sulfur impurities before burning the fossil fuel or prevent the sulfur dioxide gas formed from entering the atmosphere.
  iii Oxides of nitrogen – formed when nitrogen in the air inside a combustion engine is oxidised at very high temperatures.
5 a To condense the steam formed in the combustion of ethanol and collect it as a liquid
 b Blue cobalt chloride paper turns pink/white anhydrous copper(II) sulfate turns blue.
 c It turns cloudy as carbon dioxide is produced when the carbon atoms in ethanol are oxidised.
 d Compare the time it takes for the limewater to turn cloudy when ethanol burns with the time it takes with just air being drawn through the apparatus.
 e ethanol + oxygen → carbon dioxide + water
    $C_2H_5OH + 3O_2 \rightarrow 2CO_2 + 3H_2O$
 f Swap the spirit burner with ethanol in for a wax candle.
 g propane + oxygen → carbon dioxide + water
    $C_3H_8 + 5O_2 \rightarrow 3CO_2 + 4H_2O$
 h Carbon monoxide
6 a Ethanol, because it can be made from plants, e.g., sugar cane and sugar beet
 b The carbon dioxide released when it burns was taken in from the atmosphere as it grew.
 c Water is the only product of combustion so no more carbon dioxide is added to the atmosphere when it burns.
 d How to store the gas (large volume or very high pressures in cylinders);
    explosive with oxygen in an accident;
    how to generate the hydrogen.
 e hydrogen + oxygen → water
    $2H_2 + O_2 \rightarrow 2H_2O$
 f A fuel cell

Answers to end of chapter practice questions

1 a i (Compound containing) hydrogen and carbon *(1 mark)*
only/but no other elements *(1 mark)*
ii Only single bonds/no double bonds *(1 mark)*
b i C_nH_{2n+2} *(1 mark)*
ii C_3H_8 *(1 mark)*
iii

H H H H
| | | |
H–C–C–C–H (or displayed formula for methylpropane)
| | | |
H H H H *(1 mark)*

c i (Liquid to vapour) evaporation/boiling *(1 mark)*
(Vapour to liquid) condensation *(1 mark)*
ii Viscous *(1 mark)*
High boiling point *(1 mark)*
2 a Carbon dioxide *(1 mark)*
Water *(1 mark)*
b $CH_4 + 2O_2 \rightarrow CO_2 + 2H_2O$
Four formulae correct *(1 mark)*
Balancing *(1 mark)*
c (Solid) carbon *(1 mark)*
(Gaseous) carbon monoxide *(1 mark)*
d i High temperature *(1 mark)*
ii Acid rain *(1 mark)*
e i Remove it before fuel is burnt. *(1 mark)*
ii Remove it before it enters the atmosphere. *(1 mark)*
3 a It comes from plants. *(1 mark)*
b Sugar (cane) *(1 mark)*
Fermentation *(1 mark)*
c i Crude oil/fossil fuels are becoming depleted. *(1 mark)*
ii Less food can be grown. *(1 mark)*
d i No carbon dioxide formed *(1 mark)*
ii Storage is difficult/expensive/heavy *(1 mark)*
Risk of explosion *(1 mark)*

13 Other products from crude oil

13.1

1 a To meet our demands for fuels (petrol and diesel) from crude oil
b The hydrocarbon vapours are passed over a hot catalyst or mixed with steam at very high temperatures.
2 a i B
ii A hydrocarbon that contains at least one C=C double bond
iii Alkene
b i A **ii** Alkane
c Thermal decomposition
3 24 hydrogen atoms in $C_{12}H_{24}$
4 $C_{12}H_{26} \rightarrow C_8H_{18} + 2C_2H_4$

13.2

1 a A monomer is a small reactive molecule that joins together with other monomers to form a polymer.
A polymer is a very large molecule made up of many repeating units.
b Polymerisation
c

Ethene   Poly(ethene)

d For example, carrier bags, dustbins, clingfilm, bottles, washing-up bowls
2 Two carbon atoms are needed for a hydrocarbon to have a C=C double bond whereas in alkanes only one carbon atom is needed to form a saturated hydrocarbon as in CH_4.
3 a

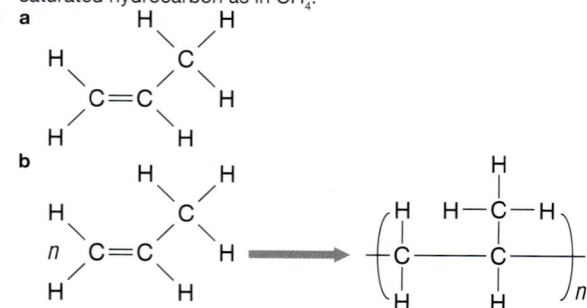

b

c For example, carpets, milk crates, ropes
d The C=C double bond 'opens up' in neighbouring propene molecules and forms single bonds, joining the monomers together in a long chain.

13.3

1 a In a tangled mess
b i Thermosoftening polymers have their individual chains held in place by relatively weak intermolecular forces whereas thermosetting polymers have strong covalent bonds (cross links) between the chains.
ii Thermosoftening plastics soften when heated and tend to be soft but thermosetting polymers are heat-resistant (eventually charring at high enough temperatures) and rigid.
2 The handles would soften if thermosoftening polymers were used as heat is conducted through the metal pan to its handle.
3 Polymer A has stronger intermolecular forces than polymer B so it takes more energy to soften it. Polymer C has strong covalent bonds holding its chains in position so it will not soften on heating.
4 a The reaction conditions are different (temperature, pressure, catalyst).
b Both HDPE and LDPE are thermosoftening polymers.
c HDPE has straighter chains which can pack more closely together so it has a higher density, whereas LDPE has randomly branched chains so the chains cannot pack regularly together so there is more space in its structure so it has a lower density.

13.4

1 a i A polymer that is developed with a specific use in mind
ii A polymer which changes in response to changes in its surroundings
b i It is light, tough, and transparent
ii For example, clothing, duvet filling
iii Thermosoftening polymer because it can be melted and re-worked.
2 a A loose network of cross-linked polymers with water trapped in its structure
b Dressings for burns, contact lenses
3 Once the top layer is peeled off, the light-sensitive polymer beneath loses its stickiness when in contact with light so can easily be removed from the skin.
4 The stitches are inserted loosely and tighten when warmed to body temperature.

13.5

1 Waste plastic mostly ends up in landfill sites where it takes many years to break down.
2 Polymers that can be broken down by microbes in the soil.
3 a Because they are biodegradable
b Lactic acid
4 a Little or not enough oxygen to aid decomposition process
b Decomposition might start before the plastic's useful life has expired.
c It can be incorporated into the structure of non-biodegradable plastics so that microbes feed on it once disposed of and speed up decomposition.
Polymers derived from cornstarch are themselves biodegradable.

Answers to end of chapter summary questions

1 a C_3H_6
b

H H
\ /
H C
\ / \
C=C H
/ \ /
H H

c Unsaturated molecule as it contains a C=C double bond
d Add bromine water, stopper, and shake – the propene will decolourise the bromine water whereas the propane has no effect/bromine water remains orange in colour.
e i Polymerisation
ii Poly(propene)
iii Monomers
iv The poly(propene) is a solid at room temperature whereas propene is a gas.
2 $n\ C_2H_4 \rightarrow -CH_2-CH_{2n}-$
3 a ethene + steam $\rightleftharpoons$ ethanol
$C_2H_4(g) + H_2O(g) \rightleftharpoons C_2H_5OH$
b 75%
c Advantage – it is a continuous process and a quicker process.
Disadvantage – it uses a non-renewable resource as its raw material (crude oil) and uses more energy in the process.

4 a Water molecules can pass out of the tiny pores in the material (as when we breathe out water vapour) so the wearer does not get clammy.
  b No, because it does not change in response to changes in its environment.
5 a The crops have been genetically modified.
  b Not in theory as it absorbs the same amount of carbon dioxide when the crop grows as it releases when it decomposes. However, processing and transporting the crop to extract the polymer will use fossil fuels which will result in some excess carbon dioxide being released.
  c Probably as the recycling bypasses the processing/fermentation stages but it would itself require some costs in terms of extracting the polymer.

Answers to end of chapter practice questions

1 a (Type 1)
    Heat   (1 mark)
    With a catalyst   (1 mark)
    (Type 2)
    Mix with steam   (1 mark)
    Very high temperature   (1 mark)
  b i (Carbon–carbon) double (covalent) bond   (1 mark)
    ii

    H $C=C$ H / H ... H (1 mark)
    iii C_4H_8 (1 mark)
  c C_4H_8 or $2C_2H_4$ (1 mark)
  d Bromine (water)   (1 mark)
    Alkanes show no change in colour.   (1 mark)
    Alkenes decolourise it.   (1 mark)
  e Ethene   (1 mark)
    Steam   (1 mark)
    Heat/catalyst   (1 mark)
2 a No double bonds/saturated   (1 mark)
    Much bigger/long chains   (1 mark)
  b i LD = low density and HD = high density   (1 mark)
    ii Melt when heated   (1 mark)
  c i Does not melt/decompose   (1 mark)
    ii Cross links/chains joined together   (1 mark)
  d Broken down by microbes   (1 mark)
    Reduce space taken up in landfill   (1 mark)
3 **0 marks** No relevant content
  Level 1 (1–2 marks) One type of polymer is correctly identified and there is a brief account of the properties, structure, and bonding of polymers.
  Level 2 (3–4 marks) Both types of polymer are correctly identified and there are some references to the properties, structure, and bonding of polymers.
  Level 3 (5–6 marks) Both types of polymer are correctly identified and there is a clear, balanced, and detailed account of the properties, structure, and bonding of polymers.
  Examples of chemistry points made in the response:
  • Polymer 2 should be made using low-density thermosoftening polymer
  • Polymer 1 should be made using thermosetting polymer
  • low-density (thermosoftening) polymer chains have many branches
  • high-density (thermosoftening) polymer chains have no/few branches
  • thermosetting polymer chains have cross links/covalent bonds between the chains
  • HD polymer chains can pack closely together
  • LD polymer chains do not pack closely together
  • thermosetting polymers cannot change shape because the cross links do not break when heated

14 Alcohols, carboxylic acids, and esters

14.1
1 a Esters   c Carboxylic acids
  b Alcohols
2 a Propanol   c Methanoic acid
  b Ethyl ethanoate
3 a
    H—C—C—O—H (with H atoms)
                                    c
                                      H—C—C—C—C (with H atoms and O, O—H)

b
$H-C-C=O$ structure / $O-C-C-H$

14.2
1 Ethanol, CH_3CH_2OH or C_2H_5OH
2 Fuels, solvents, and ethanol in alcoholic drinks
3 a sodium + methanol → sodium methoxide + hydrogen
    $2Na + 2CH_3OH → 2CH_3ONa + H_2$
  b Methanoic acid, HCOOH
  c methanol + oxygen → carbon dioxide + water
    $2CH_3OH + 3O_2 → 2CO_2 + 4H_2O$
4 Place methanol, ethanol, and propanol in separate spirit burners and weigh then accurately. Heat a known volume of water for a set time and record the rise in temperature of the water. Re-weigh the spirit burners with their remaining alcohols. Calculate the rise in temperature that would have been produced if one gram of each alcohol had burnt. The alcohol with the largest temperature rise per gram is the one that releases most energy when it burns.

14.3
1 a Carbon dioxide
  b Potassium propanoate
  c $2C_2H_5COOH + K_2CO_3 → 2C_2H_5COOK + H_2O + CO_2$
2 a   sulfuric acid catalyst
    methanoic acid + ethanol ⇌ ethyl methanoate + water
  b i Ethyl ethanoate   ii Propyl methanoate
  c They evaporate/give off vapour easily.
  d Flavourings and perfumes (polyester fabrics)
3 Propanoic acid is a weak acid because it does not ionise completely when added to water. The majority of its molecules remain intact and only a fraction form $H^+(aq)$ ions and propanoate (negative) ions. Therefore propanoic acid does not produce as high a concentration of $H^+(aq)$ ions in its solution as a solution of a strong acid of equal concentration.

Answers to end of chapter summary questions

1 a A
  b C
  c Esters
  d B
  e A is $CH_3CH_2CH_2CH_2CH_2COOH$
    C is $CH_3CH_2CH_2CH_2CH_2CH_2CH_2OH$
2 a i Sodium fizzes/effervesces/gives off bubbles of gas, and gets smaller and smaller as a solution is formed.
    ii Hydrogen
    iii Sodium ethoxide
    iv Na^+
    v $2Na + 2C_2H_5OH → 2C_2H_5ONa + H_2$
  b The reaction with lithium would be slower/gas given off more slowly/longer for lithium to dissolve into the solution.
3 a
    O
    H—C—O—H
  b i
    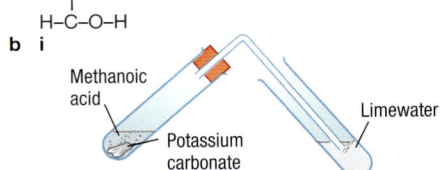
    Methanoic acid
    Potassium carbonate
    Limewater
    ii methanoic acid + potassium carbonate → potassium methanoate + water + carbon dioxide
    $2HCOOH + K_2CO_3 → 2HCOOK + H_2O + CO_2$
  c i **E** ethanol
    ii **C** methanoic acid, because ethanol is neutral so has the highest pH of 7. The rest are acids so have a pH below 7, but methanoic acid is the only weak acid. The other three being strong acids will have lower pH values.
4 a Ethyl ethanoate
  b Ethanol and ethanoic acid
  c   sulfuric acid catalyst
    ethanoic acid + ethanol ⇌ ethyl ethanoate + water
    $CH_3COOH + C_2H_5OH ⇌ CH_3COOC_2H_5 + H_2O$
  d It evaporates easily
  e Methanol and propanoic acid

5 For example – Add sodium carbonate to a solution of each liquid – only propanoic acid would fizz (giving off carbon dioxide gas); then add sodium to the other two remaining liquids, only propanol would fizz (giving off hydrogen); the remaining liquid is ethyl ethanoate (smell to check).

6 a There are 0.5 mol of solute in 1 dm^3 of solution.

b Methanoic acid will have a higher pH value than nitric acid.

c Nitric acid is a strong acid. The HNO_3 molecules all ionise in water:

$$HNO_3(aq) \rightarrow H^+(aq) + NO_3^-(aq)$$

However, methanoic acid is a weak acid so most of its molecules stay un-ionised. Only a few molecules will ionise and split up in their solutions in a reversible reaction.

$$HCOOH(aq) \rightleftharpoons HCOO^-(aq) + H^+(aq)$$

So there are fewer $H^+(aq)$ ions in a given volume of methanoic acid solution compared with the same volume of nitric acid solution.

Answers to end of chapter practice questions

1 a (Any two of)
Same general formula
Same functional group
Same/similar chemical reactions *(2 marks)*

b i CH_3OH/CH_4O *(1 mark)*

ii

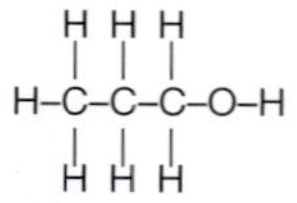

H H H
| | |
H–C–C–C–O–H
| | |
H H H
(1 mark)

c (Any two of)
Bubbles
Sodium disappears/gets smaller
Solid moves around *(2 marks)*

d Carbon dioxide *(1 mark)*
Water *(1 mark)*

e $CH_3CH_2OH + 3O_2 \rightarrow 2CO_2 + 3H_2O$
All formulae correct *(1 mark)*
Balancing *(1 mark)*

f i Ethanoic acid *(1 mark)*
ii Vinegar *(1 mark)*

2 a (Ethanol) – no change *(1 mark)*
(Ethanoic acid) – bubbles *(1 mark)*

b i 5 *(1 mark)*
ii 1 *(1 mark)*

c Ethanoic acid is weaker because the ionisation is not complete/fewer H^+ ions are formed. *(1 mark)*
Nitric acid is stronger because ionisation is complete/more H^+ ions are formed. *(1 mark)*

3 a i Methanoic acid *(1 mark)*
Propanol *(1 mark)*

ii

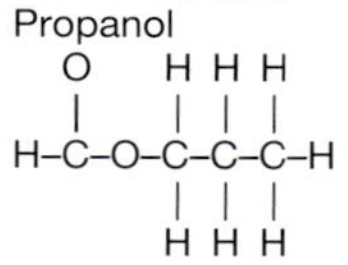

O   H H H
| | | |
H–C–O–C–C–C–H
| | |
H H H
(1 mark)

b Ethyl ethanoate *(1 mark)*

c (Food) flavourings *(1 mark)*
Perfumes *(1 mark)*

Index